COST Action MP0806 'Particles in Turbulence': International Conference on Fundamentals, Experiments, Numeric and Applications 2011

Journal of Physics: Conference Series Volume 333

Potsdam, Germany
16-18 March 2011

ISBN: 978-1-62276-680-2
ISSN: 1742-6588

Printed from e-media with permission by:

Curran Associates, Inc.
57 Morehouse Lane
Red Hook, NY 12571

Some format issues inherent in the e-media version may also appear in this print version.

Printed by Curran Associates, Inc. (2013)

Institute of Physics
Dirac House, Temple Back
Bristol BS1 6BE UK

Phone: 44 1 17 929 7481
Fax: 44 1 17 920 0979

techtracking@iop.org

Additional copies of this publication are available from:

Curran Associates, Inc.
57 Morehouse Lane
Red Hook, NY 12571 USA
Phone: 845-758-0400
Fax: 845-758-2634
Email: curran@proceedings.com
Web: www.proceedings.com

TABLE OF CONTENTS

International Conference on *Fundamentals*, *Experiments*, *Numeric* and *Applications*

Particles in turbulence

March 16, 2011 - March 18, 2011 University of Potsdam (Germany)

Organizing committee

Markus Abel, Eberhard Bodenschatz, Federico Toschi

Keynote speakers

Markus **Abel**	Claudio **Gheller**	Martin **Kissane**
Jeremie **Bec**	Wojciech **Grabowski**	Mike **Reeks**
Luca **Biferale**	Rahul **Pandit**	Jörg **Schumacher**
Eberhard **Bodenschatz**	Alain **Pumir**	Federico **Toschi**
Ulrike **Feudel**	Detlef **Lohse**	Christos **Vassilicos**
	Beat **Luethi**	

International Advisory Committee

Prof. Daniele CARATI, Prof. Nikolay VITANOV, Dr Soren OTT, Prof. Jakob MANN, Dr Jacob BERG, Prof. Mogens Hogh JENSEN, Dr Ivan MAMMARELLA, Dr Paolo MURATORE GINANNESCHI, Dr Alain PUMIR, Dr Jeremie BEC, Dr Emmanuel LEVEQUE, Dr Jean-Francois PINTON, Dr Holger SIEBERT, Prof. Eberhard BODENSCHATZ, Prof. Jörg SCHUMACHER, Prof. Joachim PEINKE, Dr Ioannis SARRIS, Mr Tamas TEL, Dr Imre M. JANOSI, Dr K. Gabor SZABO, Prof. Armann GYLFASON, Prof. Luca BIFERALE, Dr Claudio GHELLER, Dr Alessandra Sabina LANOTTE, Dr Algis DZIUGYS, Prof. Pranas BALTRENAS, Dr Robertas NAVAKAS, Dr Saulius VASAREVICIUS, Prof. Detlef LOHSE, Prof. Herman CLERCX, Prof. Gijsbert OOMS, Prof. Helge ANDERSSON, Dr Konrad BAJER, Dr Jacek POZORSKI, Prof. Tomasz KOWALEWSKI, Dr Jose PALMA, Prof. Javier JIMENEZ, Dr Beat LUETHI, Dr Markus HOLZNER, Prof. Christos VASSILICOS, Prof. Michael REEKS, Prof. Yannis HARDALUPAS, Dr David SWAILES, Prof. Stavros KASSINOS, Prof. Miroslav JICHA, Prof. Tov ELPERIN, Prof. Gregory FALKOVICH, Prof. Itamar PROCACCIA Prof. Bernhard MEHLIG

in collaboration with ICTR - International Collaboration for Turbulence Research

More information on the conference website http://mp0806.cineca.it/potsdam/
or on the **COST MP0806** website http://mp0806.cineca.it Contact email CostMP0806.phys@tue.nl

Program of the conference Particles in turbulence 2011 International Conference on Fundamentals, Experiments, Numeric and Applications
16-18 March 2011, Potsdam, Germany http://mp0806.cineca.it/potsdam

Program version - post conf

Registration will take place on 15 March 2011 from 17:00 to 19:00

16 March 2011 - Conference day 1				
Starting hour	Ending hour	Speaker	Talk title	#
8:00	8:15	**Toschi, Abel, Bodenschatz**	Welcome address	
8:15	8:45	**Bodenschatz**	*Results from the Göttingen Turbulence Facility*	1
8:45	9:00	**Hardalupas**	Preferential concentration of water droplets in homogeneous and isotropic turbulence	2
9:00	9:15	**Soldati**	Particle deposition and re-entrainment in turbulent channel flow with wall roughness	3
9:15	9:30	**Picano**	Particle Clustering in Turbulent Premixed Flames	4
9:30	9:45	**Sun**	Bubbly drag reduction and local bubble distribution in bubbly turbulent Taylor-Couette flow	5
9:45	10:00	**Pozorski**	Towards an anisotropic stochastic dispersion model for LES of particle-laden turbulent flows	6
10:00	10:15	**Brändle de Motta**	Direct numerical simulation of sustained homogeneous turbulence interacting with finite size resolved particles.	7
10:15	10:45	Coffee break		
10:45	11:15	**Pumir**	*Statistical properties of the coarse-grained velocity gradient tensor in turbulence*	8
11:15	11:30	**Skartlien**	Particles and droplets in turbulent pipe flow: Kinetic modelling and added mass effects	9
11:30	11:45	**Zimmermann**	intermittency and turbulence induced lift experienced by large particles in a turbulent flow	10
11:45	12:00	**Calzavarini**	Impact of trailing wake drag on the statistical properties and dynamics of finite-sized particle in turbulence	11
12:00	12:15	**Fouxon**	Perspectives in the study of behavior of inertial particles in real turbulence	12
12:15	12:30	**Perlekar**	Discrete population dynamics in presence of turbulence	13
12:30	13:00	**Kissane**	*Postulated Nuclear Accidents: Evaluating Radioactive Releases with Special Emphasis on Particulates*	14
13:00	14:30		**Lunch break / Lunch MC meeting (MP0806 MC members)**	
14:30	15:00	**Feudel**	*Aggregation and fragmentation of marine aggregates in random flows*	15
15:00	15:15	**Jin**	Applying a new drag and lift correlation for particle deposition	16
15:15	15:30	**Knorps**	Statistics of dispersed phase in LES of heated channel flow	17
15:30	15:45	**Andersson**	Breakup of particles in turbulent flows with application to spray painting	18
15:45	16:00	**Pakhomov**	Flow and heat transfer in steady and intermittent turbulent impinging mist spray	19
16:00	16:15	**Gualtieri**	The effects of back-reaction on turbulence modulation in shear flows	20
16:30	17:00	Coffee break		
17:00	17:30	**Reeks**	*Particle Resuspension Modeling in Turbulent Flows*	22
17:30	17:45	**Arponen**	Stratified turbulence ensemble	23
17:45	18:00	**Salvetti**	A-priori reconstruction and characterization of an ideal correction for particle motion in LES velocity fields	24
18:00	18:15	**La Mantia**	New experimental set-up to analyse cryogenic flows by visualisation	25
18:15	18:30	**Gibert**	Flow around finite-size neutrally buoyant Lagrangian particles in fully developed turbulence	26
18:30	18:45	**Zhao**	Two-way coupled simulations of particle-laden wall turbulence	27
18:45	19:15	**Vassilicos**	*Particles in Turbulence*	28

17 March 2011 - Conference day 2

Starting hour	Ending hour	Speaker	Talk title	#
8:15	8:30	**Biferale**	*Turbulent pair dispersion of inertial particles*	#REF!
8:45	8:45	**Ireland**	Entrainment studies in a shearless mixing layer through direct numerical simulation	#REF!
9:00	9:00	**van Hout**	TR-PIV measurements of polystyrene beads entrained in a turbulent water channel flow	#REF!
9:15	9:15	**Xi**	Turbulence of dilute polymer solution: the Weissenberg Number effect	#REF!
9:30	9:30	**Homonnai**	Experimental model studies of passive tracer dispersal in geostrophic turbulence	#REF!
9:45	9:45	**Swailes**	The Fully-Mixed (Zero-Drift) Limit in PDF Models for Disperse Particle Flows	#REF!
10:00	10:00	**Mukin**	Nonlinear Algebraic Reynolds Stress Model for Two-Phase Turbulent Flows Laden with Small Heavy Particles	#REF!
10:15	10:30	Coffee break		
10:45	11:00	**Pandit**	*Time scales, persistence, and dynamic mutiscaling in two-dimensional, homogeneous, isotropic fluid turbulence*	37
11:15	11:15	**van Hinsberg**	An efficient, second order method for the approximation of the Basset history force	38
11:30	11:30	**Zamansky**	Solid particle acceleration in a high Reynolds number channel flow : DNS versus LES with stochastic modelling of subgrid acceleration	39
11:45	11:45	**Scepanskis**	The simulation of lagrangian particles in EM driven recirculated turbulent flows	40
12:00	12:00	**Lavezzo**	LBM of turbulent convection laden with inertial particles	41
12:15	12:15	**Dejoan**	DNS experiments on the two-way fluid-particle interaction in the settling of heavy particles in homogeneous turbulence including Reynolds number effects	42
12:30	13:00	**Lüthi**	*Velocity gradient dynamics*	43
13:00	14:30	**Lunch break**		
14:30	15:00	**Grabowski**	*Growth of cloud droplets and raindrops in turbulent clouds*	44
15:00	15:15	**Vollmer**	Rain in the test tube?	45
15:15	15:30	**Connaughton**	The inverse Smoluchowski problem for cluster-cluster aggregation	46
15:30	15:45	**Warncke**	Particle concentrations in turbulent flows in the low- τ_P limit	47
15:45	16:00	**Benczik**	Droplet dynamics in phase separation of binary mixtures under slow cooling	48
16:00	16:15	**Mehlig**	Distribution of relative velocities in turbulent aerosols	49
16:15	16:30	**Clercx**	Lagrangian velocity, acceleration and vorticity autocorrelations in rotating turbulence	50
16:30	17:00	Coffee break		
17:00	17:30	**Toschi**	*Droplets dynamics and breakup in turbulent flows*	51
17:30	17:45	**Haszpra**	Advection of tracer particles in the free atmosphere	52
17:45	18:00	**Vitanov**	Investigation of the bounds on the transport of heat and momentum in two turbulent fluid flows	53
18:00	18:15	**Ooms**	DNS analysis of local flow topology in a particle-laden turbulent channel flow	54
18:15	18:45	**Schumacher**	*Phase relaxation of a cloud water droplet ensemble undergoing turbulent mixing*	55

18 March 2011 - Conference day 3

Starting hour	Ending hour	Speaker	Talk title	#
8:15	8:30	**Lohse**	*Torque scaling in turbulent Taylor-Couette flow with co- and counter-rotating cylinders*	56
8:45	8:45	**Scagliarini**	Vortex filaments in turbulence: a Lagrangian viewpoint	57
9:00	9:00	**Viikmäe**	Technology for Finding Optimum Fairways for Environmental Management in the Baltic Sea	58
9:15	9:15	**Tagawa**	Three-dimensional Voronoi analysis for inertial particle clustering in turbulence	59
9:30	9:30	**Nagendra Prakash**	Lagrangian statistics of light particles in turbulence	60
9:45	10:00	**Bajer**	*Clear air turbulence - the DELICAT project*	61
10:00	10:30	Coffee break		
10:30	11:00	**Abel**	*Particle advection by an intermittent deterministic field*	66
11:00	11:15	**Chibbaro**	The LES/PDF method for turbulent two phase flows	67
11:15	11:30	**Bragg**	Non-local closure model for particle dispersion tensors in a turbulent boundary layer	68
11:30	11:45	**Musacchio**	Elastic collisions of inertial particles	69
11:45	12:00	**Lalescu**	Virtual particle transport in turbulent MHD Kolmogorov flow	70
12:00	12:30	**Voth**	*Measurements of accelerations of large spheres and rotation rates of rods*	71
13:00	14:30	**Lunch break**		
14:30	15:00	**Bec**	*Finite-size effect in the dynamics of neutrally buoyant particles in turbulent flow*	73
15:00	15:15	**Janosi**	Batchelor scaling in the atmosphere	74
15:15	15:30	**Cencini**	Intermittency in the velocity distribution of heavy particles in turbulence	75
15:30	15:45	**Babler**	Breakup of small particles driven by local dissipation	76
15:45	16:00	**Vasarevicius**	Experimental investigation of particle movement in turbulent flow through rectangular elbow	77
16:00	16:15	**Monchaux**	Voronoï analysis of inertial particles clustering in turbulent flows	78
16:15	16:30	**De Lillo**	Synchronization of hydrodynamically coupled colloidal particles	79
16:30	17:00	Coffee break		
17:00	17:15	Parallel WGs meetings	To be annonced by MP0806 WG leaders	80
17:15	17:30			81
17:30	17:45			82
17:45	18:00			83
18:00	18:15	Joint WG meeting	To be annonced by MP0806 WG leaders	84
18:15	18:30			85
18:30	19:00			86
19:00	19:15	Toschi, Abel, Bodenschatz	Closing	87

A generalized Kraichnan like model for simulating stratified flows

Heikki Arponen
University of Helsinki, Dept. of Mathematics and Statistics,P.O. Box 68 (Gustaf Hällströmin katu 2b) FI-00014 University of Helsinki

E-mail: heikki.arponen@helsinki.fi

Abstract. A generalization to the Kraichnan gaussian, white noise in time model of turbulent velocity field is proposed. The generalization is designed to mimic the effects of stratification in the atmospheric boundary layer by introducing different scaling exponents for horizontal and vertical velocity components, resulting in strong vertical anisotropy. The inevitable compressibility of the model leads one to take into account also the density statistics of the fluid, which will immediately lead to a boundary layer like structure in the flow statistics and non zero time correlations. Possible applications for passive scalar, tracer and inertial particle statistics and their anomalous scaling behavior will be discussed.

1. Introduction

Compared to homogeneous, isotropic turbulence, perhaps the most significant difference in a stably stratified atmospheric boundary layer is the strong vertical anisotropy manifesting as a $\propto k^{-3}$ mesoscale spectrum (corresponding to $\propto r^2$ spatial behavior in the structure function) of the vertical velocity components, as opposed to the usual Kolmogorov-Obukhov $\propto k^{-5/3}$ spectrum of the horizontal components (see e.g. [4] and references therein). The vertical velocities are "smoothed" by the effect of gravity pulling the atmosphere toward the ground. Recently, much understanding on the nature of intermittency and anomalous scaling (or multiscaling) has been acquired by the so called Kraichnan model, where one assumes the turbulent velocity statistics to be known and completely determined by a gaussian, white in time random velocity field (see e.g. [5] for a review). The purpose of the present and future work is to modify the Kraichnan model to mimic the behavior of stratified flows and to ultimately apply the model to the passive scalar and tracers[2].

2. Kraichnan model

The Kraichnan model (see e.g. [5, 6]) is a gaussian, mean zero, white in time random field which can be completely defined via the pair correlation function

$$\langle v_\mu(t,\mathbf{x}) v_\nu(t',\mathbf{x}')\rangle \doteq \delta(\Delta t) D_{\mu\nu}(\Delta\mathbf{x}), \tag{1}$$

where $\delta(\Delta t) \doteq \delta(t-t')$ is the Dirac delta function and $D_{\mu\nu}$ is an incompressible tensor field, which may be defined as

$$D_{\mu\nu}(\mathbf{r}) \doteq D_0 \int d^3\mathbf{p} \frac{\delta_{\mu\nu} - p_\mu p_\nu / p^2}{(p^2 + L^{-2})^{3/2+\xi/2}} e^{i\mathbf{p}\cdot\mathbf{r}}. \tag{2}$$

The constant L is the integral scale of the flow and $0 < \xi < 2$ describes the spatial roughness of the velocity field and appears as a scaling exponent in the velocity structure function in the limit of infinite L as

$$\langle |\mathbf{v}(t,\mathbf{x}) - \mathbf{v}(t',\mathbf{x}')|^2 \rangle \propto \delta(\Delta t)|\Delta \mathbf{x}|^\xi. \tag{3}$$

One should note that although the infinite L limit of the structure function is finite, the tensor field of (2) diverges as L^ξ.

The behavior of a passive scalar density (e.g. dye in fluid) under the Kraichnan flow can be described by the equation

$$\partial_t \theta - \kappa \Delta \theta + \mathbf{v} \cdot \nabla \theta = f. \tag{4}$$

The forcing $f \doteq f(t,\mathbf{r})$ is usually defined as a gaussian, white in time random field with a pair correlation function

$$\langle f(t,\mathbf{r})\, f(t',\mathbf{r}') \rangle \doteq \delta(\Delta t)\mathcal{C}(\Delta \mathbf{r}/L_f) \tag{5}$$

where L_f is the scale of forcing or dye injection[5]. The role of the forcing is to counter molecular dissipation and to thus maintain a non-equilibrium steady state. The name "passive" refers to an approximation where the backreaction of the scalar on the velocity is disregarded, which is an acceptable assumption in many realistic cases.

In the case of an isotropic, large scale forcing, i.e. $\mathcal{C} \to \mathcal{C}(0)$ as $L_f \to \infty$, the passive scalar equal time structure functions scale as

$$\langle [\theta(t,\mathbf{x}+\mathbf{r}) - \theta(t,\mathbf{x})]^N \rangle \propto r^{\frac{N}{2}(2-\xi)-\Delta_N} \tag{6}$$

with $\Delta_N = \frac{N(N-2)}{2(d+2)}\xi + \mathcal{O}(\xi^2)$ for small ξ (see [5] and references therein). The simple gaussian model of turbulence has therefore lead to an intermittency effect manifesting as multiscaling of the passive scalar structure functions.

3. Modified Kraichnan model with stratification

The idea of the present work is to study the effect of stratification in the context of a Kraichnan like turbulence model. For example, one would like to determine if and how the stratification affects the anomalous scaling exponents Δ_N above. We define a stratified generalization of the Kraichnan model as

$$D_{i\nu} \doteq D_0 \int d^3\mathbf{p} \frac{P_{i\nu}(\hat{p})e^{-lp}}{(p^2 + L^{-2})^{3/2+\xi/2}} e^{i\mathbf{p}\cdot\mathbf{r}} = D_{\nu i} \tag{7}$$

for the non-vertical components with $i \in 1, 2$, $\mu \in 1, 2, 3$ and r_3 is the vertical component. The vertical component is taken as

$$D_{33} \doteq D_0 L^{\xi-\eta} \int d^3\mathbf{p} \frac{P_{33}(\hat{p})e^{-lp}}{(p^2 + (\alpha L)^{-2})^{3/2+\eta/2}} e^{i\mathbf{p}\cdot\mathbf{r}}. \tag{8}$$

Here $P_{\mu\nu}(\hat{p}) = \delta_{\mu\nu} - \hat{p}_\mu \hat{p}_\nu$ is the usual incompressibility tensor, $0 \le \xi < \eta \le 2$ are the horizontal and vertical scaling exponents, $\alpha \le 1$ is the aspect ratio of vertical to horizontal integral scales and the term $L^{\xi-\eta}$ in the prefactor ensures correct dimensionality. In an atmospheric boundary layer one would have $\eta \approx 2$, but we will keep η as a free parameter. Also included is an ultraviolet cutoff term $\propto e^{-lp}$, where the length scale l corresponds to a viscous dissipative scale. It is often omitted in the Kraichnan model, since one is usually interested in scales larger than the viscous dissipation and because the model is finite as l is taken to zero. Here the viscous scale l must be kept finite (for now) because the model is not incompressible due to the different horizontal and vertical scaling. Specifically, the quantity $\langle(\nabla \cdot \mathbf{v})^2\rangle$ diverges as l approaches zero. The diffusivity ratio behaves as $D_{33}/D_{ii} \propto \alpha^{\eta}$, which implies vertically inhibited diffusion for small aspect ratios, as is known to occur in real stratified turbulence (see e.g. [7]).

Consider now the passive scalar equation in the presence of inhomogeneous density[7],

$$\partial_t (\rho\theta) + \nabla \cdot (\mathbf{v}\rho\theta) - \kappa\nabla \cdot (\rho\nabla\theta) = 0. \tag{9}$$

As usual, the above equation should be interpreted as a stochastic PDE of Stratonovich (mid point) type [8]. Taking this carefully into account and going from Stratonovich to Itô prescription[1], using the mass continuity equation

$$\partial_t\rho + \nabla \cdot (\mathbf{v}\rho) = 0 \tag{10}$$

and finally taking the limit of vanishing molecular diffusivity κ, we obtain

$$\partial_t\theta + [v_\mu + D_{\mu\nu}(0)\partial_\nu\chi]\,\partial_\mu\theta - \frac{1}{2}\widehat{D}_0\theta = 0, \tag{11}$$

where $\chi \doteq \log(\rho/\rho_0)$ (with a reference density ρ_0), henceforth referred to as log-density, and notation $\widehat{D}_0 \doteq D_{\mu\nu}(0)\partial_\mu\partial_\nu$. The passive scalar is therefore advected by an effective velocity field $\mathbf{u} \doteq \mathbf{v} + D(0) \cdot \nabla\chi$. Although the turbulent velocity field $\mathbf{v}$ is compressible, the effective velocity field $\mathbf{u}$ is in fact incompressible[2]. It is easy to understand this physically: the effect of compressibility is that fluid flowing into a given region does not match the outgoing flow all the time. This results in temporary regional increase or decrease of the flow density, as dictated by the mass continuity equation (10). The advection of the passive scalar however depends also on the gradient of the density, which cancels the compressible effects of the turbulent velocity field $\mathbf{v}$.

The density field is therefore another random field in the model which plays an important role in e.g. passive scalar and inertial particle behavior. We will rewrite also the mass continuity equation (10) in terms of the log-density χ. The equation for the log-density follows by similar use of the rules of stochastic calculus, resulting in the equation

$$\partial_t\chi + \nabla \cdot \mathbf{v} + \mathbf{v} \cdot \nabla\chi - \frac{1}{2}\widehat{D}_0\chi + \gamma = 0, \tag{12}$$

with notation $\gamma \doteq \langle(\nabla \cdot \mathbf{v})^2\rangle = -D_{\mu\nu,\mu\nu}(0)$ (symbols after a comma denote derivatives). In addition to the velocity field pair correlation functions, we therefore need to consider also the correlators $\langle\chi\rangle$, $\langle\mathbf{v}\chi\rangle$, $\langle\chi\chi'\rangle$ etc. One should note that as the log-density is determined by a first order stochastic differential equation in time, it is evident that the above correlators will exhibit finite time correlations. This will also imply that the effective velocity field $\mathbf{u}$ is not decorrelated.

Using the above SPDE, we write the steady state equations for the mean log-density $g(z) \doteq \langle\chi\rangle$ (which we assume to depend only on the vertical coordinate $z = x_3$),

$$\widehat{D}_0 g = 2\gamma, \tag{13}$$

[1] Sketchily, $\mathbf{v}(t) \doteq \frac{d\mathbf{V}(t)}{dt}$, where $\mathbf{V}(t)$ is an infinite dimensional Wiener process satisfying $d\mathbf{V}(t)d\mathbf{V}(t') = D(t - t')dt$. Then we can go from Stratonovich to Itô prescription by $dV_\mu(t)\theta\,(t + dt/2) - dV_\mu(t)\theta\,(t) = \frac{1}{2}dV_\mu(t)d\theta(t) = -\frac{1}{2}dV_\mu(t)\nabla \cdot (d\mathbf{V}(t)\theta(t)) = \frac{1}{2}dt\partial_\nu\,(D_{\mu\nu}(0)\theta(t))$.

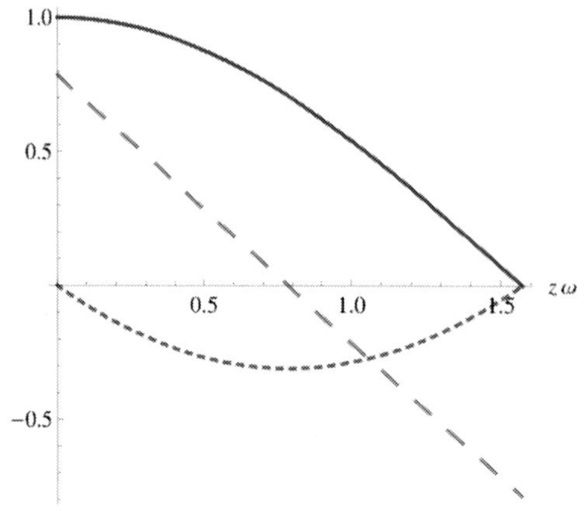

Figure 1. Plots of mean density (solid), mean log-density (small dashes) and vertical mean effective velocity (large dashes).

which can be readily solved with suitable boundary conditions, yielding

$$\langle \chi \rangle = g(z) = \frac{1}{2}\omega z \left(\omega z - \frac{\pi}{2}\right), \tag{14}$$

where $\omega^2 \doteq 2\langle(\nabla\cdot\mathbf{v})^2\rangle/\langle\mathbf{v}^2\rangle \propto \alpha^{-\eta}\,(l/L)^{\xi}\,l^{-2}$ is a compressibility like parameter. The usual definition of compressibility $p = \langle(\nabla\cdot\mathbf{v})^2\rangle/\langle(\nabla\mathbf{v})^2\rangle$ would here give $p = const.$ for $r \ll l$ and $p \propto (l/r)^{2-\xi}$ for $r \gg l$. Moreover, it turns out that the effects of compressibility cancel with contributions from the density fluctuations in e.g. the passive scalar equation[2]. The solution will therefore tend to zero as L approaches infinity. It is perhaps more instructive to consider the mean density rather than the mean log-density. The solution for mean density is

$$\langle \rho \rangle = \rho_0 \cos(\omega z), \tag{15}$$

which leads one to consider the "height" of the boundary layer as $h \doteq \pi/2\omega \propto L^{\xi/2}$. The gradient of the mean log-density (14) implies that the mean effective velocity has a nonzero vertical component, pointing towards the mean height. This implies that passive quantities will be concentrated at height $h/2$. The situation is illustrated in Fig. (1). Clearly this is not a completely realistic density profile, as one might expect since the velocity correlation function does not exhibit any height dependence in this approximation.

The equation for the flux like correlator $\mathcal{F}_\mu(t-t', \mathbf{x}-\mathbf{x}') \doteq \langle \chi(t,\mathbf{x}) v_\mu(t',\mathbf{x}')\rangle$ is

$$\partial_t \mathcal{F}_\mu = -\delta(t-t')\left[D_{\mu\nu,\nu} + g'(z) D_{\mu 3}\right] + \frac{1}{2}\widehat{D}_0 \mathcal{F}_\mu. \tag{16}$$

Since it is the gradient of the log-density that appears e.g. in the passive scalar equation, we take a derivative of the solution, which at leading order as $L \to \infty$ is approximately

$$\langle \partial_\mu \chi(t,\mathbf{x}) v_i(t',\mathbf{x}')\rangle = -\theta(\Delta t) e^{\frac{1}{2}\Delta t \widehat{D}_0} \partial_\mu \left[D_{3i} g'(z) + D_{i\nu,\nu}\right] \tag{17}$$

$$\approx \theta(\Delta t)\left[c^1_{\mu\nu} + c^2_{\mu\nu}(r/L)^2 + c^3_{\mu\nu}\Delta t/L^{2-\xi}\right], \tag{18}$$

where the c^i are some anisotropic unit tensors and $r = |\Delta \mathbf{x}|$. If we further assume that $\Delta t/L^{2-\xi} \gg 1$, we obtain $\langle \chi v_i' \rangle \propto \theta(\Delta t)(\Delta t)^{-(5-\xi)/2}$, i.e. the long time time correlation is a power law. Similar result for the vertical velocity component can be derived but with ξ replaced by η.

For the equal time two point function $G_0(\mathbf{x},\mathbf{x}') \doteq \langle \chi(t,\mathbf{x})\chi(t,\mathbf{x}') \rangle$ we have

$$\frac{1}{2}D_{\mu\nu}(0)\left(\partial_\mu\partial_\nu + 2\partial_\mu\partial'_\nu + \partial'_\mu\partial'_\nu\right)G_0 - d_{\mu\nu}\partial_\mu\partial'_\nu G_0 - D_{\mu\nu,\mu\nu} - \gamma\left(g(z) + g(z')\right) = 0. \tag{19}$$

At leading order the last term vanishes. We can then assume homogeneity, which results also in the vanishing of the first term, yielding the leading order solution (again with gradients of χ)

$$\langle \partial_\mu \chi(t,\mathbf{x}) \partial_\nu \chi(t,\mathbf{x}') \rangle \approx -\partial_\mu\partial_\nu \widehat{d}^{-1} D_{\alpha\beta,\alpha\beta}(\mathbf{x}-\mathbf{x}') \tag{20}$$

$$\approx c^4_{\mu\nu} r^{-2}, \tag{21}$$

with another anisotropic unit tensor and notation $\widehat{d} = d_{\mu\nu}\partial_\mu\partial_\nu = (D_{\mu\nu}(0) - D_{\mu\nu})\,\partial_\mu\partial_\nu$. We've also assumed $r \gg l$, hence the divergence at $r = 0$.

The equation for steady state unequal time two point function $G(t-t',\mathbf{x},\mathbf{x}') \doteq \langle \chi(t,\mathbf{x})\chi(t',\mathbf{x}') \rangle$ is in turn,

$$\partial_t G = \frac{1}{2}\widehat{D}_0 G - \gamma g(z'), \qquad t > t'. \tag{22}$$

The solution, again at leading order, is

$$\langle \partial_\mu \chi(t,\mathbf{x}) \partial_\nu \chi(t',\mathbf{x}') \rangle \approx -e^{\frac{1}{2}|\Delta t|\widehat{D}_0} \partial_\mu\partial_\nu G_0 \tag{23}$$

$$\approx c^5_{\mu\nu} L^{-\xi} \frac{1}{|\Delta t|} \tag{24}$$

with another anisotropic unit tensor.

4. Concluding remarks and future perspectives

The next step in the process would naturally be to consider e.g. the passive scalar and tracer or inertial particles in a flow defined by the above model. From the approximative expressions above, one can already gain some insight into some properties of these passive quantities. First, since there is now a mean effective velocity field pointing toward the height $h/2$, this is where one may expect the particles and passive scalar fields to be concentrated. Second, it was demonstrated that the log-density gradient field exhibits a time-like correlation, as opposed to the white noise in time correlation of the Kraichnan model. The effect of finite time correlation on clustering of inertial particles can be understood intuitively as follows: the orbits of a turbulent (incompressible) flow consist of closed, nonintersecting curves around the eddies, along which tracer particles would traverse. Inertial particles however deviate from these orbits due to centrifugal forces, and therefore tend to cluster on the boundaries of eddies. A delta correlation in time, such as in the Kraichnan model, corresponds to eddies changing so rapidly in time that the inertial particles simply do not have enough time to cluster on the boundaries of eddies. It should be noted however that clustering phenomena are possible also for decorrelated, compressible models (see e.g. [3] and references therein).

It is therefore foreseeable that such clustering occurs for this model, simply due to the anisotropic density fluctuations. Also, a common critique toward extensions of the Kraichnan model to models with time like correlations is that they violate Galilean invariance. Here such a problem does not arise since the time correlation originates from the density field in the mass continuity equation, which of course is Galilean invariant by construction.

The extent to which the model can be considered a realistic description of stably stratified turbulent flows is afflicted by the same pathologies as the Kraichnan model, namely time decorrelation, gaussianity and (therefore) lack of intermittency. Although the presence of density fluctuations introduces an effective time correlation, the velocity field itself is still decorrelated. At large spatial and timelike scales the time decorrelation is a decent approximation, but gaussianity really is not. However, as evidenced by the Kraichnan passive scalar, such non-intermittent models may still successfully describe some interesting behavior of advection and transport in real turbulent flows[5]. The presence of compressibility seemingly poses another challenge, especially from numerical perspective. However, as mentioned above, although we have $\gamma \doteq \langle(\nabla\cdot\mathbf{v})^2\rangle > 0$, it turns out that the effective velocity $\mathbf{u}$ is incompressible, i.e. $\langle(\nabla\cdot\mathbf{u})^2\rangle = 0$ [2].

Acknowledgments
The author would like to thank Antti Kupiainen and Paolo Muratore-Ginanneschi for useful discussions concerning the project. This work was supported by the Magnus Ehrnrooth foundation.

References
[1] H. Arponen (2009). Anomalous scaling and anisotropy in models of passively advected vector fields, Phys. Rev. E 79, pp 056303, doi:10.1103/PhysRevE.79.056303
[2] H. Arponen. Passive scalar in a stably stratified modification of the Kraichnan model (to be published)
[3] J. Bec, M. Cencini, R. Hillerbrand, K. Turitsyn, Stochastic suspensions of heavy particles, Physica D: Nonlinear Phenomena, Volume 237, Issues 14-17, Euler Equations: 250 Years On - Proceedings of an international conference, 15 August 2008, Pages 2037-2050, ISSN 0167-2789, DOI: 10.1016/j.physd.2008.02.022. (http://www.sciencedirect.com/science/article/pii/S0167278908000638)
[4] G. Brethouwer, P. Billant, E. Lindborg and J.-M. Chomaz (2007). Scaling analysis and simulation of strongly stratified turbulent flows. Journal of Fluid Mechanics, 585, pp 343-368, doi:10.1017/S0022112007006854
[5] G. Falkovich, G. Gawędzki, K. and Vergassola, M. (2001) Particles and fields in fluid turbulence. Rev. Mod. Phys., Vol. 73, pp 913-975, doi: 10.1103/RevModPhys.73.913
[6] Robert H. Kraichnan. Anomalous scaling of a randomly advected passive scalar. Phys. Rev. Lett., 72(7):10161019, Feb 1994.
[7] A. Vincent, G. Michaud, and M. Meneguzzi (1996) On the turbulent transport of a passive scalar by anisotropic turbulence. Phys. Fluids 8, 1312, doi:10.1063/1.868912
[8] J. Zinn-Justin (1996), Quantum Field Theory and Critical Phenomena 3rd ed. Oxford University Press

Particle clustering in turbulent premixed flames

Battista F.[1], Picano F.[1], Troiani G.[2], & Casciola C.M.[1]

[1] Dipartimento di Ingegneria Meccanica e Aerospaziale, "La Sapienza" University of Rome, Italy

[2] C.R. ENEA Casaccia, Rome, Italy

E-mail: francesco.picano@uniroma1.it

Abstract. Transport of inertial particles in turbulent reacting flows is frequent in a number of engineering and natural systems. Aim of this work is to illustrate the effect of the fluctuating instantaneous flame front on the particle spatial distribution. To this purpose a Direct Numerical Simulation of a Bunsen premixed flame seeded with small inertial particles is performed. The flamelet Stokes number St_{fl}, defined as the ratio between the particle relaxation time and the flame front time scale, is found to be the proper parameter to characterize the particle dynamics in a premixed flame. Clustering of inertial particles is apparent, especially beyond the flame front. The amount of particle segregation is here quantified by the clustering index and two distinct contributions are found to interplay. The first is independent of the particle inertia and affects also tracers. Actually it is associated to the abrupt variation of the particle concentration induced by the fluid expansion across the flame front. The second effect is mainly due to the time lag associated to the particle inertia that, in proximity of the front, affects both the mean and the fluctuation of the particle number in a fixed volume. The global effect results in an intense clustering of the inertial particles in the flame brush region with a maximum for particles with flamelet Stokes number: $St_{fl} = \mathcal{O}(1)$.

1. Introduction

Particles with different inertia and dimension dispersed in reacting flows are frequently found in many engineering or natural applications, e.g. solid fueled rocket engines or soot formation.

The inertial particle dynamics has been addressed in several incompressible flow configurations, e.g. homogeneous flows, pipe and channel flows, see e.g. [1, 2] for recent reviews. Peculiar transport phenomena have been discussed, such as small-scale clustering [3, 4, 5, 6, 7] which consists of a loss of homogeneity of the particle distribution that leads to the formation of void regions and particle clusters, turbophoresis in the wall bounded flows [8, 9, 10, 11] which amounts to a mean drift of the particles towards the wall and preferential accumulations in free jets [12, 13, 14, 15].

The inertial particle spatial distribution in presence of temperature fluctuations is addressed in [16]. The turbulent thermal diffusion in stably stratified flows induces a non uniform mean particle distribution which presents preferential segregation near the minimum of the mean temperature. The authors demonstrate that the *tangling* clustering occurring in presence of thermal fluctuations, via buoyancy effects, is more effective than the clustering observed in isothermal conditions.

In this framework, as far as the authors know, the interaction between inertial particle dynamics and the large temperature variations typical of reactive flows has not been investigated in details. This issue is relevant for particle collisions and coalescence in reactive flows, e.g. in problems concerning soot growth.

In turbulent premixed combustion, new phenomenologies arises, e.g. interactions with thermal expansion and intense temperature gradients. As a matter of fact, the flame front induces sudden fluid accelerations that could be not accommodated by finite inertia particles.

Aim of the present work is to discuss the effects induced by the simultaneous presence of finite inertia and flame front fluctuations by addressing a DNS of premixed turbulent flame endowed with inertial particles. In order to isolate the effects of inertia from concurrent effects, thermophoresis and gravitational forces are neglected. The main effect we observe is the clustering of the particles in the region spanned by the instantaneous turbulent flame front, the so called flame brush region. Clustering is quantified in terms of clustering index, see e.g. [17]. Also particles with negligible inertia present the intermittent spatial distribution typical of clustering. This feature is explained by a simple model based on the Bray-Moss-Libby assumptions for flamelets [18]. Additional intermittency is found for particle with finite inertia which present an increased clustering especially in flame brush.

2. The model for inertial particle dynamics

The basic model to address inertial particle transport consist of a very diluted suspension of small spherical particles is here considered where inter-particle collisions and force feedback on the fluid are negligible. The particle mass density is assumed much larger than the fluid one, $\rho_p/\rho_f \gg 1$ so that the particle dynamic is forced by the only Stokes drag [19, 20],

$$\begin{cases} \dfrac{d\boldsymbol{x}_p}{dt} = \boldsymbol{v} \\[2ex] \dfrac{d\boldsymbol{v}}{dt} = \dfrac{\boldsymbol{u}(\boldsymbol{x},t)|_{\boldsymbol{x}=\boldsymbol{x}_p} - \boldsymbol{v}}{\tau_p} \end{cases} \tag{1}$$

where $\boldsymbol{x}_p$ and $\boldsymbol{v}$ are the particle position and velocity, respectively, while $\boldsymbol{u}(\boldsymbol{x},t)|_{\boldsymbol{x}=\boldsymbol{x}_p}$ is the fluid velocity at particle position. $\tau_p = d_p^2/(18\,\nu)(\rho_p/\rho_f)$ is the particle relaxation time quantifying the time lag of the particle in following fluid velocity fluctuations (here d_p is the particle diameter, ρ_p and ρ_f the particle and fluid density, respectively, and ν the fluid kinematic viscosity). We neglect gravity and the thermophoretic force which may affect the motion of small particles in reacting flows. Thermophoresis acts on very small particles (not considered in the present study) producing a drift towards the reactants. Actually, it is induced by unbalanced Brownian collisions across the steep temperature gradients of the flame front.

The proper parameter controlling particle dynamics is a suitably defined Stokes number, ratio of particle relaxation time to the relevant time scale of the flow. In turbulent premixed combustion the strong fluid acceleration due to the abrupt thermal expansion across the flame front provides the smallest fluid dynamics time scale, $\tau_{fl} = \delta_{fl}/\Delta u_{fl}$, where δ_{fl} is the thermal thickness of the laminar flame and Δu_{fl} the velocity jump across the front. The thermal thickness can be estimated as $\delta_{fl} = (T_b - T_u)/|\nabla T|_{sup}$, where T_b and T_u are the temperature of burned and unburned mixtures respectively, with $|\nabla T|_{sup}$ the norm of maximum temperature gradient. Across the flame front, thermal expansion drives a velocity jump $\Delta u_{fl} = S_{fl}\,(T_b/T_u - 1)$, with S_{fl} the laminar flame speed. In this framework the Stokes number controlling particle dynamics reads [21]

$$St_{fl} = \frac{\tau_p\, S_{fl}\,(T_b/T_u - 1)}{\delta_{fl}} \,. \tag{2}$$

We stress that the flamelet Stokes number defined above depends on the particle relaxation time τ_p and on the thermo-chemical properties of the flame, namely the temperature ratio T_b/T_u and

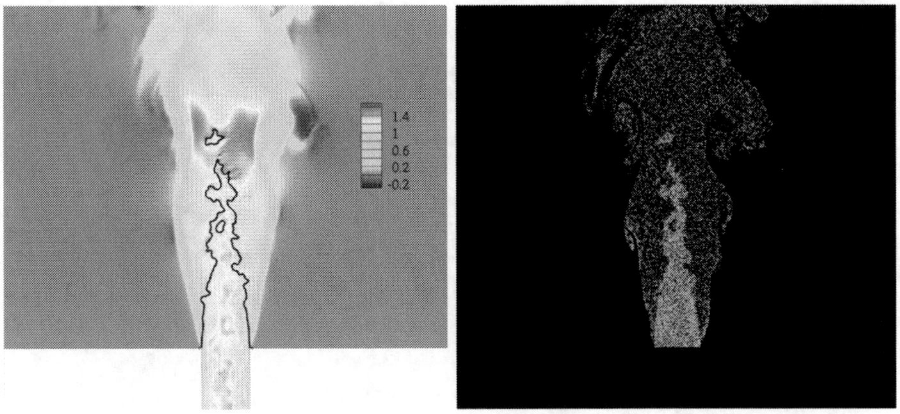

Figure 1. Right panel: instantaneous field of axial velocity fluid with contours and iso-level of reactants at $Y_R/Y_R^0 = 0.5$. Right panel: lighter particles $St_{fl} = 0.022$ instantaneous position.

the flamelet speed S_{fl} and thickness δ_{fl}. It should be remarked that usually τ_{fl} is smaller than the Kolmogorov time scale τ_η, e.g. here $\tau_\eta \simeq 2\,\tau_{fl}$.

3. Methodology

The problem is addressed by means of an Eulerian Direct Numerical Simulation (DNS) of a turbulent Bunsen burner coupled with a Lagrangian solver for particle evolution. The Eulerian algorithm discretizes the Low-Mach number formulation of the Navier-Stokes equations [22] in a cylindrical domain, which describes a low Mach number flow with arbitrary flame-induced density variations neglecting acoustics effects,

$$\frac{\partial \rho}{\partial t} + \boldsymbol{\nabla}\cdot(\rho \boldsymbol{u}) = 0 \tag{3}$$

$$\frac{\partial \rho \boldsymbol{u}}{\partial t} + \boldsymbol{\nabla}\cdot(\rho \boldsymbol{u}\boldsymbol{u}) = \frac{1}{Re}\boldsymbol{\nabla}\cdot\left[\mu\left(\boldsymbol{\nabla}\boldsymbol{u} + \boldsymbol{\nabla}\boldsymbol{u}^T\right)\right] - \boldsymbol{\nabla}p + \frac{1}{Fr^2}\rho \boldsymbol{f} \tag{4}$$

$$\frac{\partial \rho Y_R}{\partial t} + \boldsymbol{\nabla}\cdot(\rho Y_R \boldsymbol{u}) = \frac{1}{Re Sc_R}\boldsymbol{\nabla}\cdot(\mu \boldsymbol{\nabla} Y_R) - \omega_R \tag{5}$$

$$\boldsymbol{\nabla}\cdot\boldsymbol{u} = \frac{1}{\mathcal{P}}\left[\frac{1}{Re Pr}\boldsymbol{\nabla}\cdot(\mu\boldsymbol{\nabla}T) + \frac{\gamma-1}{\gamma} Ce\,\omega_R\right] \tag{6}$$

$$T = \frac{\mathcal{P}}{\rho} \tag{7}$$

with ρ, $\mathbf{u}$ and p the density, the velocity and the dynamic pressure, respectively- T, $\mathcal{P}$ Y_R and ω_R are the temperature, the thermodynamic pressure, the reactant concentration and its global reaction rate, respectively. μ is the dynamic viscosity of the mixture, $Re = \rho_0 U_0 L_0/\mu_0$ is the Reynolds number and $Pr = \mu_0/(\rho_0 \alpha)$ is the Prandtl number giving the ratio between thermal and viscosity of the mixture (here $Pr = 0.6$). The ratio between the heat capacity coefficient is denoted by γ (here $\gamma = 1.3$) and the ratio between mass and momentum transport reactant coefficient is given by the Smidth number $Sc_R = \mu_0/(\rho_0 D_\alpha)$ (here $Sc_R = 0.6$). The chemical kinetics is given by a simple one-step irreversible reaction transforming premixed fresh mixture R into combustion products P, with an Arrhenius model for the reaction rate: $\omega_R = Da\,(\rho Y_R)e^{T/T_a}$ (Da the nominal Damkohler number, T_a the activation temperature, Ce the non dimensional

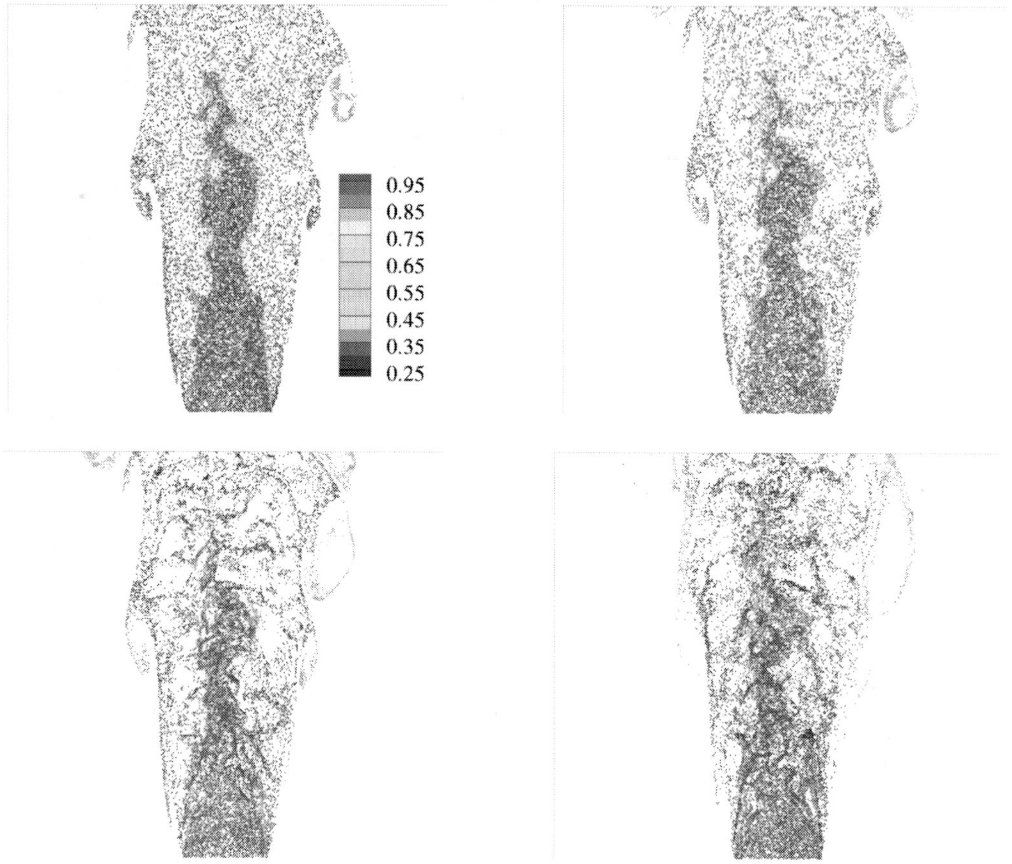

Figure 2. Thin slice of width $R/20$ in the axial-radial plane of the instantaneous particle configuration for different Stokes time. Top-left: $St_{fl} = 0.022$; Top-right: $St_{fl} = 0.54$; Bottom-left $St_{fl} = 2.16$; Bottom-right $St_{fl} = 8.65$. Colors denote the density of the fluid ρ/ρ_u at the particle position.

heat release). The model is tuned to reproduce a lean premixed methane-air Bunsen flame with equivalence ratio $\phi \sim 0.7$ ($T_b/T_u = 5.3$ imposed by the parameters $T_a/T_u = 30$, $Da = 45000$, $Ce = 18.5$), with temperature dependent viscosity according to Sutherland's law, $\mu \propto T^{1/2}$. The ratio of laminar flame speed to the bulk velocity at jet exit is $S_{fl}/U_0 \simeq 0.05$, with laminar flame-thickness $\delta_{fl} \simeq 0.019\,D$.

Spatial discretization is based on central second order finite differences in conservative form on a staggered grid. The convective terms of the relevant reaction-advection-diffusion equations for the scalars are discretized by a bounded central difference scheme designed to avoid spurious oscillations [23]. This non-linear scheme imposes the boundness property characteristic of the convection terms without introducing a large extra-dissipation as in classical TVD schemes. Temporal evolution is performed by a low-storage third order Runge-Kutta scheme [24].

Velocity is prescribed (Dirichlet condition) at the inflow section, by using a cross-sectional plane of a periodic turbulent pipe flow, provided by a companion time-evolving DNS. A convective outflow condition is adopted together with a traction-free condition at the side boundary which mimics the open environment surrounding the flame. The DNS reproduces a premixed Bunsen flame with Reynolds number based on diameter D equal to $Re_D = U_0 D/\nu_\infty = 6000$, with U_0 the bulk velocity, ν_∞ the kinematic viscosity of the cold flow.

The computational domain, $[\theta_{max} \times R_{max} \times Z_{max}] = [2\pi \times 12.34 \times 14]R$, is discretized by $N_\theta \times N_r \times N_z = 128 \times 201 \times 560$ nodes with radial mesh stretching for accurate resolution of the shear layer. The grid size in the jet region is about two/three times the Kolmogorov scale allowing also for an accurate description of the instantaneous flame front. More details on the code can be found in [15, 14] for incompressible jets and in [21] for the reactive case.

Particles evolve by a Lagrangian tracking method which integrates equation (1) by means of the same Runge-Kutta method used for the fluid phase. Interpolation of fluid velocity at particle positions is done by using mixed second/third order Lagrangian polynomials. Particles are assumed to be diluted and do not counteract on the flow (one-way coupling). In this limit the dynamical effect of inter-particle collisions can be safely neglected [2]. Nonetheless it should be remarked that also results obtained in the one-way coupling regime may be useful to model collision kernels, see e.g. [3]

Four particle populations are considered to mimic a laboratory Bunsen premixed flame seeded with alumina particles ($\rho_p = 4000\ kg/m^3$) with diameters $d_p = 1\mu m$, $d_p = 5\mu m$, $d_p = 10\mu m$, $d_p = 20\mu m$. In the four cases the flamelet Stokes number is respectively $St_{fl} = 0.022$, $St_{fl} = 0.54$, $St_{fl} = 2.16$, $St_{fl} = 8.65$, while the inlet Stokes number based on the Kolmogorov time scale is about half the flamelet Stokes number $St_\eta \simeq St_{fl}/2$.

Particles are introduced in the field at a fixed rate with homogeneous distribution at the jet inlet section with velocity matching the local fluid. About six million particles are evolved in the simulation.

The simulation runs for $30\,D/U_0$ to achieve a statistical steady state before collecting about one hundred fields, separated by $0.125\,D/U_0$, for statistical analysis.

The left panel of figure 1 provides a slice of the instantaneous fluid axial velocity field (contours) and of the position of the flame front localized by the reactant iso-level $Y_R/Y_R^0 = 0.5$ (black solid line). The turbulent pipe flow originates a fully turbulent inflow for the Bunsen jet so that the instantaneous flame front appears highly wrinkled. The right panel of figure 1 reports the instantaneous configuration of the smallest particles ($St_{fl} = 0.022$) where the effect of the abrupt expansion across the flame front appears.

4. Particle clustering

The instantaneous configuration of the four particle populations is presented in figure 2. Colors denote the fluid density at particle positions and allow to detect the flame front. The huge variations in the particle concentration across the flame front is induced by the sudden expansion due to heat release. According to their inertia, the particles can differently follow the abrupt acceleration experienced by the fluid. Actually for larger particles, the effect is not restricted just beyond the flame front, but the particle distribution is influenced well into the burned gas region.

The clustering, apparent in figure 2, takes place at smaller and smaller scales as the Stokes number is reduced. The clusters in the burned region are particularly well defined for intermediate weight particles, $St_{fl} = 0.54$ and 2.16, with the typical length scale of the voids large enough to be clearly detected. The lightest particles are more evenly distributed both in the burned and unburned regions recovering a tracer-like behavior characterized by a sharp jump of the particle density across the front.

Particle segregation can be measured by the deviation of the actual particle distribution with respect to randomly distributed independent positions. These reference conditions are expressed by the Poisson distribution giving the probability $p(n)$ to find n particles in a certain domain Ω of volume ΔV with characteristic length $\ell = \sqrt[3]{\Delta V}$. Once the average particle number $\bar{n}$ is known, the probability density function (pdf) reads:

$$P(n, \bar{n}) = \frac{\bar{n}^n e^{-\bar{n}}}{n!}. \tag{8}$$

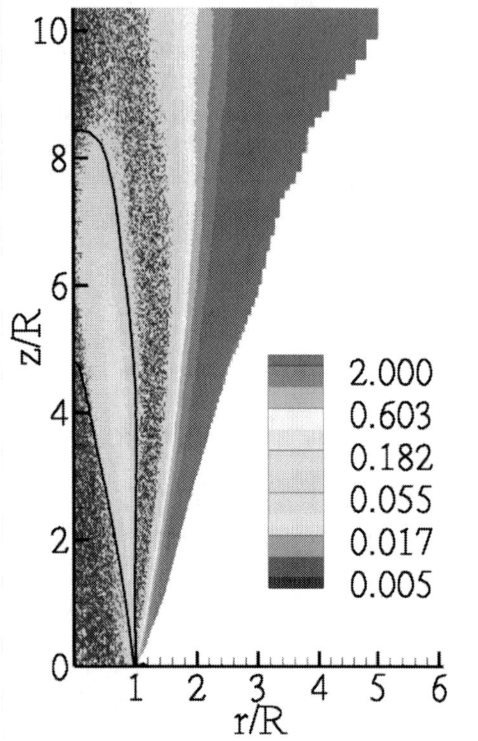

Figure 3. Clustering Index field K (contour) for the smallest particles ($St_{fl} = 0.022$). The control volumes correspond to the computational cells allowing to exploit the azimuthal homogeneity for the statistics, so the typical dimension becomes $\ell \propto r^{1/3}$. The solid lines indicate the boundaries of the flame brush, here identified by the isolevels of the mean progress variable $\bar{c} = 1 - Y_R/Y_R^0 = 0.05$ and $\bar{c} = 1 - Y_R/Y_R^0 = 0.95$.

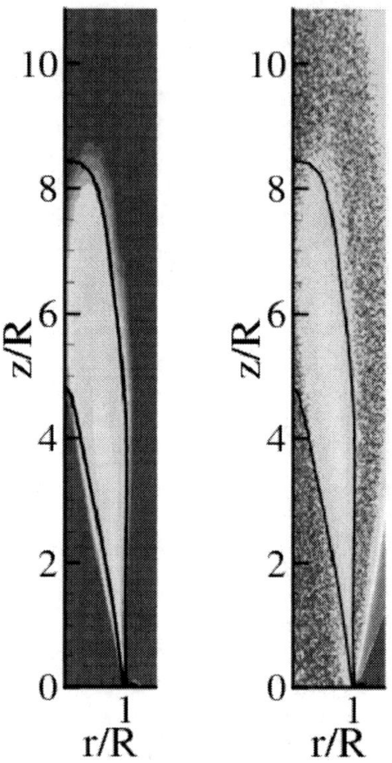

Figure 4. Clustering index K: model (13), left panel, and numerical results, right panel. The control volumes correspond to the computational cells allowing to exploit the azimuthal homogeneity for the statistics, so the typical dimension becomes $\ell \propto r^{1/3}$. The solid lines indicate the boundaries of the flame brush, here identified by the isolevels of the mean progress variable $\bar{c} = 1 - Y_R/Y_R^0 = 0.05$ and $\bar{c} = 1 - Y_R/Y_R^0 = 0.95$.

Since the variance of Poissonian processes $\sigma_n^2 = \overline{(n - \bar{n})^2} = \overline{(\delta n)^2}$ equals the mean value $\bar{n}$ a measure of particle clustering is provided by the clustering index K, see e.g. [17], which is based on the departure from the Poissonian distribution,

$$K = \frac{\overline{(\delta n)^2}}{\bar{n}} - 1. \tag{9}$$

Obviously a Poissonian distribution implies $K = 0$, due to the absence of any clustering or preferential localization effect. A negative clustering index $K < 0$ instead corresponds to a more uniformly distributed particle configuration characterized by a smaller ratio of the variance with respect to the mean value. On the contrary, if particles preferentially aggregate in clusters, the variance exceeds the mean value, yielding a positive clustering index, $K > 0$. Since the clustering index depends on the length scale ℓ, $K(\ell)$, via $\bar{n}(\ell)$ and $\overline{(\delta n)^2}(\ell)$, all the above features are scale dependent. Additionally for inhomogeneous processes the clustering index becomes a field since it depends on the position of the control domain $\Omega(\boldsymbol{x})$, $K = K(\boldsymbol{x}, \ell)$.

The clustering index field can be exploited to quantify the amount of particle segregation induced by the fluctuating turbulent flame front. An example is reported in figure 3 for particles with flamelet Stokes number 0.022 and control volumes corresponding to the computational mesh cells. The region spanned by the flame front is called the flame brush. It can be defined in terms of the average progress variable $\bar{c}$, where the instantaneous progress variable $c = 1 - Y_R/Y_R^0$ ranges from 0 in the reactants to 1 in the products. The flame brush is then the region where $0 < \bar{c} < 1$ whose boundaries are shown as thick lines in figure 3. The clustering index K, shown by colors in the same figure, is positive in the flame brush and in the outer region of the jet. In the outer region the huge value of the clustering index is induced by the intermittency of the boundary between the hot jet seeded with particles and the external, particle-free, cold environment. Such process leads to a particle distribution characterized by a large value of K. More interesting for our purpose is the flame brush region which is spanned by the instantaneous flame front. Away from these two regions, the clustering index almost vanishes suggesting an almost purely Poissonian arrangement of these quasi-Lagrangian particles at $St_{fl} \ll 1$.

As shown in the figure, the clustering observed in the flame brush is remarkable also for very light particles. To try and explain this effect of the flame we resort to the Bray-Moss-Libby (BML) description of the flamelet [18] and to the assumption of a Poissonian distribution for the particles in the two nearby regions, burned and unburned gases respectively. We remark that in order to apply the BML formalism, the control volume where we count the particle number should contain only a single state of the mixture, either burned or unburned gas. These assumptions, which are exact for tracers, are reasonable also for sufficiently small particles ($St_{fl} \ll 1$) whenever the instantaneous flame front is thin.

In the flamelet context the state of the mixture is either $c = 0$ (unburned) or $c = 1$ (burned), and the probability distribution is $p(\boldsymbol{x}, c) = [\alpha(\boldsymbol{x})\delta(c) + \beta(\boldsymbol{x})\delta(1-c)]$, where α and β are the probability to find the unburned or burned gases at $\boldsymbol{x}$. By evaluating $\bar{c}$ from the above pdf it follows $\beta(\boldsymbol{x}) = 1 - \alpha(\boldsymbol{x}) = \bar{c}(\boldsymbol{x})$, see [18, 25]. Let us now address the joint probability distribution of particle number in the volume V and instantaneous progress variable c, $p(\boldsymbol{x}; c, n) = p(\boldsymbol{x}, c)\, p(n|c)$. We shall assume that the conditional probability of the particle number conditioned to the state c of the mixture is Poissonian both for the burned and unburned state. In these conditions the conditional probability is entirely determined by the average particle number conditioned to the state of the mixture, $\langle n|c\rangle$, $p(n|c) = P(n; \langle n|c\rangle)$ where P is the Poissonian distribution, eq. 8. It follows

$$p(\boldsymbol{x}; c, n) = \qquad (10)$$
$$\{[1 - \bar{c}(\boldsymbol{x})]\, \delta(c) + \bar{c}(\boldsymbol{x})\delta(1-c)\}\ P(n; \langle n|c\rangle).$$

Starting from 10 we can express the first and second order moments of the particle number pdf,

$$\bar{n}(\boldsymbol{x}) = [1 - \bar{c}(\boldsymbol{x})]\overline{n_u} + \bar{c}(\boldsymbol{x})\overline{n_b} \qquad (11)$$

$$\overline{n^2}(\boldsymbol{x}) = [1 - \bar{c}(\boldsymbol{x})]\overline{n_u^2} + \bar{c}(\boldsymbol{x})\overline{n_b^2} \qquad (12)$$

where $\bar{n}_{u/b}$ is the mean particle number in the volume V corresponding to fully unburned and burned conditions. Under our assumptions these two quantities are strictly related to the density a purely Lagrangian tracer would have in corresponding conditions. It is than clear that, according to our model, the ratio $\bar{n}_u/\bar{n}_b$ equals the gas density ratio across the thin flame front which is estimated by the expansion ratio $\tau = T_b/T_u$, $\bar{n}_u/\bar{n}_b = \tau$. Considering equations (9), (11) and (12), the clustering index reads,

$$K_0(\boldsymbol{x}) = \frac{\overline{n}_u}{\tau}\frac{\bar{c}(\boldsymbol{x})[1 - \bar{c}(\boldsymbol{x})]\,(\tau - 1)^2}{\bar{c}(\boldsymbol{x}) + \tau[1 - \bar{c}(\boldsymbol{x})]}. \qquad (13)$$

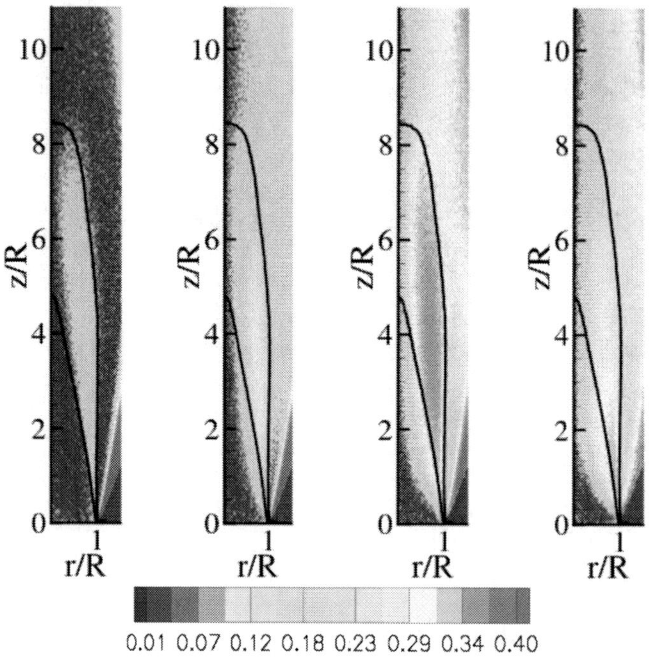

Figure 5. Contours of the Cluster Index for all particle populations; from left: $St_{fl} = 0.022$, $St_{fl} = 0.54$, $St_{fl} = 2.16$ and $St_{fl} = 8.65$, respectively. The control volumes correspond to the computational mesh cells. The lines indicate two isolevels of the mean progress variable $\bar{c} = 1 - Y_R/Y_R^0 = 0.05$ and $\bar{c} = 1 - Y_R/Y_R^0 = 0.95$.

Hence the clustering index deviates from zero only in the flame brush where $0 < \bar{c} < 1$, to vanish in the burned and unburned region where the statistics comes back to Poissonian.

Relation (13) highlights that for light particles, a positive clustering index, $K_0 > 0$, is mainly due to the fluctuation of the thin flame front. A Poissonian distribution is obviously recovered in the limit of $\tau \to 1$. The model proposed for the flame brush region can be easily extended to model the behavior of the particles in the intermittent region separating the burned gas and the external particle-free environment. To this purpose we shall assume that the instantaneous interface between hot gas and environment is sufficiently thin, as expected for high Reynolds number flows. Across the instantaneous interface the average particle number jumps from $\bar{n}_b$ to $\bar{n}_e = 0$. Denoting by $\tau_e = \bar{n}_b/\bar{n}_e \to \infty$, the empty external region is formally characterized by a limiting Poissonian process with vanishing average particle number. According to the model the clustering index pertaining to this outer part of the jet is described by relation (13) in the limit $\tau \to \infty$ with $\bar{n}_u$ replaced by $\bar{n}_b$, i.e. $K_0 = \bar{n}_b \bar{c}_e(\boldsymbol{x})$. In full analogy with the flame brush, the instantaneous indicator function $c_e(\boldsymbol{x})$ has a singular bi-modal distribution taking the value 0 in the fully burned region and 1 in the external fluid while its average $\bar{c}_e$ goes smoothly from zero to one. The huge value of K, seen in figure 3, in the outer part of the jet can actually be interpreted in this framework, after noting that the relevant Stokes number should be interpreted in terms of the local turbulent time scale which is typically larger than the flamelet characteristic time.

Focusing on the flame brush region, we show in figure 4 the capability of the model to excellently reproduce the numerical results by comparing the DNS data for the smallest particles ($St_{fl} = 0.022$) with the model (13). The vanishing of the clustering index on the axis both in the model and in the DNS is explained by taking in consideration the radial variation of the

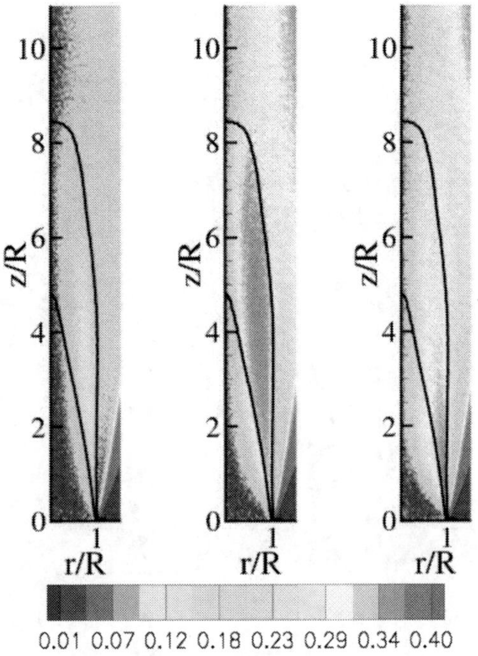

Figure 6. Estimation of the clustering index K using the right hand side of equation (15). From left to right: data concerning particles with $St_{fl} = 0.54$, $St_{fl} = 2.16$ and $St_{fl} = 8.65$, respectively.

coarsening length, which following the grid size is proportional to the cubic root of the radial coordinate, $\ell \propto r^{1/3}$. Consistently the average particle number $\bar{n}_u$ drops to zero towards the axis entailing the vanishing of the clustering index.

Relation (13) and figure 4 show that the quasi-Lagrangian particles, whose inertia is negligible, present a positive clustering index in the flame brush region. This phenomenon suggests that a positive clustering index is induced not only by the interaction between particle inertia and flame front fluctuations or turbulence, but also by the mere dynamics of the flame front that induces a strong decrease in the particle concentration due to the fluid thermal expansion. This peculiar behavior promotes a substantial instantaneous fluctuation of the particle concentration with respect to the local average, which is detected by the clustering index.

The influence of the finite inertia on flame-induced particle clustering is addressed in figure 5 where the clustering index is reported for each particle population. Two features are immediately apparent, the increase of the clustering intensity with increasing inertia and the extension of the clustering region towards the burned gas well outside the flame brush. For the range of parameters we have investigated, the maximum deviation always occurs in the flame brush and we find an overall maximum clustering intensity for particles with $St_{fl} = 2.16$. The observation that peak clustering occurs at flamelet Stokes numbers order one confirms that this is actually the parameter controlling particle localization effects in turbulent premixed flames.

Since inertia increases the clustering index over the already positive value it has for tracers one needs to carefully separate the two effects. To this purpose it is convenient to define the

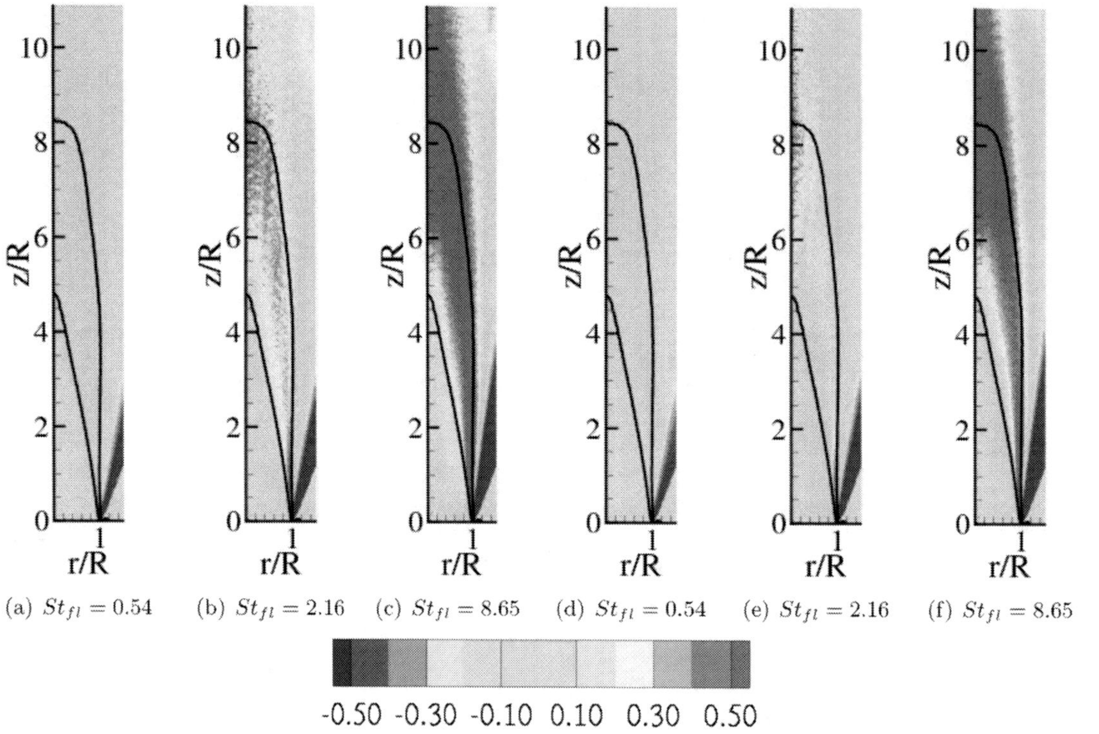

(a) $St_{fl} = 0.54$ (b) $St_{fl} = 2.16$ (c) $St_{fl} = 8.65$ (d) $St_{fl} = 0.54$ (e) $St_{fl} = 2.16$ (f) $St_{fl} = 8.65$

Figure 7. Panels (a)-(b)-(c) represent $\Delta\overline{\delta n^2}/\bar{n}_0$; panels (d)-(e)-(f) represent $(K_0 + 1)\Delta\bar{n}/\bar{n}_0$. See equation (15) and the related explanation for more details.

mean particle number and its variance as,

$$\begin{aligned} \bar{n} &= \bar{n}_0 + \Delta\bar{n} \\ \overline{\delta n^2} &= \overline{\delta n_0^2} + \Delta\overline{\delta n^2} \end{aligned} \tag{14}$$

where the subscript 0 is referred to tracer-like particles ($St \to 0$) and the two additional terms on right-hand-side quantify the correction due to departure from the reference state of tracers. Inserting relations (14) in definition (9), to first order in the corrections the Clustering index $K(\ell)$ reads,

$$K(\ell) \simeq K_0 + \frac{\Delta\overline{\delta n^2}}{\bar{n}_0} - (K_0 + 1)\frac{\Delta\bar{n}}{\bar{n}_0} \tag{15}$$

where $K_0 = \overline{\delta n_0^2}/\bar{n}_0 - 1$ is the Clustering index for tracers, eq. (13). In these approximations the effect of inertia are condenced to the last two terms in right-hand-side of equation (15). The first term, $\Delta\overline{\delta n^2}/\bar{n}_0$, takes care of the inertia-induced fluctuation of the particle number in the control volumes. The second term, $-(K_0 + 1)\Delta\bar{n}/\bar{n}_0$, accounts for the variation of the mean particle number. In fact inertia prevents the particles from quickly responding to the gas expansion implying that their mean number in a fixed volume exceeds that of tracers.
The right-hand-side of equation (15) is represented in figure 6. The excellent agreement between the approximate clustering index estimated by equation (15) and the exact one reported in figure 5 4demonstrates the reliability of the approximations. Figure 7 reports $\Delta\overline{\delta n^2}/\bar{n}_0$ and

$(K_0+1)\Delta\bar{n}/\bar{n}_0$ for particles with $St_{fl} = 0.54; 2.16; 8.65$. Concerning the most segregating particles, $St_{fl} = 2.16$, we observe that $\Delta\overline{\delta n^2}/\bar{n}_0$ is large in the flame brush region, especially near the burned gas side. We infer that the interaction between the corrugated flame front and the particle inertia is a source of particle clustering via an increase of the particle number fluctuation. Since, though small, $(K_0+1)\Delta\bar{n}/\bar{n}_0$ is positive it tends to mitigate the clustering. For this kind of particles ($St_{fl} = 2.16$), $\Delta\overline{\delta n^2}/\bar{n}_0$ is larger than $(K_0+1)\Delta\bar{n}/\bar{n}_0$, resulting in a quite large clustering in the flame brush. For particles with larger inertia, $St_{fl} = 8.65$, both terms increase. In particular $(K_0+1)\Delta\bar{n}/\bar{n}_0$ becomes comparable with $\Delta\overline{\delta n^2}/\bar{n}_0$ resulting in a less intense clustering than before. It is clear that a maximum clustering level should be achieved at intermediate Stokes numbers. The appropriate selection of the physically relevant time scale, the flamelet time scale, normalizes the Stokes number to be order one.

5. Final Remarks

Several issues concerning the behavior of inertial particles in a turbulent reacting jet has been discussed. We found that the flamelet Stokes number which compares particle relaxation time and flame front characteristic time, controls the particle dynamics, see also [21].

Inertial particles aggregate in clusters which are especially evident in the hot gas region. The clustering index was exploited to quantify the amount of clustering as a function of particle inertia.

Since the gas expands the concentration of Lagrangian tracers changes abruptly across the front, in these conditions the front oscillations produce local fluctuations in the tracer concentration which result in a positive clustering index in the flame brush region. This effect is well described by a model based on the Bray-Moss-Libby formalism for flamelets (negligible thickness of the flame front).

Inertia leads to the increase of clustering with respect to this base level. The increase is due to two contributions operating in opposite directions. One is due to the increment of the particle number fluctuation induced by the interaction between fluctuating wrinkled flame front and the particle inertia. This contribution increases with inertia. The other is associated with the increment of the mean particle concentration in the flame brush. Actually inertial particles respond slowly to the expansion of the flow, resulting in a more gentle reduction of the mean particle concentration across the front. This term operates to reduce the clustering.

The competition between the two effects gives rise to a maximum clustering intensity for particle with $St_{fl} = 2.16$. For larger inertia the two effects become comparable resulting in a less intense clustering. The fact that maximum clustering is achieved for $St_{fl} = \mathcal{O}(1)$ confirms that the flow time scale controlling the particle segregation in the flame is the flamelet time scale $\tau_{fl} = \delta_{fl}/\Delta u_{fl}$, where δ_{fl} is the flame thickness and Δu_{fl} is the jump velocity across the instantaneous flame front.

6. Acknowledgements

We want to thank the COST Action MP0806 Particles in Turbulence for supporting the present work.

7. References

[1] Toschi F and Bodenschatz E 2009 *Annual Review of Fluid Mechanics* **41** 375–404
[2] Balachandar S and Eaton J 2010 *Annual Review of Fluid Mechanics* **42** 111–133 ISSN 0066-4189
[3] Sundaram S and Collins L 1997 *Journal of Fluid Mechanics* **335** 75–109 ISSN 0022-1120
[4] Balkovsky R, Falkovich G and Fouxon A 2001 *Physical Review Letters* **86** 2790
[5] Bec J, Biferale L, Cencini M, Lanotte A, Musacchio S and Toschi F 2007 *Physical Review Letters* **98** 084502
[6] Gualtieri P, Picano F and Casciola C 2009 *J. Fluid Mech.* **629** 25
[7] Coleman S and Vassilicos J 2009 *Physics of Fluids* **21** 113301

[8] Reeks M 1983 *J. Aerosol Sci.* **14** 729–739
[9] Rouson D and Eaton J 2001 *J. Fluid Mech.* **428** 149
[10] Marchioli C, Soldati A, Kuerten J, Arcen B, Taniere A, Goldensoph G, Squires K, Cargnelutti M and Portela L 2008 *International Journal of Multiphase Flow* **34** 879–893
[11] Picano F, Sardina G and Casciola C 2009 *Phys. Fluids* **21** 093305
[12] Hardalupas Y, Taylor A and Whitelaw J 1989 *Proceedings of the Royal Society of London. A. Mathematical and Physical Sciences* **426** 31 ISSN 1364-5021
[13] Longmire E and Eaton J 1992 *Journal of Fluid Mechanics* **236** 217–257
[14] Picano F, Sardina G, Gualtieri P and Casciola C 2010 *Physics of Fluids* **22** 051705
[15] Casciola C, Gualtieri P, Picano F, Sardina G and Troiani G 2010 *Physica Scripta* **T142** 014001
[16] Eidelman A, Elperin T, Kleeorin N, Melnik B and Rogachevskii I 2010 *Physical Review E* **81** 56313
[17] Kostinski A and Shaw R 2001 *J. Fluid Mech.* **434** 389–398
[18] Bray K, Libby P A and Moss J 1985 *Combustion and Flame* **61** 87 – 102
[19] Maxey M and Riley J 1983 *Physics of Fluids* **26** 883
[20] Stella A, Guj G, Kompenhans J, Raffel M and Richard H 2001 *Experiments in Fluids* **30** 167–180
[21] Picano F, Battista F, Troiani G and Casciola C 2011 *Experiments in Fluids* **50**(1) 75–88
[22] Majda A and Sethian J 1985 *Combustion Sci. and Tech.* **42** 185
[23] Waterson N and Deconinck H 2007 *J. Computational Physics* **224** 182–207
[24] Lundbladh A, Henningson D and Johansson A 1992 Tech. rep. FFA-TN 1992-28,KTH, Stockholm, Sweden
[25] Troiani G, Marrocco M, Giammartini S and Casciola C 2009 *Comb. and Flame* **156** 608–620

Spatial and velocity statistics of inertial particles in turbulent flows

J. Bec[1], L. Biferale[2,5], M. Cencini[3], A.S. Lanotte[4], F. Toschi[5,6]
[1]Université de Nice-Sophia Antipolis, CNRS, OCA, Lab. Cassiopée, B.P. 4229, 06300 Nice, France
[2]Dept. Physics and INFN, University of Rome "Tor Vergata", Via della Ricerca Scientifica 1, 00133 Rome, Italy
[3]CNR-ISC, Via dei Taurini 19, 00185 Rome, Italy
[4]CNR-ISAC & INFN-Sez. di Lecce, Str. Prov. Lecce-Monteroni, 73100 Lecce, Italy
[5]Department of Physics Eindhoven University of Technology, 5600 MB Eindhoven, The Netherlands
[6]Department of Mathematics and Computer Science, Eindhoven University of Technology, 5600 MB Eindhoven, The Netherlands and CNR-IAC, Via dei Taurini 19, 00185 Rome, Italy

Abstract. Spatial and velocity statistics of heavy point-like particles in incompressible, homogeneous, and isotropic three-dimensional turbulence is studied by means of direct numerical simulations at two values of the Taylor-scale Reynolds number $Re_\lambda \sim 200$ and $Re_\lambda \sim 400$, corresponding to resolutions of 512^3 and 2048^3 grid points, respectively. Particles Stokes number values range from $St \approx 0.2$ to 70. Stationary small-scale particle distribution is shown to display a singular –multifractal– measure, characterized by a set of generalized fractal dimensions with a strong sensitivity on the Stokes number and a possible, small Reynolds number dependency. Velocity increments between two inertial particles depend on the relative weight between smooth events - where particle velocity is approximately the same of the fluid velocity-, and *caustic* contributions - when two close particles have very different velocities. The latter events lead to a non-differentiable small-scale behaviour for the relative velocity. The relative weight of these two contributions changes at varying the importance of inertia. We show that moments of the velocity difference display a quasi bi-fractal-behavior and that the scaling properties of velocity increments for not too small Stokes number are in good agreement with a recent theoretical prediction made by K. Gustavsson and B. Mehlig *arXiv:1012.1789v1 [physics.flu-dyn]*, connecting the saturation of velocity scaling exponents with the fractal dimension of particle clustering.

1. Introduction

Particle suspensions of dust, droplets, or bubbles advected by incompressible turbulent flows are commonly encountered in many natural phenomena (see e.g. [1–6]). The problem is more complicated than in the case of fluid tracers, i.e. small particles with the same density as the carrier fluid. Suspended particles with a finite size and/or a density ratio different from that of the fluid do not follow exactly the flow because of inertial forces. As a consequence, it is crucial to understand the correlations between particle positions and structures of the underlying flow. For instance, heavy/light particles are expelled/attracted from/into vortical structures, a phenomenum that can be exploited to evaluate temporal properties of coherent turbulent structures [7]. As a result, particle *preferentially concentrate* in specific regions of the flow, and

	N	Re_λ	η	δx	ε	ν	τ_η	$t_{\rm dump}$	δt	T_L
Run I	512	185	0.01	0.012	0.9	0.002	0.047	0.004	0.0004	2.2
Run II	2048	400	0.0026	0.003	0.88	0.00035	0.02	0.00115	0.000115	2.2

Table 1. Eulerian parameters for the Runs analysed here: Run I and Run II in the text. N is the number of grid points in each spatial direction; Re_λ is the Taylor-scale Reynolds number; η is the Kolmogorov dissipative scale; $\delta x = \mathcal{L}/N$ is the grid spacing, with $\mathcal{L} = 2\pi$ denoting the physical size of the numerical domain; $\tau_\eta = (\nu/\varepsilon)^{1/2}$ is the Kolmogorov dissipative time scale; ε is the average rate of energy injection; ν is the kinematic viscosity; $\tau_{\rm dump}$ is the time interval between two successive dumps along particle trajectories; δt is the time step; $T_L = L/U_0$ is the eddy turnover time at the integral scale $L = \pi$, and U_0 is the typical large-scale velocity. Stokes number available from the two simulations are: Run I:(0; 0.16; 0.27; 0.37; 0.48; 0.59; 0.69; 0.80; 0.91; 1.01; 1.12; 1.34; 1.60; 2.03; 2.67; 3.31). Run II: (0.0; 0.16; 0.6; 1.0; 3.0; 5.0; 10; 20; 30; 40; 50; 70)

strong inhomogeneities in the particle spatial distribution develop, depending on the relative importance between inertial forces and turbulent advection.

Recent theoretical [8–10]) and numerical [11–14] work have focused on these aspects. Advances in the understanding of the statistical characterization of small particle aggregates have been obtained by studying heavy particles advected by stochastic flows [15–18], and/or in two-dimensional turbulent flows [19]. Temporal properties along single trajectory have been addressed both numerically and experimentally for small, heavy particles and light particles (see, e.g., [20–28]), and for large particles, where inertial effects combine with finite size ones (see e.g. [29–31]).

In our simulations, particles are organised in different families according to the values of their *Stokes number*, St. The Stokes number is defined as $St = \tau_s/\tau_\eta$, i.e. the ratio between the particle response time τ_s and the flow Kolmogorov timescale $\tau_\eta = (\nu/\varepsilon)^{1/2}$, where ν is the flow kinematic viscosity and ε the average rate of energy injection. The response time $\tau_s = 2a^2\rho_f/(9\nu\rho_p)$ depends on the particle radius a, which is assumed to be much smaller than the Kolmogorov scale η, and the ratio between the fluid ρ_f and particle density ρ_p.

In this paper we present a new data analysis concerning a series of high-resolution Direct Numerical Simulations (DNS). In particular, we intend to discuss (i) non-homogeneous effects in the particle spatial distribution, characterized by the whole set of generalized fractal dimensions $D_q(St)$ at changing both the order of the moment q and the Stokes number St; (ii) the connection between scaling properties of high-order moments of particle-particle relative velocity increments and the (multi)fractal spatial properties. Concerning the latter point, we show that our numerical results agree with a recent theoretical prediction [32], at least for not too small Stokes numbers.

2. Equation of motion and numerical details

We present results from a $3D$ DNS of a turbulent flows seeded with inertial particles. The flow phase is described by the Navier-Stokes equations for the velocity field $\boldsymbol{u}(\boldsymbol{x}, t)$

$$\partial_t \boldsymbol{u} + \boldsymbol{u} \cdot \boldsymbol{\nabla} \boldsymbol{u} = -\frac{1}{\rho_f}\boldsymbol{\nabla} p + \nu \Delta \boldsymbol{u} + \boldsymbol{f}, \quad \boldsymbol{\nabla} \cdot \boldsymbol{u} = 0\,. \tag{1}$$

The statistically homogeneous and isotropic external forcing $\boldsymbol{f}$ injects energy in the first low wave number shells, by keeping constant their spectral content [33]. Kinematic viscosity is chosen such that the Kolmogorov length scale $\eta \approx \delta x$, where δx is the grid spacing: this choice ensures that small-scale velocity dynamics is well resolved. The numerical domain is cubic and 2π-periodic in the three directions of space. We use a fully dealiased pseudospectral algorithm with 2nd order Adam-Bashforth time-stepping (for details see [20, 21]). We performed two series of DNS: Run I with numerical resolution of 512^3 grid points, and the Reynolds number at the Taylor scale $Re_\lambda \approx 200$; Run II with 2048^3 resolution and $Re_\lambda \approx 400$. Details of the runs can be found in Table 1.

The dispersed phase is constituted by millions of heavy particles, with different Stokes number values. Particle sizes are assumed to be much smaller than the Kolmogorov scale of the flow and with a negligible Reynolds number relative to the particle size. These assumptions allow us to consider point particles which evolve according to the dynamics

$$\dot{\boldsymbol{X}} = \boldsymbol{V}\,, \qquad \dot{\boldsymbol{V}} = -\frac{1}{\tau_s}\left[\boldsymbol{V} - \boldsymbol{u}(\boldsymbol{X},t)\right]\,, \tag{2}$$

where the dots denote time derivatives. The particle position and velocity are $(\boldsymbol{X}(t), \boldsymbol{V}(t))$, respectively, while $\boldsymbol{u}(\boldsymbol{X}(t),t)$ is the Eulerian fluid velocity evaluated at the particle position.

Equation (2) has been derived under the assumption of very dilute suspensions, where particle-particle interactions (collisions) and hydrodynamic coupling to the flow can be neglected (see Ref.[34] for a discussion of the complete equation of motion of a small, spherical particle in a non-uniform flow).

For each Stokes number we follow 7×10^6 (Run I) and 10^8 (Run II) particles, for 5 and 2 large-scale eddy-turnover times, respectively. Particles are injected uniformly in the box with initial velocity equal to that of the fluid, and measurements are started only after a steady state statistics is reached (see, e.g., Ref.[21] for details).

3. Mass Distribution

Preferential concentration is responsible for a highly non trivial spatial distribution of particles at changing inertia. Due to its connection with collision kernels, much work has been done to measure and characterize the two-point particle-particle distribution, i.e. the probability to find a pair of particles at a distance $r \ll \eta$, in an ensemble of inertial particle in homogeneous turbulence [11, 13]. The dissipative dynamics of particles is responsible for the convergence of their trajectories towards a dynamically evolving attractor in the position-velocity phase space. As a consequence, the probability density function (PDF) of the inter-particle distances behaves at scales smaller than the Kolmogorov dissipative scale η, as a power law, i.e. $p_2(r) \sim r^{D_2(St)-1}$, where the exponent $D_2(St)$ is usually called the *correlation dimension*. In general, there is no reason to expect a complete statistical self-similarity of the mass distribution, which would give equal generalized dimension $D_q(St)$ for any order q. Thus the full characterization of the particle density field requires the knowledge of the full spectrum of generalized dimensions. This has already been argued theoretically and found numerically for the case of paticles in random flows and/or smooth flows [35, 36]. We present here the first measurements of the generalized fractal dimensions $D_q(St)$ as a function of the Stokes number and give some insight on the large mass fluctuations.

The formal definition of generalized dimensions generally relies on box-counting techniques. For that, let us cover at a fixed time t, for each given St, the particle set with $N_r(St)$ balls of size r. We define the instantaneous local mass distribution $m_r(\boldsymbol{x}_i, t)$ inside the ith ball centered at $\boldsymbol{x}_i$ as the fraction of particles inside this ball. Then, one can fully characterize the particle distribution by studying the behaviour of the moments of order q of m_r at changing the ball radius, r. By definition of Hausdorff dimension we have that $\langle N_r(St)\rangle \sim r^{-D_0(St)}$, where $\langle\cdot\rangle$ denote ensemble averages (equivalent to time averages when ergodicity is assumed). The whole spatial particle distribution can then be characterized by the behaviours of the moments of m_r

for different orders:

$$M_r^{(q)}(St) = \left\langle \sum_{i=1}^{N_r(St)} m_r^q(\boldsymbol{x}_i, t) \right\rangle = \left\langle \overline{m_r^{q-1}} \right\rangle, \tag{3}$$

where the over-bar average is a Lagrangian average; it is made on the attractor so that it is locally weighted with the mass of particles itself. For a multifractal distribution we expect for $r \ll \eta$:

$$M_r^{(q)}(St) \sim r^{(q-1)D_q(St)}. \tag{4}$$

The generalized dimensions $D_q(St)$ fully determines the spatial mass distribution for each Stokes number.

In the stationary regime, for almost all balls centered on the attractor, the local dimension $h = \ln m_r / \ln r$ tend to a time-constant, deterministic value $D_1(St)$, which is usually called the *information dimension*. For finite length scales r, the fluctuations with respect to $D_1(St)$ of the local dimension h on the attractor are described by a large-deviation principle, which implies that they take the form $\overline{p}_r(h; St) \propto r^{f(h;St)}$, where the density probability $\overline{p}_r$ is evaluated on balls centered on the attractor. The function $f(h)$ is generally called the multifractal (or singularity) spectrum. The generalized dimensions are related to it via

$$(q-1)\, D_q(St) = \inf_h \left[(q-1)\, h + f(h; St)\right]. \tag{5}$$

Note that if the particles were to be distributed with a spatial density $\rho(\boldsymbol{x}, t)$, the coarse-grained mass would simply read

$$m_r(\boldsymbol{x}_i, t) = \int_{|\boldsymbol{y}|<r} d\boldsymbol{y}\, \rho(\boldsymbol{x}_i + \boldsymbol{y}, t), \tag{6}$$

so that its moment of order q would relate to the integral over balls of size r of the q-th order correlator of the density:

$$M_r^{(q)}(St) = \int_{|\boldsymbol{y}_2|<r} \cdots \int_{|\boldsymbol{y}_q|<r} d\boldsymbol{y}_2 \cdots d\boldsymbol{y}_q \, \langle \rho(0,t)\rho(\boldsymbol{y}_2, t) \cdots \rho(\boldsymbol{y}_q, t) \rangle \,. \tag{7}$$

A measurement and evaluation of the generalized fractal dimensions that is based on one of their properties described above requires having a good small-scale approximation of the mass distribution. This is generally not the case in numerical simulations since the finite number of particles is responsible for very strong effects of discreteness at small scales.

In order to cope with this difficulty, it is much easier to deal with the q-point particle distribution. The probability $P_q(r; St)$ to observe q particles randomly chosen on the attractor, all lying within a distance r, behaves at small scales as $P_q(r; St) \propto r^{(q-1)D_q(St)}$. For the case $q = 2$, one indeed gets $\mathrm{d}P_2(r)/\mathrm{d}r = p_2(r) \propto r^{D_2(St)-1}$. We made use of this technique based on the q-point cumulative probabilities P_q in order to evaluate the fractal dimensions of inertial particles in our numerical simulations.

Figure 1 (left) represents the generalized fractal dimensions as a function of Stokes from Run I. The statistics shows a clear signature of multifractality, confirming the results already obtained in random flows [36]. A straightforward method for evaluating the scaling properties of $P_q(r)$ is to look at the behavior of its logarithmic derivative $(\mathrm{d} \ln P_q(r))/(\mathrm{d} \ln r)$. Such a quantity defines a local slope, which tends to $(q-1)D_q$ as $r/\eta \to 0$. To extract a scaling exponent, a choice is to average this local slope over an interval where it is reasonably constant; error bars can then be defined as maximal deviations of the actual local slope from this average. Following such procedure error bars may become large, order %10 for the highest moment shown in the left panel of Fig. (1). Here we made use of a different strategy. As seen from the right panel of Fig. 1,

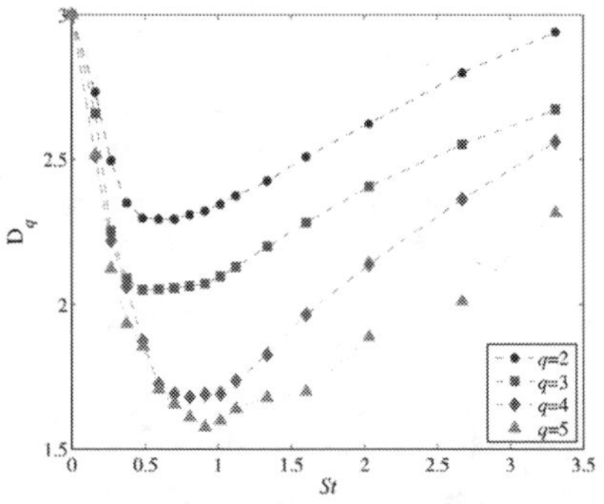

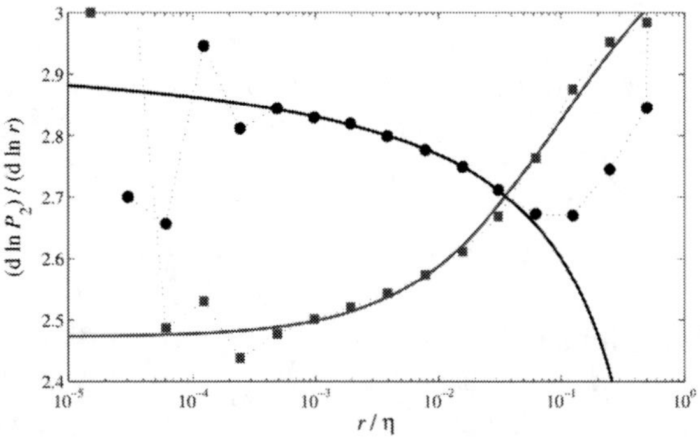

Figure 1. Left: generalized fractal dimensions $D_q(St)$ measured from Run I, as a function of the Stokes number for moment of order $q = 2$ to $q = 5$. Results from Run II do not show any significative dependency (not shown). Right: local slope of the two-point cumulative probability $P_2(r; St)$ as a function of scale for two different values of the Stokes number (lower blue: $St = 0.48$ and upper black: $St = 3.31$); the two solid lines represent the fit obtained from accounting for expected subleading terms.

there is no clear scaling range larger than a fraction of a decade where the local slope observed from our numerical data is decently constant. However, there is a sufficient statistical accuracy for resolving a smooth behaviour of the local slope over more than two decades. Making use of this requires accounting for subleading terms and thus formulating the following Ansatz for the small-scale behaviour of $P_q(r)$:

$$P_q(r) \approx r^{(q-1)D'_q}\left[\alpha_0 + \alpha_1 r + \alpha_2 r^2 + \cdots\right] + r^{3(q-1)}\left[\beta_0 + \beta_1 r + \beta_2 r^2 + \cdots\right], \quad (8)$$

The decomposition in two separate contributions follows from [37]. The first term involves the phase-space dimension D'_q, which describes the particle distribution in the full position-velocity space. The second term comes from the projection of this set onto the position space and relates to the development of caustics (i.e. of spatial locations with more than one particle velocity that become singular upon projection). This Ansatz suggests that $D_q = \min(D'_q, 3)$. The form of the first-order subleading term at a given scale r strongly depends on the values of the coefficients α_i and β_i. At small values of the Stokes number, $D'_q < 3$ and the β_i's are very close to 0. This relates the particle clustering properties, and to a very small probability to observe caustics, which is expected to be of the form $\propto \exp(-c/St)$ (see [38] and next section). At larger Stokes numbers, the caustics give the main subleading contribution.
Hence, two different possible fits for the q-point probability can be considered

$$P_q(r) \approx \alpha_0\, r^{(q-1)D'_q} + \alpha_1\, r^{(q-1)D'_q+1} \text{ for } St \le St_\star, \quad (9)$$

$$P_q(r) \approx \alpha_0\, r^{(q-1)D'_q} + \beta_0\, r^{3(q-1)} \quad \text{for } St \ge St_\star, \quad (10)$$

where $St_\star$ approximately corresponds to the value of the Stokes number where the minimum of the dimension D_q is attained, and small-scale clustering is maximal. When applied to the logarithmic derivative, such forms allow one to fit the data with two parameters (D'_q and the ratio of the two constants) over more than two decades, as seen from Fig. 1 (right).

The measurement of fractal dimensions obtained from this technique are shown on Fig. 1 (left). They show two very different behaviours at very small and moderate Stokes. On the left-hand side of the minimum, the dimensions D_q seem to saturate very quickly to a fixed limit

D_∞ when q increases. The saturation is much slower at larger values of the Stokes number. This suggests that, eventhough two particle sets with different Stokes numbers located below and above $St_\star$ can display the same correlation dimension D_2, the fluctuations of their mass at the same scale will be of very different nature. Such a consideration, which still requires to be understood in a quantitative way, might be very important when evaluating instantaneous collision rates between heavy particles.

4. Statistical Properties of the Relative Velocity

In this section, we discuss the statistics of the relative velocity between heavy particle pairs, having the same Stokes number, conditioned on their separation. This is an important subject relevant to the build up of models for particle collisions [3, 17, 39]. We define the relative velocity between a particle pair as

$$\delta_R V_{St}(t) = |\boldsymbol{V}_1(\boldsymbol{X}_1(t), t) - \boldsymbol{V}_2(\boldsymbol{X}_2(t), t)|, \tag{11}$$

with $R = |\boldsymbol{X}_1 - \boldsymbol{X}_2|$, the pair distance vector. The statistical properties of $\delta_R V_{St}$ can be characterized in terms of the conditional moments

$$S_p(r; St) = \langle |\delta_R V_{St}|^p \mid R{=}r \rangle\,. \tag{12}$$

Since we work in stationarity conditions, with an homogeneous and isotropic velocity field, the dependence of S_p on time, space and direction of $\boldsymbol{R}$ drops out. In the following, we will discuss the scaling behaviour of moments of relative velocity at varying the degree of inertia and the distance between the particles.

The qualitative behavior of $\delta_R V_{St}$ can be readily understood as follows. If the particle response time τ_s is much smaller than the characteristic time $\tau(R)$ of fluid velocity fluctuations on scale R, particle velocities will be very close to fluid velocities at particle position so that $\delta_R V(St) \approx \delta_R u$. On the other hand, if the particle response time is larger than $\tau(R)$, particle velocities will be uncorrelated with the fluid and among each other, thus one expects that $\delta_R V(St) \approx const$. This leads to introduce the notion of scale dependent Stokes number [40–42] $St(r) = \tau_s/\tau(R)$, where $\tau(R) \approx \epsilon^{-1/3} R^{2/3}$ if $R > \eta$, i.e. for inertial range scales, and $\tau(R) = \tau_\eta$ for $R \le \eta$, i.e. for scales belonging to the dissipative range.

Hence, for what concerns the statistics of relative velocity, at a given scale R particle inertia is small or large depending on $St(R)$. In Fig. 2, we observe that such a qualitative behavior is well reproduced by the second order structure function $S_2(r, St)$, as a function of the scale and for different values of $St = \tau_s/\tau_\eta$. For larger scales, $S_2(r, St)$ tends to recover the fluid behavior characterized by a Kolmogorov like $r^{2/3}$ scaling [42]. For larger Stokes values, the scale R at which the fluid velocities are recovered becomes larger (see Fig.5 in [43]). However, at scales within the dissipative range, we observe a strong dependence of $S_2(r; St)$ on the Stokes number suggesting that

$$S_p(r, St) \sim r^{\xi_p(St)} \tag{13}$$

i.e. within the available scale range accessible by our DNS, we observe a scaling behavior with an exponent $\xi_p(St)$ which non-trivially depends both on the order of the statistics and on the Stokes number. it is interesting to discuss such dependency and to connect it with particle statistical dynamics. In Figure 3, we show the behavior of the fitted exponent $\xi_p(St)$ at varying both the order p and the Stokes number St [43]. Two features can be readily identified. First we note that, at fixed St, for increasing the order p, the exponent $\xi_p(St)$ changes from a linear dependence on p ($\xi_p(St) = p$) to a p-independent value $\xi_\infty(St)$, showing a tendency to saturation, similarly to what happens in Burgers turbulence [44]. In particular, the scaling exponents are rather well approximated by a bi-fractal behavior:

$$\xi_p(St) = \min\{p, \xi_\infty(St)\}\,. \tag{14}$$

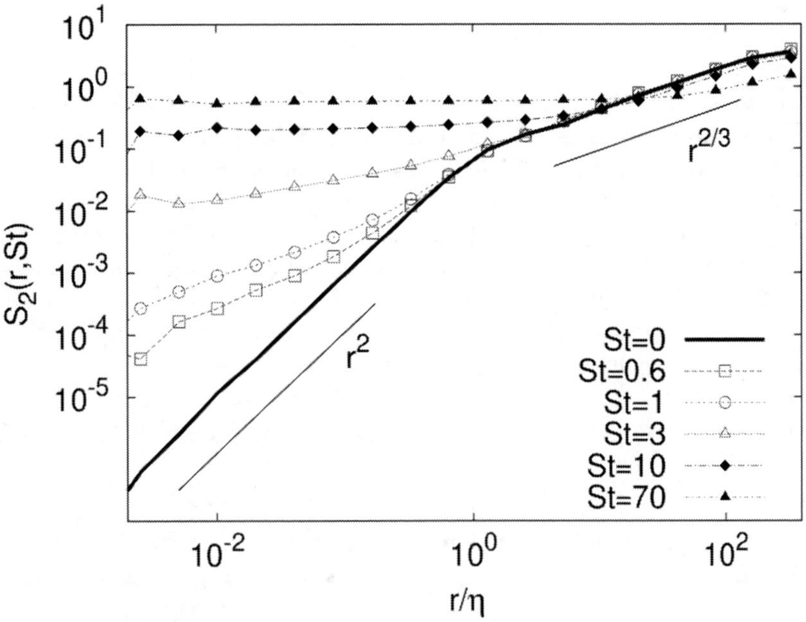

Figure 2. Log-log plot of the second order moment of the conditional velocity difference $S_2(r; St)$, as a function of r/η for Run II (from [42]).

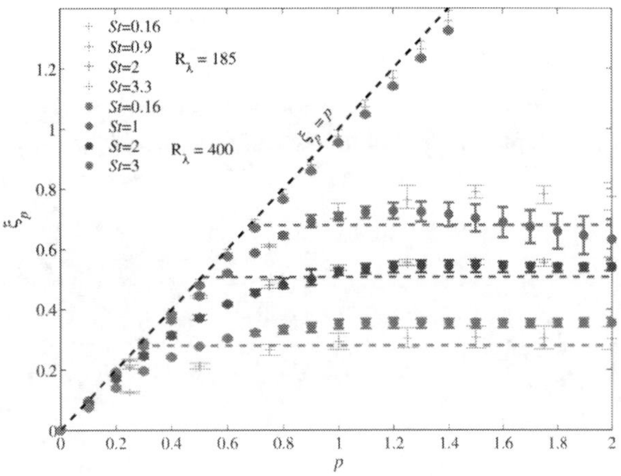

Figure 3. The scaling exponents ξ_p vs the order p, for different Stokes numbers and for both Runs (see also [43]).

Note that to disentangle a *pure* bi-fractal behavior from a multifractal one in presence of a saturation of scaling exponents is not simple, and requires a statistical convergence not easily achievable for 3D flows. However, as discussed below, there are good reasons to favor the former type of scaling behavior.
Second we note that the saturation exponent $\xi_\infty(St)$ dramatically depends on St. In particular, it decreases from values larger than 1, for $St < 1$, to values smaller and smaller at increasing St. Data suggest that $\xi_\infty(St) \approx 0$ when St reaches values $\approx 7 - 8$. Therefore, velocity increments

become rougher as St increases, departing more and more from the smooth differentiable behaviour.

These observations can be interpreted in terms of *caustic* contributions [38, 45], or *sling effect* [3, 10]. The physical observation is that, due to inertia, relative velocity differences $\delta_R V_{St}$ may not go smoothly to zero when the particle separations decreases. Indeed, particle dynamics in phase space is dissipative and the set where particle trajectories evolve may fold in the velocity direction leading to the formation of such caustics [17]. This effect becomes dramatic when inertia is such that particle response time is larger than any turbulent flow time scale: in such case, nearby particles will move with uncorrelated velocities [37, 46], and caustics are dense in space. The statistical importance of caustics does change with the Stokes number.

Recently, theoretical arguments have been put forward about the scaling behaviour for conditioned structure functions $S_p(r; St)$ in the dissipative range of the 3D turbulence, based on results obtained in stochastic flows [32, 47]. In particular, for scales $r/\eta \to 0$, it is expected that,

$$S_p(r; St) \sim A_p(St)(r/\eta)^p + B_p(St)(r/\eta)^{3-D_2(St)}\,. \tag{15}$$

The first term in (15) corresponds to the smooth (differentiable) part of the particles' velocity distribution where $A_p(St)$ is a $O(1)$ constant. The second term represents the contribution from caustics. Notice that the exponent $D_2(St)$ is the correlation dimension of the set in position space where particles concentrate (see previous section). The above prediction was obtained by solving in the limit $r \to 0$, the equation that rules the joint probability to find a particle pair at separation $R = r$ with a relative velocity $\delta_R V_{St}$. Such equation can be solved for a one dimensional stochastic flow. Extending the same ideas to flows in two and three dimensions, one gets to (15) as shown recently in [32]. In equation 15, the constant in front of the caustic contribution is suggested to be

$$B_p(St) \sim \exp(-C_p/St)\,, \tag{16}$$

and it is related to the rate at which caustics are formed [38, 48]. Note that while $A_p(St)$ is an $O(1)$ constant, possibly dependent on St, the activation factor B_p goes extremely fast at zero when $St \to 0$.

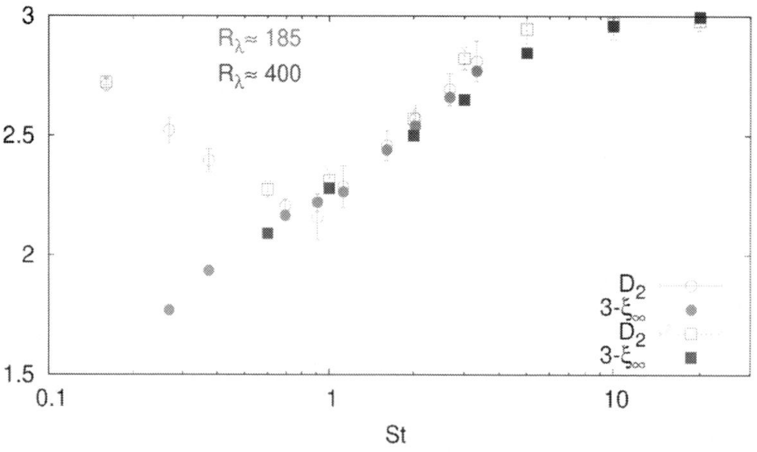

Figure 4. Comparison of the exponents $D_2(St)$ and $3 - \xi_\infty(St)$ vs St for Run I (red data, circles) and Run II (blue data, squares).

From equation (15) a simple bi-fractal behaviour is predicted for asymptotically small distances i.e., $S_p(r, St) \sim r^{\xi_p(St)}$ with

$$\xi_p(St) = \min\{p, 3 - D_2(St)\}. \tag{17}$$

Comparing eq. (17) with eq. (14), obtained on the basis of the observed behavior (Fig. 3), we get a prediction for the saturation exponent

$$\xi_\infty(St) = 3 - D_2(St)\,, \tag{18}$$

that can be checked with the DNS data. Figure 4 shows in the same plot $D_2(St)$ and $3 - \xi_\infty(St)$ as measured in both Runs.
For Stokes numbers larger than that for which $D_2(St)$ is minimal ($St \simeq 0.7$), the prediction (18) is rather well verified by our numerical data. While for smaller Stokes values, a discrepancy is observed between the value od the correlation dimension and the saturation exponent.

Two different comments can be made. By matching the statistical weigh of the two terms in (15), we observe that the asymptotic regime is expected only for scales $r < r_c(p, St) \sim B_p(St)^{1/(p-3+D_2)}$, which become exponentially small when $St \sim 0$. In our numerics, the range of scales allowing to fit the scaling properties is limited to an interval $[r_{min}, \eta]$. Clearly, the bifractal behavior (17) can be observed only if $r_{min} \ll r_c(p, St)$, otherwise only the smooth scaling would be observed $\xi_p(St) = p$. Hence, in general, it may be very difficult to test the prediction (18) for small Stokes number. On the other hand, we notice that for small Stokes values our data do not suggest a pure smooth scaling $\xi_p(St) = p$, but are compatible with a saturation of high order scaling exponents ξ_p to a value different from that predicted by eq. (18). In conclusion, it is fair to say that while for $St > 1$ the picture derived from (15) works rather well, for $St \ll 1$ there is still some difficulty to validate it. Either the saturation we observe is the result of some spurious effect due to the limited range of avalaible scales, or there is still something to understand in the limit of very small inertia.

5. Conclusions

We studied the spatial and velocity statistics of point-like heavy particles in stationary, incompressible, homogeneous and isotropic three-dimensional turbulence at two values of the Reynolds number, $Re_\lambda \sim 200$ and $Re_\lambda \sim 400$. We considered the behavior of tracers ($St = 0$), and heavy particles with Stokes number from $St = 0.16$ to 70.
Stationary small-scale particle distribution display a singular –multifractal– measure, characterized by a set of generalized fractal dimensions with a strong sensitivity on the Stokes number and a small –if any– Reynolds number dependency. Velocity increments between a pair of particles conditioned to be at a given distance depend on the relative weight of *smooth* events, where the particle velocity difference is approximately the equal to the fluid velocity difference, and *caustics* contributions, i.e. associted to close particles having very different velocities. The latter events lead to a non-differentiable small-scale behaviour for the relative velocity. The relative weight of these two mechanisms changes at varying Stokes number. We showed that moments of the velocity difference enjoy a quasi bi-fractal-behavior, and that the scaling properties of velocity increments for not too small Stokes numbers are in good agreement with a recent theoretical prediction made by K. Gustavsson and B. Mehlig [32].

ACKNOWLEDGEMENTS. This study benefited from fruitful discussions with G. Falkovich, B. Mehlig, and M. Wilkinson. JB and AL acknowledge support from NSF under grant No. PHY05 51164 for their stay at the Kavli Institute for Theoretical Physics in the framework of the 2008 Physics of Climate Change program. Part of this work was supported by ANR under grant No. BLAN07-1 192604. Simulations were performed at CASPUR and CINECA (Italy), and in the framework of the DEISA Extreme Computing Initiative supported by the DEISA Consortium (co-funded by the EU, FP6 project 508830).

References

[1] Csanady, G. 1980 *Turbulent diffusion in the environment.* Geophysics and Astrophysics Monographs Vol. 3 D. Reidel Publishing Company.
[2] Eaton, J.K.& Fessler, J.L. 1994 Preferential concentrations of particles by turbulence. *Intl. J. Multiphase Flow* **20**, 169–209.
[3] Falkovich, G., Fouxon, A. & Stepanov, M. 2002 Acceleration of rain initiation by cloud turbulence. *Nature* **419**, 151–154.
[4] Post, S. & Abraham, J. 2002 Modeling the outcome of drop-drop collisions in Diesel sprays. *Intl. J. Multiphase Flow* **28**, 997–1019.
[5] Shaw, R.A. 2003 Particle-turbulence interactions in atmospheric clouds. *Ann. Rev. Fluid Mech.* **35**, 183–227.
[6] Toschi, F. & Bodenschatz, E. 2009 Lagrangian Properties of Particles in Turbulence. *Ann. Rev. Fluid Mech.* **41**, 375–404 .
[7] Biferale, L., Scagliarini, A. & Toschi, F. 2010 On the measurement of vortex filament lifetime statistics in turbulence. *Phys Fluids* **22** 065101.
[8] Balkovsky, E., Falkovich, G. & Fouxon, A. 2001 Intermittent distribution of inertial particles in turbulent flows. *Phys. Rev. Lett.* **86**, 2790–2793.
[9] Zaichik, L.I., Simonin, O. & Alipchenkov, V.M. 2003 Two statistical models for predicting collision rates of inertial particles in homogeneous isotropic turbulence. *Phys. Fluids* **15**, 2995–3005.
[10] Falkovich, G. & Pumir, A. 2004 Intermittent distribution of heavy particles in a turbulent flow. *Phys. Fluids* **16**, L47–L51.
[11] Collins, L.R. & Keswani, A. 2004 Reynolds number scaling of particle clustering in turbulent aerosols. *New J. Phys.* **6**, 119.
[12] Chun, J., Koch, D.L., Rani, S., Ahluwalia, A. & Collins, L.R. 2005 Clustering of aerosol particles in isotropic turbulence. *J. Fluid Mech.* **536**, 219–251.
[13] Bec, J., Biferale, L., Boffetta, G., Cencini, M., Lanotte, A.S., Musacchio, S. & Toschi, F. 2007 Heavy particle concentration in turbulence at dissipative and inertial scales. *Phys. Rev. Lett.* **98**, 084502.
[14] Goto, S. & Vassilicos, J.C. 2008 Sweep-stick mechanism of heavy particle clustering in fluid turbulence. *Phys. Rev. Lett.* **100** 054503.
[15] Sigurgeirsson H. & Stuart A.M. 2002 A Model for Preferential Concentration. *Phys. Fluids* **14**, 4352–4361.
[16] Mehlig, B. & Wikinson, M. 2004 Coagulation by random velocity fields as a Kramers problem. *Phys. Rev. Lett.* **92**, 250602.
[17] Bec, J., Celani, A., Cencini, M. & Musacchio, S. 2005 Clustering and collisions of heavy particles in random smooth flows. *Phys. Fluids* **17**, 073301.
[18] Olla, P. 2002 Transport properties of heavy particles in high Reynolds number turbulence. *Phys. Fluids* **14**, 4266–4277.
[19] Boffetta, G., De Lillo, F. & Gamba, A. 2004 Large scale inhomogeneity of inertial particles in turbulent flows. *Phys. Fluids* **16**, L20–L24.
[20] Bec, J., Biferale, L., Boffetta, G., Celani, A., Cencini, M., Lanotte, A.S., Musacchio, S., & Toschi, F. 2006 Acceleration statistics of heavy particles in turbulence. *J. Fluid Mech.* **550**, 349–358.
[21] Cencini, M., Bec, J., Biferale, L., Boffetta, G., Celani, A., Lanotte, A.S., Musacchio, S. & Toschi, F. 2006 Dynamics and statistics of heavy particles in turbulent flows. *J. Turb.* **7**, 36, 1.
[22] Gylfason, A., Ayyalasomayajula, S., Bodenschatz, E. & Warhaft, Z. 2006 Lagrangian measurements of inertial particle accelerations in grid generated wind tunnel turbulence. *Phys. Rev. Lett* **97**, 144507.
[23] Gerashchenko, S., Sharp, N.S., Neuscamman, S. & Warhaft, Z. 2008 Lagrangian measurements of inertial particle accelerations in a turbulent boundary layer. *J. Fluid Mech.* **617**, 255–281.
[24] Zaichik, L.I. & Alipchenkov, V.M. 2008 Acceleration of heavy particles in isotropic turbulence. *Intl. J. Multiphase Flow* **34**, 865–868.
[25] Ayyalasomayajula, S., Warhaft, Z. & Collins, L.R. 2008 Modeling inertial particle acceleration statistics in isotropic turbulence. *Phys. Fluids* **20** 095104.
[26] Volk, R., Mordant, N., Verhille, G. & Pinton, J.-F. 2008 Laser Doppler measurement of inertial particle and bubble accelerations in turbulence, *Europhys. Lett* **81**, 34002.
[27] Volk, R., Calzavarini, E.,Verhille, G., Lohse, D., Mordant, N., Pinton, J.-F. & Toschi, F. 2009 Acceleration of heavy and light particles in turbulence: comparison between experiments and direct

numerical simulations, *Phys. D* in press.

[28] Benzi, R., Biferale, L., Calzavarini, E., Lohse, D. & Toschi, F. 2009. Velocity gradients statistics along particle trajectories in turbulent flows: the refined similarity hypothesis in the Lagrangian frame. *Phys. Rev. E* **80**, 066318.

[29] Qureshi, N.M., Bourgoin, M., Baudet, C., Cartellier, A., & Gagne, Y., 2007 Turbulent transport of material particles: An experimental study of finite size effects. *Phys. Rev. Lett* **99**, 184502.

[30] Xu H. & Bodenschatz E. 2008 Motion of inertial particles with size larger than Kolmogorov scale in turbulent flows. *Physica* D **237**, 2095–2100.

[31] Qureshi, N.M., Arrieta, U., Baudet, C., Cartellier, A., Gagne, Y. & Bourgoin, M. 2008 Acceleration statistics of inertial particles in turbulent flow. *Eur. Phys. J. B* **66**, 531–536.

[32] Gustavsson, K. & Mehlig, B. 2010. Distribution of relative velocities in turbulent aerosols. arXiv:1012.1789.

[33] Chen, S., Doolen, G.D., Kraichnan, R.H. & She, Z.-S. 1993 On statistical correlations between velocity increments and locally averaged dissipation in homogeneous turbulence. *Phys. Fluids A* **5**, 458–463.

[34] Maxey, M.R. & Riley, J. 1983 Equation of motion of a small rigid sphere in a nonuniform flow. *Phys. Fluids* **26**, 883–889.

[35] Bec, J. 2003 Fractal clustering of inertial particles in random flows *Phys. Fluids* **15**, L81–L84.

[36] Bec, J., Gawedzki, K. & Horvai, P. 2004. Multifractal clustering in compressible flows. *Phys. Rev. Lett.* **92**, 224501.

[37] Bec, J., Cencini, M. & Hillerbrand, R. 2007 Heavy particles in incompressible flows: the large Stokes number asymptotics. *Physica* D **226**, 11–22.

[38] Wilkinson M. & Mehlig B. 2005 Caustics in turbulent aerosols. *Europhys. Lett* **71**, 186–192.

[39] Wilkinson, M. , Mehlig B., & Bezuglyy, V. 2006. Caustic activation of rain showers. *Phys. Rev. Lett.* **97**, 48501.

[40] Falkovich, G. Fouxon, A. & Stepanov, P. 2003 in *Sedimentation and Sediment Transport*, edited by A. Gyr and W. Kinzelbach, Kluwer Academic, Dordrecht, pp. 155-158.

[41] Bec, J., Cencini, M. & Hillerbrand, R. 2007 Clustering of Heavy particles in random self-similar flows . *Phys. Rev. E* **75**, 025301.

[42] Bec, J., Biferale, L., Lanotte, A.S., Toschi, F. & Scagliarini, A. 2010 Turbulent pair dispersion of inertial particles *J. Fluid Mech.* **645** 497.

[43] Bec, J., Biferale, L., Cencini, M., Lanotte, A.S. & Toschi, F. 2010 Intermittency in the velocity of heavy particles in turbulence. *J. Fluid Mech.* **646** 527.

[44] Frisch, U. 1995 *Turbulence. The legacy of A.N. Kolmogorov* Cambridge University Press, Cambridge (UK).

[45] Falkovich, G. & Pumir, A. 2007 Sling effect in collision of water droplet in turbulent clouds. *J. Atm. Sci.* **64**, 4497–4505.

[46] Abrahamson J. 1975 Collision rates of small particles in a vigorously turbulent fluid. Chem. Eng. Sci. **30** 1371–1379.

[47] Cencini, M., Bec, J. 2009 Talk given at working group meeting of COST action MP0806.

[48] Derevyanko, S.A., Falkovich, G., Turitsyn, K. & Turitsyn S. 2007 Lagrangian and Eulerian descriptions of inertial particles in random flows. *J. Turb.* **8**, 16, 1–19.

Statistical properties of an ideal subgrid-scale correction for Lagrangian particle tracking in turbulent channel flow

F. Bianco[1,2], S. Chibbaro[2], C. Marchioli[1], M.V. Salvetti[3], A. Soldati[1]

[1]Energy Technology Dept.and Centro Interdipartimentale di Fluidodinamica e Idraulica, University of Udine, 33100 Udine – Italy

[2] Jean Le Rond D'Alembert Institute, University Pierre et Marie Curie and CNRS UMR7190, 75252 Paris Cedex 05 – France

[3]Aerospace Engineering Dept., University of Pisa, 56100 Pisa – Italy

E-mail: mv.salvetti@dia.unipi.it

Abstract. One issue associated with the use of Large-Eddy Simulation (LES) to investigate the dispersion of small inertial particles in turbulent flows is the accuracy with which particle statistics and concentration can be reproduced. The motion of particles in LES fields may differ significantly from that observed in experiments or direct numerical simulation (DNS) because the force acting on the particles is not accurately estimated, due to the availability of the only filtered fluid velocity, and because errors accumulate in time leading to a progressive divergence of the trajectories. This may lead to different degrees of inaccuracy in the prediction of statistics and concentration. We identify herein an ideal subgrid correction of the a-priori LES fluid velocity seen by the particles in turbulent channel flow. This correction is computed by imposing that the trajectories of individual particles moving in filtered DNS fields exactly coincide with the particle trajectories in a DNS. In this way the errors introduced by filtering into the particle motion equations can be singled out and analyzed separately from those due to the progressive divergence of the trajectories. The subgrid correction term, and therefore the filtering error, is characterized in the present paper in terms of statistical moments. The effects of the particle inertia and of the filter type and width on the properties of the correction term are investigated.

1. Introduction

Dispersion and entrainment of inertial heavy particles in turbulent flows are crucial in a number of industrial applications and environmental phenomena. Examples are mixing, combustion, depulveration, spray dynamics, pollutant dispersion or cloud dynamics. The dynamics of inertial particles in turbulent flows is characterized by large-scale clustering and preferential concentration, due to the tendency of inertial particles to distribute preferentially at the periphery of strong vortical regions and to segregate into straining regions [1–5]. In turbulent boundary layers clustering and preferential concentration also influence settling, deposition and entrainment to and from the wall [2, 5, 6].

Direct numerical simulation (DNS) together with Lagrangian particle tracking (LPT) have successfully been used to investigate and quantify the behavior of particles in wall-bounded turbulent flow. DNS studies have shown, for instance, that particle segregation and subsequent deposition phenomena are strictly related with the interaction between particles and the near-wall coherent vortical structures [2, 6, 7]. More specifically, pairs of counter-rotating quasi-streamwise vortices, which are the statistically most common near-wall structures, are able to entrain particles from the core region towards the wall on their downwash side and away from the wall on their upwash side. This transport mechanism, called turbophoresis [8], produces net particle fluxes toward the wall leading to irreversible particle accumulation at the wall.

On the other hand, DNS is still nowadays limited to low Reynolds numbers, often very far from those characterizing practical applications, and to academic flow configurations. Conversely, thanks to the huge growth of available computational resources, large-eddy simulation (LES) can tackle increasingly complex problems. In LES the basic idea is to directly simulate only the turbulence scales larger than a given dimension, while the effects of the unresolved subgrid scales (SGS) on the large-scale motion are modeled. From a formal viewpoint, the separation between scales is provided through filtering, which can be explicitly applied to the flow equations or implicitly introduced by the numerical discretization. Thus, the exact dynamics of the filtered fluid velocity field can at best be obtained from LES (*ideal LES*), this when SGS modeling and numerical errors are negligible. Let us restrict ourselves to the context of an ideal LES, which can be mimicked in the so-called a-priori tests, i.e. by assuming that the LES velocity fields are represented by the filtered DNS ones. In LPT coupled with LES of the fluid phase, the forces acting on particles can be estimated from the filtered fluid velocity fields, by simply neglecting the effects of the SGS scales. A critical issue is, however, to evaluate the impact of this approximation on the ability to capture the previously described phenomena characterizing the particle dynamics, i.e. preferential concentration and turbophoresis, and to accurately predict quantities of practical interest, such as particle concentration profiles.

In the last decades, several examples of LPT coupled with LES of wall-bounded turbulent flows can be found in the literature [9–19]. In most of the earlier studies, the fluid SGS velocity fluctuations were simply neglected in the particle motion equations [9–11, 14, 15]. This is based on the assumption that, if the particle response time is large enough compared to the smallest time scale resolved in the LES, the effects of the SGS fluid velocity fluctuations on the particle velocity statistics are negligible [4, 11]. For well-resolved LES, this assumption holds to capture satisfactorily the statistics of particle velocity. Nonetheless, more recent studies [12, 13, 16, 19] have demonstrated that neglecting the effects of the fluid SGS velocity fluctuations on particle motion leads to significant underestimation of preferential concentration and, consequently, to weaker deposition fluxes and lower near-wall accumulation. The issue is therefore how to model the SGS effects in the particle motion equations.

A few different ways have been explored in the literature. Filter inversion or approximate deconvolution have been used in LPT in homogeneous isotropic turbulence [20, 21] and in turbulent channel flow [12, 13, 22]. These techniques appeared to be successful in recovering the correct fluid energy content for the resolved scales and improved the prediction of preferential concentration and turbophoresis[12, 13, 22]. Nonetheless, these models still cannot recover the effects of the unresolved SGS scales and previous studies [22] show that the reintroduction of the correct amount of fluid velocity fluctuations does not warrant accurate quantitative prediction of particle preferential concentration and near-wall accumulation and suggest that additional information on the SGS flow field is required. Attempts of introducing information on the unrepresented scales have been made through fractal interpolation [22] or kinematic simulation[19]. Stochastic models have also been proposed for LPT coupled with LES of the carrier phase, but, they have mainly been used for homogeneous isotropic turbulence [21, 23, 24].

In order to devise efficient SGS models for LPT, the quantification and the analysis of the errors introduced by the lack of SGS scales on particle dynamics is clearly a necessary step. Even if we restrict ourselves to ideal LES, i.e. we assume that the LES of the fluid phase provides the exact dynamics of the resolved scales, this task is complicated by the fact that there are still two sources of errors. The first one, is the inaccuracy in the estimation of the forces acting on particles when they are computed only from the filtered fluid velocity, output of the LES of the carrier phase. This can be considered as the pure filtering error. The second contribution to the error comes from the fact that, due to the inaccurate estimation of forces, considering the same initial distribution of particles, the trajectories obtained in LES progressively diverge from the *exact* DNS ones and thus the forces are evaluated at different locations. When comparing particle concentrations and statistics obtained in LES to the DNS reference data, the effects of the two errors can not easily be singled out. In a recent work [25] the two contributions have been computed separately for couples of tracer particles in a-priori LES of homogeneous isotropic turbulence.

The main goal of the present paper is to compute a SGS correction term for the LES fluid velocity seen by inertial particles, which accounts for the pure filtering error alone. This correction is computed by assuming that the trajectories of particles having the same initial position tracked in a-priori LES coincide with the ones obtained in DNS of turbulent channel flow. This is clearly an idealized situation and is a stronger requirement with respect to what is usually asked to LPT coupled to LES, i.e. to correctly reproduce the particle statistics and concentration. Nevertheless, in this way, the effects of the pure filtering error can be singled out and quantified, since the divergence of trajectories is a-priori prevented. This *ideal* correction (IC) can be computed for each single particle and at each time step along its trajectory, and the properties of IC can be characterized in terms of statistics and probability distribution function. In the present paper, the correction term is characterized in terms of statistical moments, while PDFs will be presented in forthcoming studies. The aim is to obtain indications of the key features to be incorporated in SGS models for the particle motion equations. The IC term is computed here for different filter widths and types, corresponding roughly to varying amounts of resolved flow energy, and for different particle inertia. A point at issue is indeed whether SGS models for LPT should explicitly take into account inertia effects or general conclusions may be drawn. Finally, an additional question is whether the flow inhomogeneity has an impact on the key features of IC and therefore whether a SGS model should have different characteristics in the near-wall region than in the center of the channel.

2. Physical Problem and Methodology

2.1. Particle-laden turbulent channel flow and DNS methodology

The flow into which particles are introduced is a turbulent channel flow of gas. The reference geometry consists of two infinite flat parallel walls: the origin of the coordinate system is located at the center of the channel and the $x-$, $y-$ and $z-$ axes point in the streamwise, spanwise and wall-normal directions respectively. Periodic boundary conditions are imposed on the fluid velocity field in x and y, no-slip boundary conditions are imposed at the walls. In the present study, we consider air (assumed to be incompressible and Newtonian) with density $\rho = 1.3\ kg\ m^{-3}$ and kinematic viscosity $\nu = 15.7{\times}10^{-6}\ m^2\ s^{-1}$. The flow is driven by a mean dimensionless pressure gradient, such that the shear Reynolds number, $Re_\tau = u_\tau h/\nu$, based on the shear (or friction) velocity, u_τ, and on the half channel height, h, is equal to 150. The shear velocity is defined as $u_\tau = (\tau_w/\rho)^{1/2}$, where τ_w is the mean shear stress at the wall.

Particles with density $\rho_p = 1000\ kg\ m^{-3}$ are injected into the flow. The motion of particles is described by a set of ordinary differential equations for particle velocity and position. For particles much heavier than the fluid ($\rho_p/\rho \gg 1$) the most significant forces are Stokes drag and buoyancy and Basset force can be neglected being an order of magnitude smaller [26]. In the present simulations, the aim is to minimize the number of degrees of freedom by keeping the simulation setting as simplified as possible; thus the effect of gravity has also been neglected. Particles, which are assumed pointwise, rigid and spherical, are injected into the flow at concentration low enough to neglect particle collisions. The effect of particles onto the turbulent field is also neglected (one-way coupling assumption).

With the above assumptions, a simplified version of the Basset-Boussinesq-Oseen equation [27] is obtained. In vector form it writes:

$$\frac{d\mathbf{x}_p}{dt} = \mathbf{v} \ , \tag{1}$$

$$\frac{d\mathbf{v}}{dt} = \frac{\mathbf{u}_s - \mathbf{v}}{\tau_p}(1 + 0.15 Re_p^{0.687}) \ , \tag{2}$$

where $\mathbf{x}_p$ is the particle position, $\mathbf{v}$ is the particle velocity, and $\mathbf{u}_s$ is the fluid velocity at the particle position, $\mathbf{u}(\mathbf{x}_p(t), t)$. The righthand side of Eq. (2) represent the drag force per unit mass acting of the particle, D, in which τ_p is the particle relaxation time, defined as $\tau_p = \rho_p d_p^2/18\mu$, d_p being the particle diameter, and $Re_p = d_p|\mathbf{v} - \mathbf{u}|/\nu$ is the particle Reynolds number, ν being the fluid kinematic viscosity.

The fully-developed channel flow previously described has been simulated through DNS. The fluid governing equations (omitted here for the sake of brevity) are discretized using a pseudo-spectral method based on transforming the field variables into wavenumber space, using Fourier representations for the periodic streamwise and spanwise directions and a Chebyshev representation for the wall-normal (non-homogeneous) direction. A two level, explicit Adams-Bashforth scheme for the non-linear terms, and an implicit Crank-Nicolson method for the viscous terms are employed for time advancement. Further details of the method can be found in previous articles, e.g. [28].

The calculations were performed on a computational domain of size $4\pi h \times 2\pi h \times 2h$ in x, y and z respectively. The size of the computational domain in wall units is $1885 \times 942 \times 300$. Wall units are obtained combining u_τ, ν and ρ. The computational domain is discretized in physical space with $128 \times 128 \times 129$ grid points (corresponding to 128×128 Fourier modes and to 129 Chebyshev coefficients in the wavenumber space). This is the minimum number of grid points required in each direction to ensure that the grid spacing is always smaller than the smallest flow scale and that the limitations imposed by the point-particle approach are satisfied [16].

To calculate particle trajectories in the DNS flow field, 6th-order Lagrangian polynomials are used to interpolate fluid velocities at the particle position. With this velocity, Eqns. (1) and (2) are advanced in time using a 4th-order Runge-Kutta scheme.

At the beginning of the simulation, particles are distributed homogeneously over the computational domain and their initial velocity is set equal to that of the fluid at the particle initial position. Periodic boundary conditions are imposed on particles moving outside the computational domain in the homogeneous directions, and perfectly-elastic collisions are assumed at the smooth walls when the particle center is at a distance lower than one particle radius from the wall. For the simulations presented here, large samples of 10^5 particles, characterized by different response times, are considered. When the particle response time is made dimensionless using wall variables, the Stokes number for each particle set is obtained as $St = \tau_p^+ = \tau_p/\tau_f$ where $\tau_f = \nu/u_\tau^2$ is the viscous timescale of the flow. Table 1 shows all the parameters of the particles injected into the flow field.

The non-dimensional timestep size used for particle tracking was chosen to be equal to the non-dimensional timestep size used for the fluid, $\delta t^+ = 0.045$; the total tracking time in DNS was, for each particle set, $t^+ = 21150$ (in wall units).

2.2. A-priori LES methodology and computation of the ideal correction

In the *a-priori* tests the Lagrangian tracking of particles is carried out starting from the filtered fluid velocity field, $\bar{\mathbf{u}}$, obtained through explicit filtering of the DNS velocity by means of either a

Table 1. Particle parameters; τ_p is the particle relaxation time, d_p the particle diameter, V_s the particle settling velocity and Re_p the particle Reynolds number.

St	τ_p (s)	d_p^+	d_p (μm)	$V_s^+ = g^+ \cdot St$	$Re_p^+ = V_s^+ \cdot d_p^+/\nu^+$
1	$1.133 \cdot 10^{-3}$	0.153	20.4	0.0943	0.01443
5	$5.660 \cdot 10^{-3}$	0.342	45.6	0.4717	0.16132
25	$28.32 \cdot 10^{-3}$	0.765	102.0	2.3584	1.80418

cut-off or a top-hat filter. Both filters are applied in the homogeneous streamwise and spanwise directions in the wave number space:

$$\bar{u}_i(\mathbf{x},t) = FT^{-1}\begin{cases} G(\kappa_1)\cdot G(\kappa_2)\cdot \hat{u}_i(\kappa_1,\kappa_2,z,t) & \text{if } |\kappa_j| \le |\kappa_c| \text{ with } j=1,2\ , \\ 0 & \text{otherwise}\ . \end{cases} \tag{3}$$

where FT is the 2D Fourier Transform, $\kappa_c = \pi/\Delta$ is the cutoff wave number (Δ being the filter width in the physical space), $\hat{u}_i(\kappa_1, \kappa_2, z, t)$ is the Fourier transform of the ith component of the fluid velocity, namely $\hat{u}_i(\kappa_1, \kappa_2, z, t) = FT[u_i(\mathbf{x}, t)]$ and $G(\kappa_i)$ is the filter transfer function:

$$G(\kappa_j) = \begin{cases} 1 & \text{for the cut-off filter}\ , \\ \frac{sin(\kappa_j\Delta/2)}{\kappa_j\Delta/2} & \text{for the top-hat filter}\ . \end{cases} \tag{4}$$

Three different filter widths have been considered, corresponding to a grid Coarsening Factor (CF) in each homogeneous direction of 2, 4 and 8 with respect to DNS, i.e. to 64×64, 32×32 and 16×16 Fourier modes in the homogeneous directions respectively. Note that a coarsening factor of 8 leads to a very coarse resolution and is considered as an extreme case of under-resolved LES, while CF of 2 and 4 correspond to resolutions more currently used in LES. In the wall-normal direction data are not filtered, since often in LES the wall-normal resolution is DNS-like. [29]

In a-priori tests, the filtered flow velocity $\bar{\mathbf{u}}$ is used in Eq. (2) for Lagrangian particle tracking, instead of the DNS one. The SGS correction term is computed by imposing that the particle trajectories tracked in a-priori LES coincide with those in DNS, at each time step, t^n, and for each particle, k, we impose:

$$\mathbf{x}_k^{LES}(t^n) = \mathbf{x}_k^{DNS}(t^n)\ ;\ \mathbf{v}_k^{LES}(t^n) = \mathbf{v}_k^{DNS}(t^n) \tag{5}$$

in which the superscript LES denotes the particle position and velocity obtained in a-priori LES, while the superscript DNS those obtained in DNS. The correction term computed for the particle k at time t^n can be expressed as follows:

$$\delta\mathbf{u}(\mathbf{x}_k(t^n), t^n) = \mathbf{u}(\mathbf{x}_k(t^n), t^n) - \bar{\mathbf{u}}(\mathbf{x}_k(t^n), t^n) \tag{6}$$

in which $\mathbf{x}_k(t^n)$ is the position of particle k at time t^n and the superscripts LES and DNS have been dropped since the particle positions are forced to be the same in DNS and a priori LES. It is thus evident from Eq. (6) that in this way the correction term expresses the filtering error on the fluid velocity seen by the particles, while the error accumulation due to the progressive divergence of the trajectories computed in LES and DNS is eliminated by assuming the coincidence of the trajectories. As previously discussed in the Introduction, this is obviously an idealized situation, but it allows a contribution to the error in Lagrangian tracking of particles in LES, i.e. the one due to filtering of the fluid velocity, to be singled out and characterized. Note also that Eq.

(6) gives the difference between the DNS and the filtered fluid velocities *at the particle position*, i.e. computed by following the particle trajectories. As it will be shown also in the following, this may be substantially different from the Eulerian measure of the filtering error, classically computed in the literature, i.e. the difference between DNS and LES velocities at fixed points in space.

Following Eq. (6), the SGS correction term to the fluid velocity seen by the particles has been computed for each particle along the trajectories computed in the DNS fluid velocity fields, i.e. at $x_k^{DNS}(t^n)$. The computation has been carried out for the three considered particle sets characterized by different inertia and for the different filter widths and types, for a total time $\Delta t^+ \simeq 8285$ and sampled every $\delta t^+ = 4.5$. It has been checked that this leads to converged statistics of $\delta \mathbf{u}$.

3. Statistical moments of the SGS correction

The Eulerian statistical moments of the correction term (Eq. (6)) are presented and analyzed in this section. They are computed by dividing the computational domain in slabs having dimensions L_x, L_y and $\Delta_z(i)$ $(i = 1, N_z)$, $\Delta_z(i)$ being the difference between two adjacent Chebyshev collocation points and N_z the number of collocation points used in DNS. Averaging is carried out over the particles laying at a given time instant inside each slab and in time over the time interval specified in Sec. 2.2. All the quantities shown herein are non dimensional, in wall units.

3.1. Effects of particle inertia

In order to analyze first the effects of particle inertia, let us fix the filter width and type. The influence of the filter will be investigated in Sec. 3.2. We consider, in particular, the cut-off filter with a coarsening factor of 4 with respect to DNS.

Figures 1-3 show the mean value of the components of the SGS velocity correction term along the streamwise, spanwise and normal directions, computed for the considered filter and for different particle inertia, as a function of the distance from the wall. While the mean values of the spanwise component of the correction are very low along the whole channel, the spanwise and the normal components show significant (negative or positive) mean values in the near wall region, with peaks located approximately at $z^+ \simeq 15 - 20$. A first observation is that, if the correction term was computed at fixed grid points, instead of following the particle trajectories, its mean value would have been rigorously equal to zero in all the directions, since the filter does not affect the mean value of the velocity. Note that, in priori and a-posteriori tests in the literature, it is usual to analyze the effects of filtering (or, more in general, of errors) on the LES fluid velocity fields at fixed points (see e.g. [12, 16]). When dealing with Lagrangian tracking of particles, conversely, it seems to be more appropriate to study the effects of filtering on the fluid velocity seen by the particles. As it will be shown, the behavior of the mean streamwise and normal components of $\delta \mathbf{u}$ is indeed related to the effects of filtering on the turbulent structures, and in particular on near-wall structures, combined with the preferential sampling of particles.

Let us start by analyzing the streamwise component of $\delta \mathbf{u}$. As well known, the near wall region is dominated by the presence of low- and high-speed streaks and it has been shown [6, 30, 31] that inertial particles tend to preferentially sample low-speed streaks. The effect of filtering is to attenuate the velocity fluctuations and therefore to smooth the streaks. As a consequence, the correction term in mean tends to reintroduce the effect of low-speed streaks smoothened by filtering and, thus, in mean it tends to decrease the LES fluid velocity seen by the particles near the wall (negative peaks of mean δu_1 in the near wall region). As for the effects of inertia, the behavior of the mean streamwise component of the correction term is qualitatively the one previously described for all the considered particle sets. Quantitatively, the mean correction is more important for the particles having larger inertia (St=5 and 25). This is consistent with

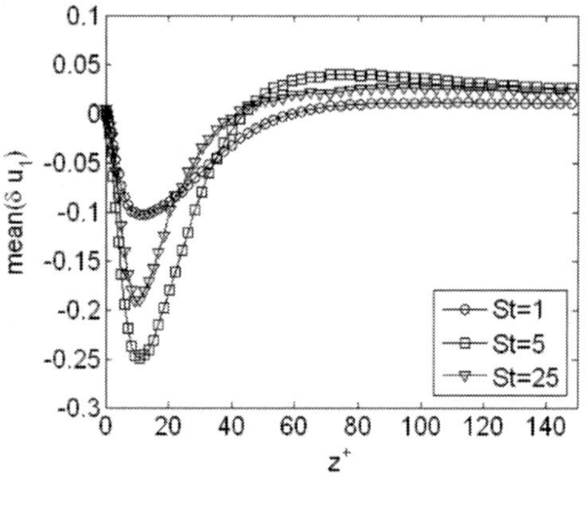

Figure 1. Mean values of the SGS velocity correction component in the streamwise direction as a function of z^+. Cut-off filter with CF=4.

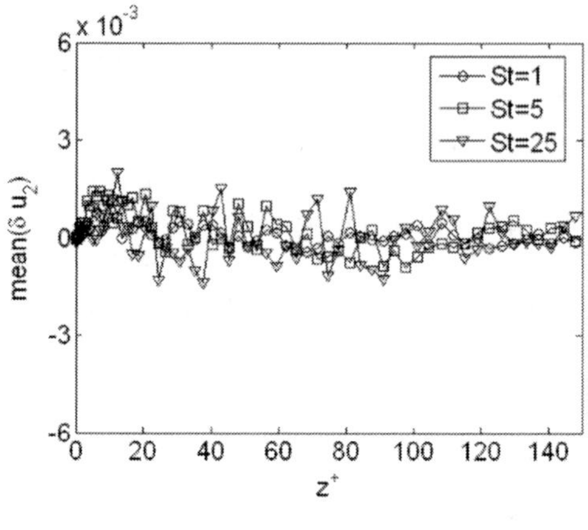

Figure 2. Mean values of the SGS velocity correction component in the spanwise direction as a function of z^+. Cut-off filter with CF=4.

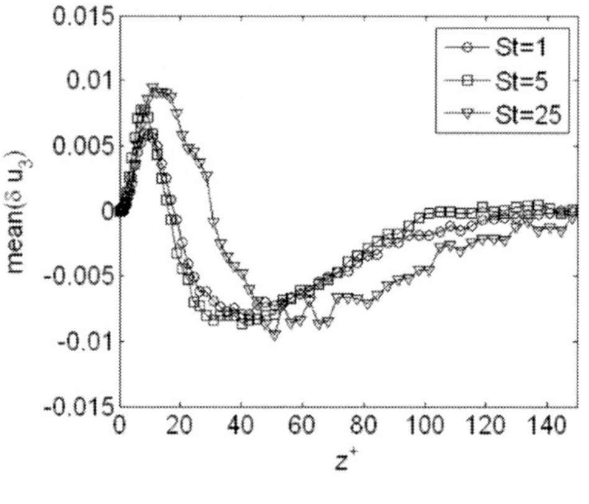

Figure 3. Mean values of the SGS velocity correction component in the wall-normal direction as a function of z^+. Cut-off filter with CF=4.

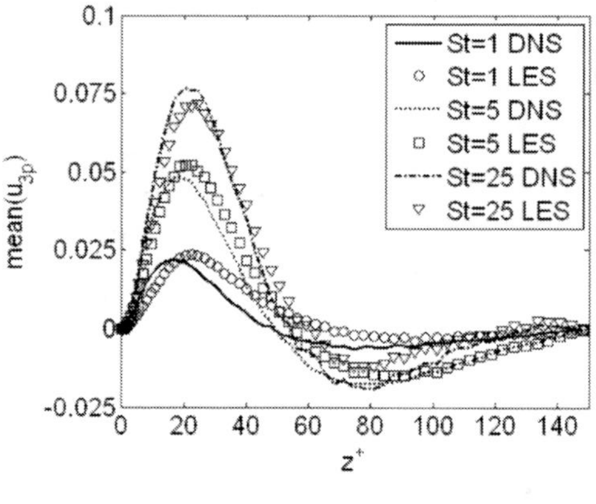

Figure 4. Effects of filtering on the fluid mean normal velocity seen by the particles. Cut-off filter with CF=4.

the previous explanation of the origin of the mean values; indeed, for the considered Reynolds number, the particles characterized by St=5 and 25 have been found in previous DNS studies [6, 30, 31] to preferentially concentrate more than particles of smaller inertia.

As for the mean wall normal component (Fig. 3), positive values of the SGS correction term are found in the near wall region, corresponding to a correction oriented away from the wall, while it becomes negative as z^+ increases, to eventually vanish moving towards the center of the channel. This behavior is again related to the effects of filtering on the normal fluid velocity seen by the particles. It has been observed in DNS that the mean fluid velocity seen by the particles is positive in the near wall region [30], due to the fact that in the near-wall region inertial particles preferentially sample ejection-like environments, in spite they have a mean drift directed towards the wall. Fig. 4 reports the mean normal velocity of the fluid seen by the different considered sets of particles in DNS, confirming the previously described behavior. Note how again the preferential sampling of the ejection-like environments is more pronounced for particles having larger inertia. The same velocity profiles as for DNS obtained in a-priori

LES are also shown in Fig. 4. The effect of filtering is to reduce the mean fluid normal velocity seen by the particles near the wall and to overestimate it moving towards the center of the channel. This is due to a smoothing of the near-wall structures, which, as for the streamiwse streaks, tends to attenuate the sweep events (see e.g. Fig. 6 in [32]). In practice, the mean correction term is given by the difference between the mean wall-normal fluid velocities seen by the particles in DNS and in LES. Thus, the correction term in mean tends to be positive near the wall, with a peak at $z^+ = 15 - 20$, and eventually becomes negative by moving towards the channel center. There is a significant effect of particle inertia on the values of z^+ at which the positive and negative peaks are located and at which the correction changes sign, being larger for the particles having $St = 25$ than for the other two considered particle sets. Note, however, that also the fluid velocity seen by the particles in DNS significantly varies with the particle inertia, this indicating a different interplay between the near-wall vortical structures, sweep and ejection events and particles, depending on particle inertia. Therefore, it is not surprising that the effects of filtering in wall-normal direction are also different for different particle inertia. It will be shown in Sec. 3.2 that the mean correction term in the wall-normal direction is also significantly sensitive to the filter width.

Note that preferential concentration and the interaction with the near-wall turbulent structures have been identified in previous DNS studies [6, 30, 31] as key ingredients to explain turbophoresis, i.e. the tendency of inertial particles to accumulate at solid walls. On the other hand, it has been observed [12, 13, 16] that LES underestimates the turbophoretic effect. The previous analysis of the mean SGS velocity correction term seems to indicate that this underestimation is related to the fact the near wall turbulent structures and their interactions with particles are incorrectly captured in LES due to errors introduced by filtering.

The r.m.s. profiles of the three components of the correction term are reported in Figs. 5-7. In all directions, the main effect is to compensate the underestimation of fluid velocity fluctuations due to filtering. This reduction of the fluid velocity fluctuations in LES is well known since it can also be found in classical error analysis carried out at fixed points, and the SGS models previously used for LPT in LES fields, such as for instance approximate deconvolution or filtering inversion, are mainly aimed at counteracting this effect, i.e. at reintroducing the correct amount of fluid velocity fluctuations [13, 16]. The r.m.s. profiles obtained for the three different sets of particles are very similar each other and thus no significant effects of particle inertia are present.

Figures 8-10 show the skewness profiles of the three components of the correction term. For the spanwise component, the skewness values remain very low along the whole channel, roughly oscillating around zero. Conversely, the streamwise component is characterized by significant values of the skewness, positive very close to the wall and negative in the largest part of the channel. Finally the skewness of the normal component of the correction is also characterized by positive significant values close to the wall, which become negative as z^+ increases, eventually become slightly positive again to vanish at the center of the channel. This behavior indicates that significant asymmetric deviations from Gaussianity are expected in the pdfs of the streamwise and normal components of the correction term, and, thus, that significant contributions to the streamwise and normal components of the correction term are given by isolated events of a given sign. This is consistent with the previous analysis, in which it has been shown that the particles mainly feel the effects of filtering on the low-speed streaks (negative velocity fluctuations) and on ejection events. The effects of particle inertia on the skewness of the streamwise component of the correction term appear to be more significant in the central part of the channel, while for the normal component larger differences are observed in the near wall region.

Finally, the flatness profiles of the three components of the correction term are reported in Figs. 11-13. In all cases large values are present very near the wall decreasing to an almost constant value of about 4-5 moving towards the center of the channel (3 being the

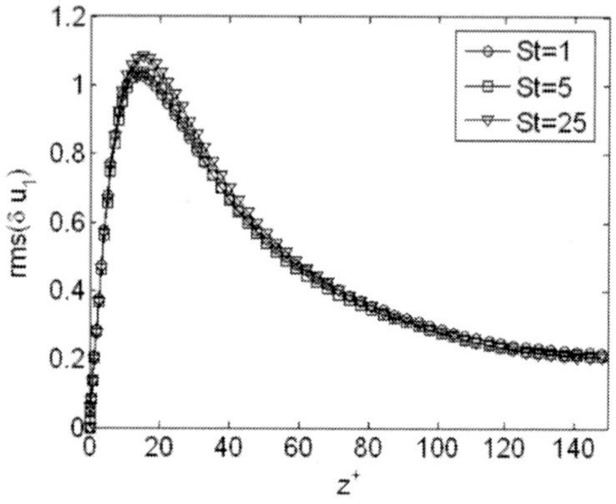

Figure 5. R.m.s values of the SGS velocity correction component in the streamwise direction as a function of z^+. Cut-off filter with CF=4.

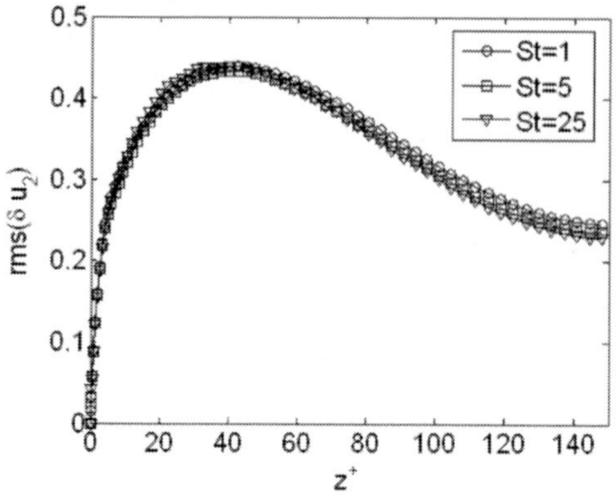

Figure 6. R.m.s values of the SGS velocity correction component in the spanwise direction as a function of z^+. Cut-off filter with CF=4.

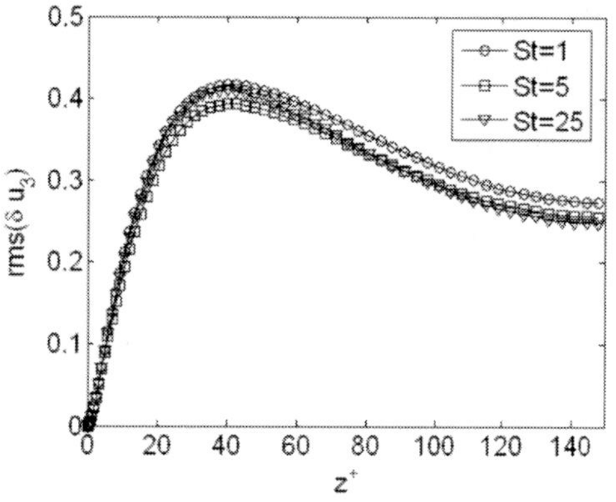

Figure 7. R.m.s values of the SGS velocity correction component in the wall-normal direction as a function of z^+. Cut-off filter with CF=4.

value corresponding to a Gaussian pdf). The effects of particle inertia are negligible.

3.2. Influence of filtering

We investigate now the effect of filter type and width on the statistics of the SGS velocity correction and we consider the particle set characterized by St=5.

As for the mean value of the streamwise component of the correction, shown in Fig. 14, the behavior is qualitatively similar for all the considered filters but the absolute value of the negative peak in the near wall region increases with the filter width. This could be expected, since, as pointed out in Sec. 3.1, this negative peak of the correction is related to the need of counteracting the attenuation of the low-speed streaks, preferentially sampled by particles, due to filtering. As for filter type, at fixed width, the top-hat filter introduces a stronger smoothing of the fluid velocity fluctuations than the cut-off one and thus the mean correction term is more important for the top-hat filter, especially for CF=2 and CF=4. The mean spanwise component of the correction term, for all the considered filters, always has very low values roughly oscillating around zero (not shown here for the sake of brevity). Figure 15 shows the mean normal component. Qualitatively, in all cases the behavior is the one described in Sec. 3.1, i.e. the mean correction is positive in the near-wall region, then becomes negative to eventually vanish at the center of the channel. However, the filter width not only affects the values of the

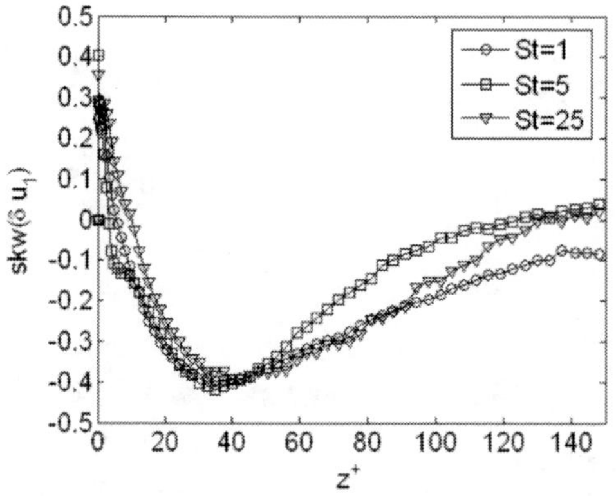

Figure 8. Skewness of the SGS velocity correction component in the streamwise direction as a function of z^+. Cut-off filter with CF=4.

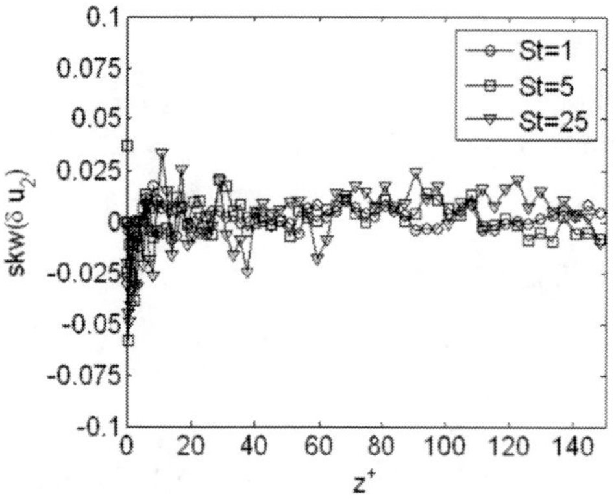

Figure 9. Skewness of the SGS velocity correction component in the spanwise direction as a function of z^+. Cut-off filter with CF=4.

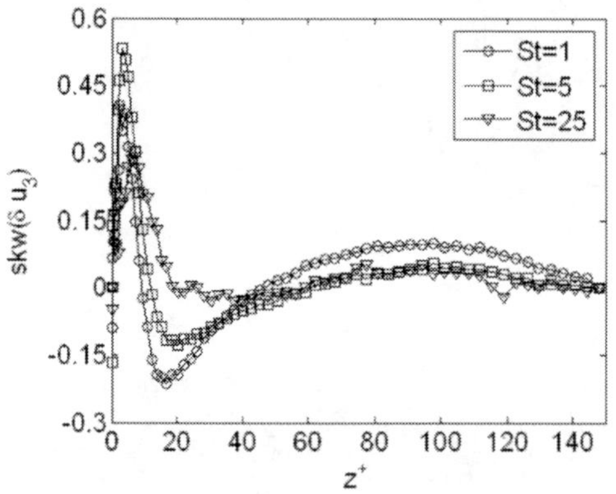

Figure 10. Skewness of the SGS velocity correction component in the wall-normal direction as a function of z^+. Cut-off filter with CF=4.

positive and negative peaks, but also their location, which both tend to move away from the wall as the filter width increases.

As for the r.m.s. of all the components of the correction term, reported in Figures 16-18, their values increase with the filter width, as expected since for larger filter widths a larger amount of fluid velocity fluctuations is damped. There is no significant effect of the filter type.

The skewness of the different components of the SGS correction term is shown in Figs. 19-21. As observed in Sec. 3.1, the values for the spanwise component are practically negligible for all the considered filters. Significant effects of the filter width, and in a less extent of the filter type, are conversely observed for both the streamwise and wall-normal components, although the qualitative behaviors remain the same.

Finally, the SGS velocity correction flatness slightly tends to decrease when increasing the filter width (Figs. 22-24), but the behavior remains qualitatively the same for all the considered cases as described in Sec. 3.1.

3.3. Concluding Remarks

The pure effect of filtering on the fluid velocity seen by inertial heavy particles has been singled out by imposing that the particle trajectories in a-priori LES of turbulent channel flow coincide with those computed in DNS. An ideal SGS correction term has therefore been computed for

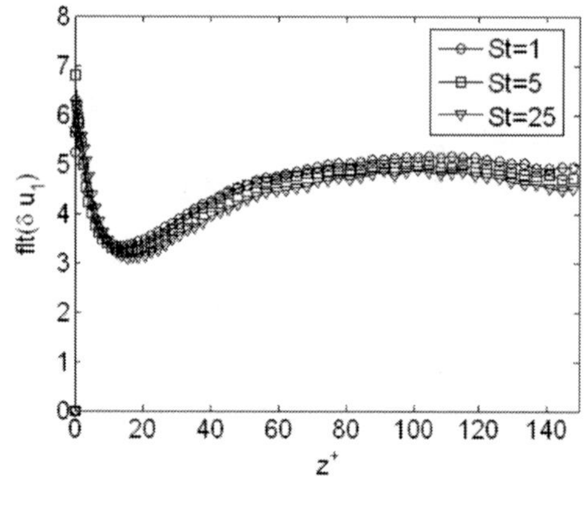

Figure 11. Flatness of the SGS velocity correction component in the streamwise direction as a function of z^+. Cut-off filter with CF=4.

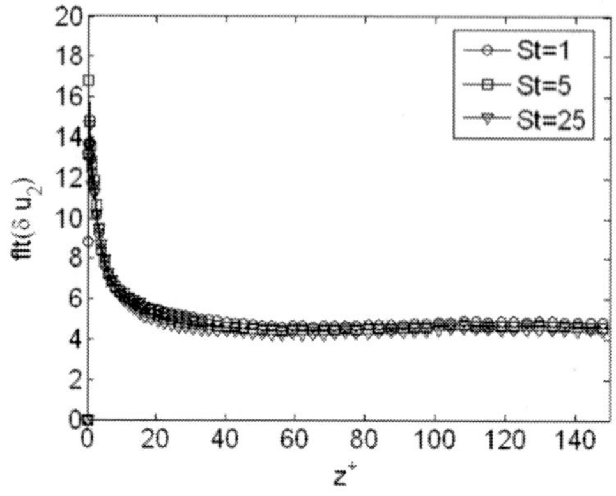

Figure 12. Flatness of the SGS velocity correction component in the spanwise direction as a function of z^+. Cut-off filter with CF=4.

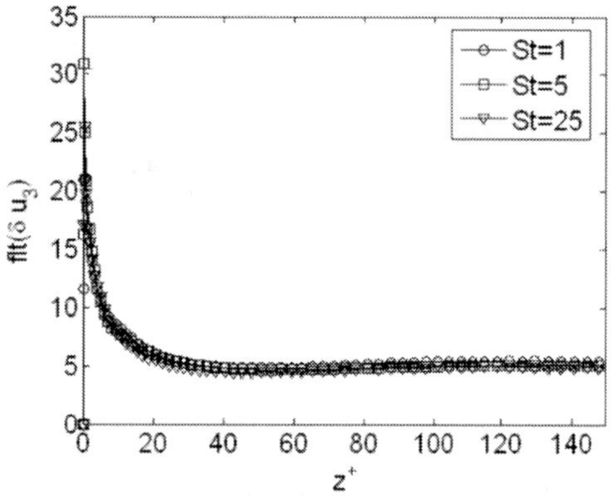

Figure 13. Flatness of the SGS velocity correction component in the wall-normal direction as a function of z^+. Cut-off filter with CF=4.

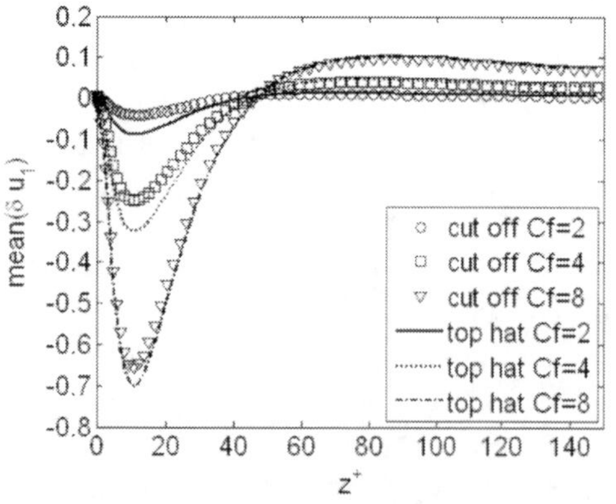

Figure 14. Mean values of the correction term component in the streamwise direction as a function of z^+. St=5.

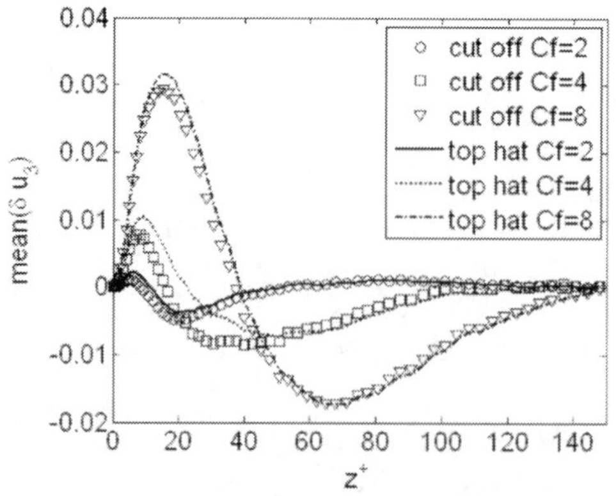

Figure 15. Mean values of the correction term component in the wall-normal direction as a function of z^+. St=5.

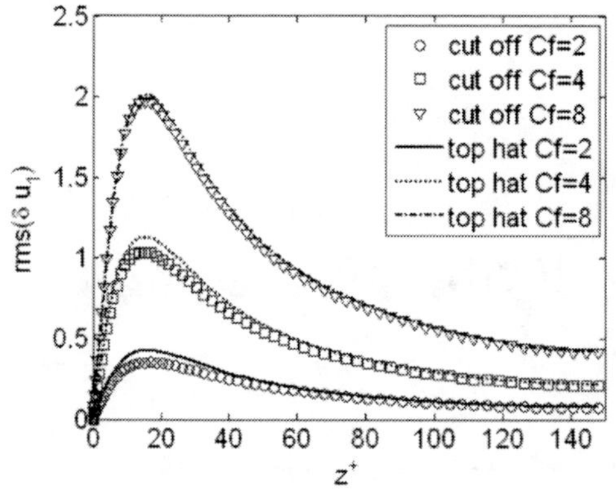

Figure 16. R.m.s. values of the correction term component in the streamwise direction as a function of z^+. St=5.

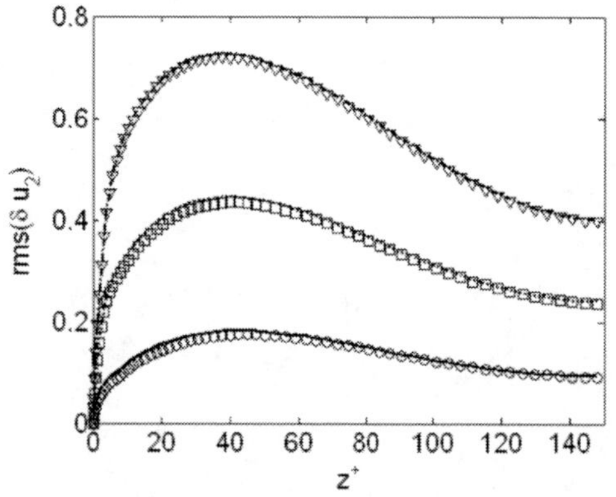

Figure 17. R.m.s. values of the correction term component in the spanwise direction as a function of z^+. St=5.

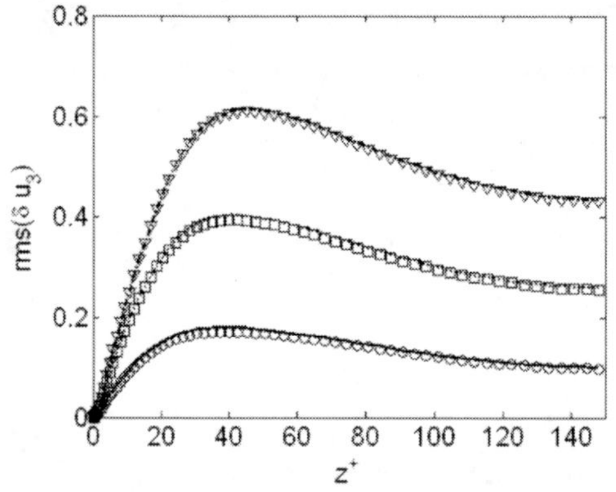

Figure 18. R.m.s. values of the correction term component in the wall-normal direction as a function of z^+. St=5.

different particle inertia and different filter types and widths. The statistical moments of this correction term have been investigated.

The first surprising result is that the mean values of the correction term components in the streamwise and wall-normal directions are not zero, as it would be expected from an Eulerian analysis, i.e. by looking at filtering effects on the mean value of fluid velocity at fixed points. Indeed, the ideal correction takes into account the effects of filtering on the fluid velocity seen by the particles, i.e. computed along the particle trajectories and the non-zero mean values depend on the effects of filtering on the near-wall turbulence structures combined with particle preferential sampling. As for the streamwise component, the correction term in mean tends to reintroduce the effects of low-speed streaks preferentially sampled by the particles and smoothed by filtering; therefore, it is characterized by a negative peak in the near wall region. The quantitative value of this near-wall peak depends on both particle inertia and filter width. As for the mean wall-normal component, the behavior of the ideal correction term is more complex, but again tends to compensate filtering effects on the near-wall structures, which in turn affect ejection events, preferentially sampled by inertial particles. Particle inertia and filter width also affect the qualitative behavior of this component of the mean SGS correction. In spite of this different and complex behavior, as a general remark, the error introduced by filtering on inertial particle motion is not only due to the reduction of fluid velocity fluctuations, which is a well known effect pointed out in classical error analysis in LES carried out for fixed points. This

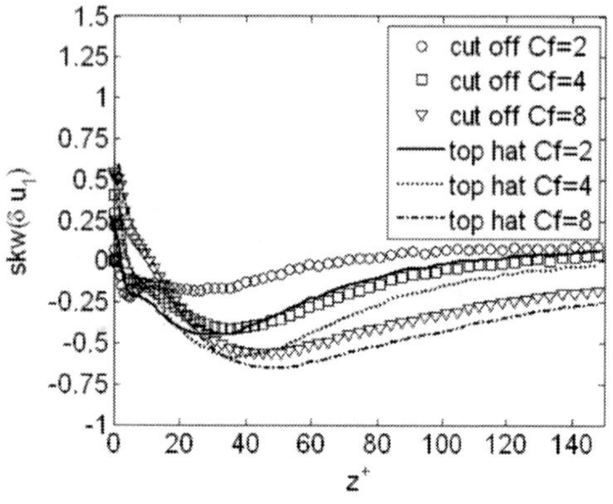

Figure 19. Skewness of the correction term component in the streamwise direction as a function of z^+. St=5.

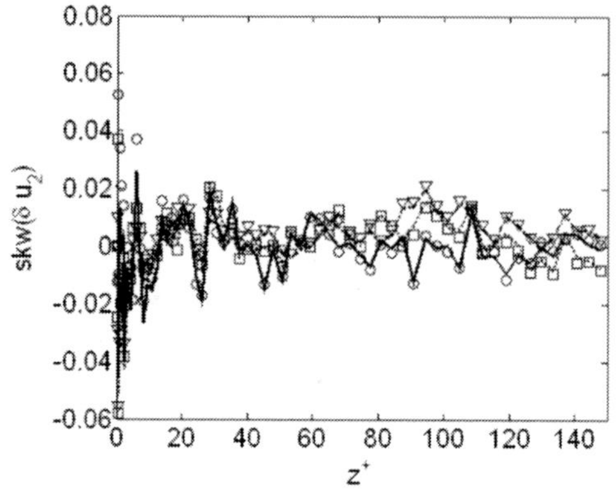

Figure 20. Skewness of the correction term component in the spanwise direction as a function of z^+. St=5.

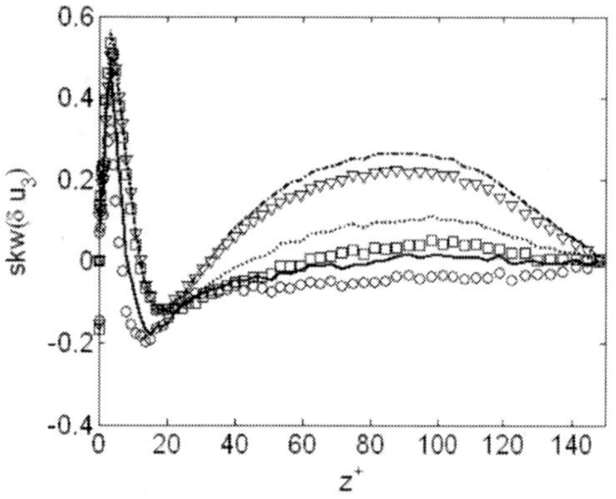

Figure 21. Skewness of the correction term component in the wall-normal direction as a function of z^+. St=5.

supports the conclusion drawn in our previous [16, 22], i.e. that the reintroduction of the correct amount of fluid velocity fluctuations by a SGS model seems not to be enough to have an accurate prediction of particle preferential concentration and near-wall accumulation.

By analyzing the higher-order moments of the correction term, the pdf of the SGS correction term is expected to significantly deviate from a Gaussian distribution. This deviation seems to be more important near the wall and the shape of the pdf is expected to be different for the different components of the ideal correction. Conversely, the effects on the pdf shape of the filter width and type, and even more of particle inertia, are expected to be small. As previousl mentioned, the Eulerian and Lagrangian pdfs of the ideal correction term will be presented and analyzed in forthcoming studies.

It would be also interesting to compute the ideal SGS correction for different values of the Reynolds number, in order to investigate whether Reynolds number effects should be taken into account.

Acknowledgments
CINECA supercomputing center (Bologna, Italy) is gratefully acknowledged for generous allowance of computer resources. The authors wish to thank Prof. B.J. Geurts for the interesting discussion which stimulated the present study.

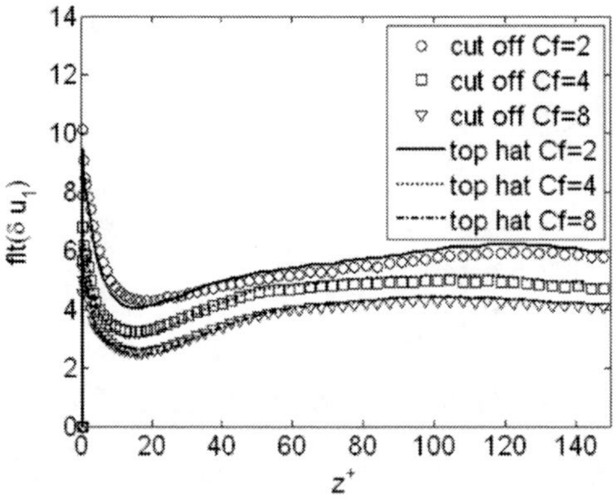

Figure 22. Flatness of the correction term component in the streamwise direction as a function of z^+. St=5.

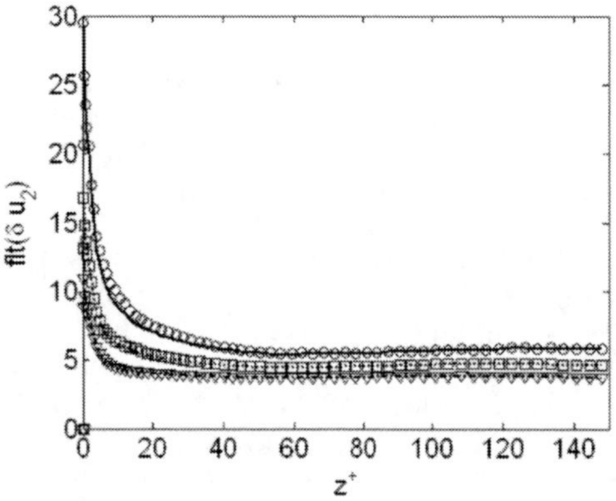

Figure 23. Flatness of the correction term component in the spanwise direction as a function of z^+. St=5.

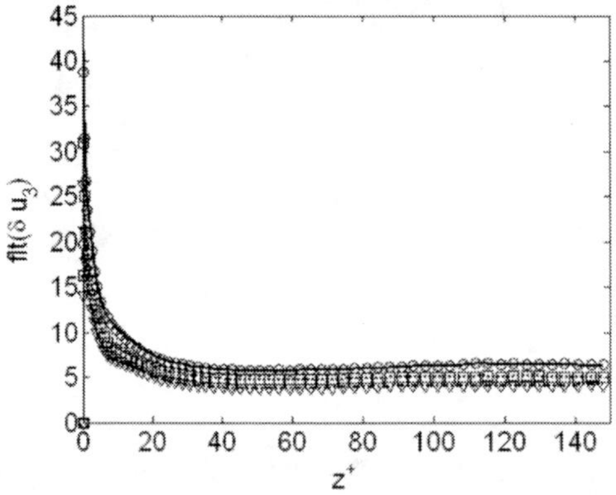

Figure 24. Flatness of the correction term component in the wall-normal direction as a function of z^+. St=5.

References

[1] Wang L P and Maxey M R 1993 *J. Fluid Mech.* **256** 27
[2] Marchioli C and Soldati A 2002 *J. Fluid Mech.* **468** 283
[3] Bec J, Biferale L, Cencini M, Lanotte A, Musacchio S and Toschi F 2007 *Phys. Rev. Lett.* **98** 084502
[4] Fede P and Simonin O 2006 *Phys. Fluids* **18** 045103
[5] Rouson D W and Eaton J K 2001 *J. Fluid Mech.* **428** 149
[6] Soldati A and Marchioli C 2009 *Int. J. Multiphase Flow* **20** 827
[7] Picciotto M, Marchioli C and Soldati A 2005 *Phys. Fluids* **17** 098101
[8] Reeks M W 1983 *J. Aerosol Sci.* **14** 729
[9] Uijttewaal W S J and Oliemans R W A 1996 *Phys. Fluids* **8** 2590
[10] Wang Q and Squires K D 1996 *Int. J. Multiphase Flow* **22** 667
[11] Armenio V, Piomelli U and Fiorotto V 1999 *Phys. Fluids* **11** 3030
[12] Kuerten J G M and Vreman A W 2005 *Phys. Fluids* **17** 011701
[13] Kuerten J G M 2006 *Phys. Fluids* **18** 025108
[14] Yamamoto Y, Potthoff M, Tanaka T, Kajishima T and Tsuji Y 2001 *J. Fluid Mech.* **442** 303
[15] Marchioli C, Armenio V, Salvetti M V and Soldati A 2006 *Phys. Fluids* **18** 025102
[16] Marchioli C, Salvetti M V and Soldati A 2008 *Phys. Fluids* **20** 040603
[17] Vreman A W, Geurts B J, Deen N G, MKuipers J A and Kuerten J G M 2009 *Flow Turb. Comb.* **82** 47
[18] Sengupta K, Shotorban B, Jacobs G B and Mashayek F 2009 *Int. J. Multiphase Flow* **35** 811
[19] Khan M A I, Luo X Y, Nicolleau F C G A, Tucker P G and Iacono G L 2010 *Int. J. Num. Meth. Biomed. Eng.* **26** 999
[20] Shotorban B and Mashayek F 2005 *Phys. Fluids* **17** 081701
[21] Gobert C 2010 *J. Turbulence* **11** 1

[22] Marchioli C, Salvetti M V and Soldati A 2008 *Acta Mech.* **201** 277
[23] Shotorban B and Mashayek F 2006 *J. Turbulence* **7** 18
[24] Pozorski J and Apte S V 2009 *Int. J. Multiphase Flow* **35** 118
[25] Calzavarini E, Donini A, Lavezzo V, Marchioli C, Pitton E, Soldati A and Toschi F 2011 *Direct and Large-Eddy Simulation 8* Ercoftac Series ed Armenio V, Geurts B, Frohlich J and Kuerten J G M (Springer)
[26] Elghobashi S E and Truesdell G C 1992 *J. Fluid Mech.* **242** 655
[27] Crowe C, Sommerfeld M and Tsuji T 1998 *Multiphase flows with droplets and particles* (New York: CRC Press)
[28] Pan Y and Banerjee S 1996 *Phys. Fluids* **17** 2733
[29] Pope S 2004 *New J. Phys.* **6** 1
[30] Picciotto M, Marchioli C, Reeks M W and Soldati A 2005 *Nuclear Eng. Design* **235** 1239
[31] Picciotto M, Marchioli C and Soldati A 2005 *Phys. Fluids* **17** 098101
[32] Marchioli C, Salvetti M V and Soldati A 2011 *Quality and Reliability of Large-Eddy Simulations II* Ercoftac Series ed Salvetti M V, Geurts B J, Meyers J and Sagaut P (Springer)

Some remarks on the inverse Smoluchowski problem for cluster-cluster aggregation

Colm Connaughton[1 2] and Peter P. Jones[2]

[1] Mathematics Institute, University of Warwick, Coventry CV4 7AL, UK
[2] Centre for Complexity Science, University of Warwick, Coventry CV4 7AL, UK

E-mail: `connaughtonc@gmail.com`

E-mail: `P.P.Jones@warwick.ac.uk`

Abstract. It is proposed to revisit the inverse problem associated with Smoluchowski's coagulation equation. The objective is to reconstruct the functional form of the collision kernel from observations of the time evolution of the cluster size distribution. A regularised least squares method originally proposed by Wright and Ramkrishna (1992) based on the assumption of self-similarity is implemented and tested on numerical data generated for a range of different collision kernels. This method expands the collision kernel as a sum of orthogonal polynomials and works best when the kernel can be expressed exactly in terms of these polynomials. It is shown that plotting an "L-curve" can provide an a-priori understanding of the optimal value of the regularisation parameter and the reliability of the inversion procedure. For kernels which are not exactly expressible in terms of the orthogonal polynomials it is found empirically that the performance of the method can be enhanced by choosing a more complex regularisation function.

1. Introduction

The effects of clouds and precipitation is one of the largest sources of uncertainty in our current attempts to simulate the Earth's climate [1]. The reason for this is that most phenomena associated with clouds happen on scales below those which are explicitly resolved by the current climate models. Their feedback on resolved scales must therefore be parameterised. Such parameterisations require a strong understanding of the underlying physical processes taking place. One such process which has attracted considerable attention in recent years is the time evolution of the droplet size distribution in clouds and its connection with precipitation. Water droplets in clouds are seeded by cloud condensation nucleii such as aerosol particles. They initially grow by condensation and subsequently by coagulation as droplets collide with each other in the cloud to produce larger droplets. The detailed micro-physics of the coagulation process is difficult to understand theoretically since turbulence within the cloud is believed to play a key role in determining the collision rate of droplets [2]. A complete theoretical description is hampered by the difficulty in describing the statistical interplay between particles and turbulence analytically. On the other hand the quality of the data available for the study of this problem is improving rapidly due to recent advances in observational techniques [3] and direct numerical simulation of droplets in turbulent flows [4, 5].

In the context of the droplet coagulation problem, much research has focused on using our improved observational understanding of the behaviour of droplets in turbulent flows to calculate

Journal of Physics: Conference Series **333** (2011) 012005 doi:10.1088/1742-6596/333/1/012005

more accurately the functional form of the droplet coagulation rate, $K(m_1, m_2)$, as a function of droplet masses, or equivalently (if droplets are assumed spherical), droplet size. In this note, we argue that the improved availability of data suggests that we should also revisit the corresponding inverse problem which can be stated as follows: given observations or measurements of the time evolution of the droplet size distribution, how much information can be extracted about the functional form of the coagulation rate? This problem was studied in the past by Wright and Ramkrishna [6] in the context of chemical mixing and has been recently revisited in the context of droplets in turbulence by Onishi et al. [7]. Although there are many difficulties associated with such inverse problems, as discussed below, there are potential rewards. The approach is data driven and could provide a guide to modeling in situations where the underlying microphysics remains incompletely understood. The inverse approach could also begin to quantify the extent to which available data and measurements can distinguish between different models. In what follows, following the original work of Wright and Ramkrishna [6] we shall assume scaling in formulating the inverse problem. We therefore require that the evolution of the cluster size distribution to be sufficiently self-similar as described below.

First a word about the usual forward problem. If the collision kernel, $K(m_1, m_2)$, is known, the evolution of the cluster size distribution, $N_m(t)$ is described by the Smoluchowski coagulation equation [8]:

$$\begin{aligned}\partial_t N_m(t) &= \int_0^m dm_1\, K(m_1, m - m_1)\, N_{m_1}(t)\, N_{m-m_1}(t)\\ &- 2N_m(t) \int_0^\infty dm_1\, K(m, m_1)\, N_{m_1}(t). \qquad (1)\end{aligned}$$

It is applicable when spatial correlations between particles are sufficiently weak that collisions between particles can be considered statistically independent. A huge amount is known theoretically about the solutions of the Smoluchowski equation. See [9] for a modern review.

2. Homogeneous collision kernels and self-similarity

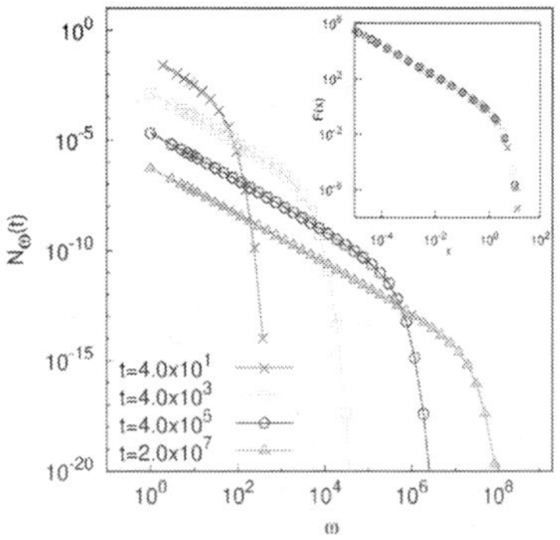

Figure 1. Snapshots of the self-similar evolution of the cluster size distribution, $N_m(t)$, obtained from a numerical simulation of Eq. (1) with the collision kernel $K(m_1, m_2) = (m_1, m_2)^{\frac{1}{4}}$. The inset shows the scaling function obtained when the data is collapsed according to Eq. (2) with the typical scale, $s(t)$, obtained as in Eq. (3) with $n = 2$.

In some applications (see [10] for some discussions), the collision kernel is a homogeneous function of its arguments whose degree we shall denote by γ:

$$K(am_1, am_2) = a^\gamma\, K(m_1, m_2).$$

For such kernels, the evolution of the cluster size distribution is often self-similar. That is to say we can write:

$$N_m(t) \sim s(t)^{-2} f(z) \qquad z = \frac{m}{s(t)} \qquad (2)$$

where $s(t)$ is the typical cluster size. This can be defined intrinsically as a ratio of moments of the size distribution:

$$s(t) = \frac{M_{n+1}(t)}{M_n(t)} \qquad M_p(t) = \int_0^\infty m^p\, N_m(t)\, dm. \tag{3}$$

The scaling function, $f(z)$, determines the shape of the cluster size distribution. An example of this self-similar evolution obtained from numerical simulation of Eq. (1) with $K(m_1, m_2) = (m_1\, m_2)^{1/4}$ is shown in Fig. 1. The inset shows the scaling function, $f(z)$, obtained by collapsing the data according to Eq. (2) with the typical scale, $s(t)$, obtained as in Eq. (3) with $n = 2$. All numerical simulations of Eq. (1) reported in this paper were done using the pairwise binning method described in [11].

In what follows, our focus on self-similar problems simply stems from the fact that they provide the simplest class of coagulation problems on which we can begin to study the inverse problem described above. For kernels having $\gamma > 1$, the typical cluster size, $s(t)$, diverges in a finite time, t_*, a phenomenon known as gelation [9] and self-similarity is only observed in the vicinity of this singularity. This introduces technical problems which we wish to avoid. We therefore restrict our discussion to kernels having $\gamma < 1$ in order to maintain our focus on the inverse problem. A homogeneous collision kernel and self-similar time evolution are not plausible assumptions for the droplet coagulation problem in clouds which has motivated this study [12]. If one restricts attention to the gravitational coagulation-dominated regime, one could perhaps argue that the kernel is approximately homogeneous. It turns out, however, that the resulting differential sedimentation kernel,

$$K(m_1, m_2) = \left(m_1^{\frac{1}{3}} + m_2^{\frac{1}{3}}\right)^2 \left|m_1^{\frac{2}{3}} - m_2^{\frac{2}{3}}\right|,$$

is homogeneous of degree $\gamma = 4/3$, and might be expected to exhibit gelation. Closer investigation [13] reveals, however, that the solution of the Smoluchowski equation with this kernel (at least if taken literally) is ill-defined since the gelation is complete (all clusters merge into an infinite cluster at $t = t_*$) and instantaneous ($t_* = 0+$) . For details the interested reader is referred to [13, 14].

3. The inverse Smoluchowski problem

Let us introduce the cumulative cluster mass distribution:

$$\mathcal{F}_m(t) = \int_0^m m_1\, N_{m_1}(t) dm_1. \tag{4}$$

The original cluster size distribution, $N_m(t)$ is recovered from $\mathcal{F}_m(t)$ by differentiation:

$$N_m(t) = \frac{1}{m}\frac{\partial \mathcal{F}_m(t)}{\partial m}. \tag{5}$$

In terms of $\mathcal{F}_m(t)$, Eq. (1) can be written in the compact form:

$$\partial_t \mathcal{F}_m(t) = -\int_0^m d\mathcal{F}_{m_1}(t) \int_{m-m_1}^\infty \frac{d\mathcal{F}_{m_2}(t)}{m_2} K(m_1, m_2) \tag{6}$$

If we assume scaling as in Eq. (2) then $\mathcal{F}_m(t) = F(z)$ and the scaling function of the cumulative mass distribution, $F(z)$, satisfies the following scaling equation:

$$z\frac{dF}{dz} = -\int_0^z dF(z_1) \int_{z-z_1}^\infty \frac{dF(z_2)}{z_2} K(z_1, z_2). \tag{7}$$

This is a complicated nonlinear integro-differential equation if we are required to determine $F(z)$ given $K(z_1, z_2))$ (the forward problem). It is, however, a linear equation if we are required to determine $K(z_1, z_2)$ given $F(z)$ (the inverse problem). This inverse problem is, however, ill-posed. This is most easily seen by considering what happens if we discretise Eq. (7) on N z-points. We obtain a linear system,

$$\mathbf{g} = S\,\mathbf{k},$$

consisting of N equations for the N^2 values of $K(z_1, z_2)$ on the discretisation points. The entries of the vector $\mathbf{g}$ are obtained from a centred difference approximation to the left-hand side of Eq. (7) and the entries of the matrix S are obtained from approximating the integrals on the right-hand side of Eq. (7) with Riemann sums. Actually we can reduce the number of unknowns almost by a factor of 2 since $K(z_1, z_2) = K(z_2, z_1)$ but the conclusion remains unchanged. This system is enormously under-determined, which implies that one can find many solutions but they are typically over-fitting the measurements of $F(z)$.

One way of dealing with under-determined systems is via Tikhonov Regularisation (Ridge regression). The idea is to solve a minimization problem with a regularisation term:

$$\mathbf{k}_{\text{est}} = \arg\min_{\mathbf{k}} \left\{ \|S\,\mathbf{k} - \mathbf{g}\|^2 + \lambda \|\mathbf{k}\|^2 \right\}. \tag{8}$$

Noise-dominated solutions have to compete against the regularization term $\lambda\|\mathbf{k}\|^2$ in the minimisation. This approach was pioneered by Wright and Ramkrishna [6]. It has the advantage that, in principle, one does not need to know the functional form of the collision kernel a-priori. The trick is to choose the "best" value of the regularization parameter, λ. Before discussing the selection of the value of λ, we discuss briefly how to set up the minimization problem.

In this article we formulate the inverse problem in the way suggested in [6] and focus on the determination of the most appropriate value of λ. We assume that $K(m_1, m_2)$ can be expanded in terms of Laguerre polynomials, $L_i(x)$, up to order p in the m_1 and m_2 directions:

$$K(m_1, m_2) = \sum_{n=1}^{p^2} a_n\, L_n(m_1, m_2) \tag{9}$$

where

$$L_n(x, y) = L_i(x)\, L_j(y)$$

where $n = (i-1)\,p + j$ is a compound index which rolls the indices, i and j, in the m_1 and m_2 directions into one. With Eq. (9), the minimisation problem Eq. (8) is transformed into a different minimisation which aims to determine the coefficients, a_n in Eq. (9). This has a number of advantages. It ensures that the minimisation problem automatically finds kernels which are fairly smooth functions of m_1 and m_2. It also enormously reduces the size of the problem since the number of polynomials, p, required in each direction is typically rather small. Furthermore, this choice allows the entries of the matrix S in Eq. (8) to be calculated semi-analytically as described in [6]. It has the disadvantage, however, of constraining the solution to those kernels which can be expressed in the form of Eq. (9) which effectively requires a-priori knowledge of the class of kernels which we are seeking, as was done in [7].

As mentioned above, the tricky part of this procedure is to determine the most apprpriate value of λ to use in Eq. (8). If λ is too small, the result will over-fit the observations. If λ is too large then $\mathbf{k}_{\text{est}}$ will be pushed to zero by the weight of the second term in Eq. (8) and will retain little information about the Smoluchowski dynamics encoded in the matrix S. A rational approach to determining λ is provided by plotting an "L-curve" (for a clear review see [15]). This is a plot of the norm of the solution, $\|\mathbf{k}\|$, as a function of the norm of the residual, $\|S\,\mathbf{k} - \mathbf{g}\|$, for a range of values of λ. If there is a clear transition in the minimisation problem

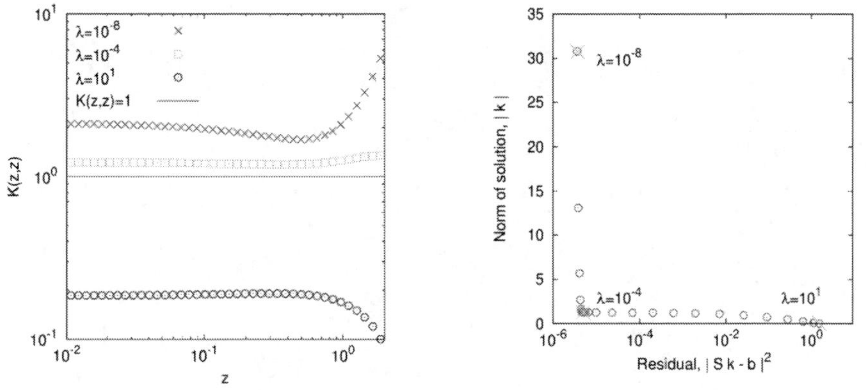

Figure 2. Results for the constant kernel, $K(m_1, m_2) = 1$. The right panel shows the L-curve obtained by performing the minimisation Eq. (8) over a range of values of λ. The left panel shows the diagonal, $K(z, z)$, of the reconstructed kernels compared to the theoretical curve for values of λ in the upper left, lower right and the "elbow" of the L-curve.

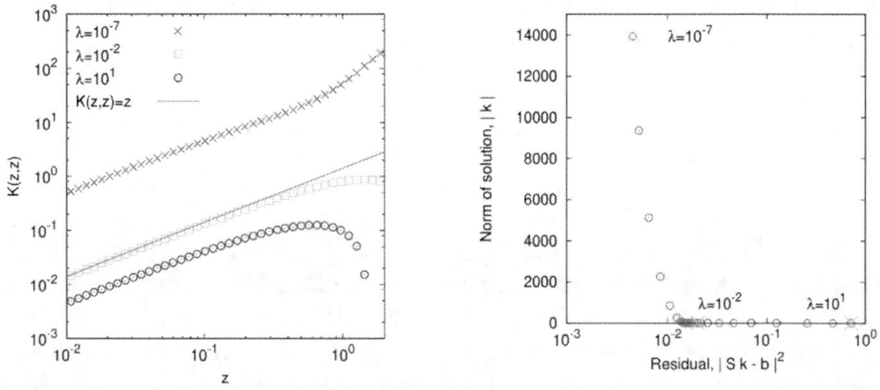

Figure 3. Results for the sum kernel, $K(m_1, m_2) = \frac{1}{2}(m_1 + m_2)$. The right panel shows the L-curve obtained by performing the minimisation Eq. (8) over a range of values of λ. The left panel shows the diagonal, $K(z, z)$, of the reconstructed kernels compared to the theoretical curve for values of λ in the upper left, lower right and the "elbow" of the L-curve.

from a regime dominated by overfitting to a regime dominated by the regularisation then the L-curve will have the distinctive L-shape and the best values of λ are those in the "elbow" of the curve.

4. Results

The regularised least squares method described above was applied to several sets of data obtained by the numerical integration of Eq. (1) with monodisperse initial data with different model kernels. We used 40 discretisation points in the interval $z \in [0.01, 2.0]$ and chose $p = 3$ Laguerre polynomials in each direction. In each case, the data collapse and extraction of the scaling function was done using the method described in [16] and then fitted to a function of the form $C\, z^{\alpha-1}\, e^{-\beta z}$ as suggested in [6]. These fitted scaling functions were then used as the

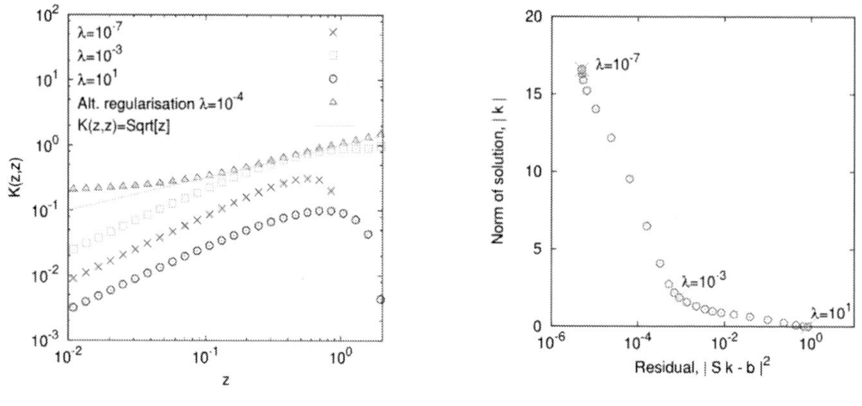

Figure 4. Results for the generalised sum kernel, $K(m_1, m_2) = \frac{1}{2}(\sqrt{m_1} + \sqrt{m_2})$. The right panel shows the L-curve obtained by performing the minimisation Eq. (8) over a range of values of λ. The triangles show the best results obtained by experimentation with empirical regularisation term Eq. (11). The left panel shows the diagonal, $K(z, z)$, of the reconstructed kernels compared to the theoretical curve for values of λ in the upper left, lower right and the "elbow" of the L-curve.

"observations", $F(z)$, in the discretisation of Eq. (7).

Figs. 2 and 3 show some results of the regularised least squares for the constant kernel, $K(m_1, m_2) = 1$, and the sum kernel, $K(m_1, m_2) = \frac{1}{2}(m_1 + m_2)$, respectively. These are good test cases to begin with since the solution of Eq. (1) with monodisperse initial data is known explicitly for each of these kernels allowing the numerical aspects of the calculation to be validated. The right panels of the figures show the L-curves obtained by performing the minimisation (8) over a range of values of λ. In both cases, a clear elbow is visible in the resulting curve indicating an appropriate range of values for λ. The left panel shows, for ease of visualisation, the diagonal, $K(m, m)$, of the kernels obtained by regularised least squares as a function of m for different values of λ. The λ curves shown correspond to values of λ which are too big, too small and "just right" (meaning a value in the elbow of the L-curve). It is clear that the method does a reasonable job of extracting the basic shape of the kernel given the scaling function, $F(z)$. Surprisingly, we found that it was not necessary to explicitly enforce the positivity of $K(m_1, m_2)$ by constrainting the minimisation as was done in [6]. The method seemed to do equally well, and in some cases, better without these constraints. For the sum kernel, we did enforce the constraint that the kernel should vanish for particles of zero mass which seems physically reasonable - the results were less convincing without this constraint.

Fig. 4 shows the corresponding results for the generalised sum kernel $K(m_1, m_2) = \frac{1}{2}(\sqrt{m_1} + \sqrt{m_2})$. The results are clearly less convincing. This is probably related to the fact that this kernel cannot be expressed in terms of the Laguerre polynomials chosen to represent the kernel in Eq. (9). It is worth pointing out that the corresponding L-curve does not have a sharp elbow indicating that the method does not find a clear "best" value of λ. This seems to illustrate a point in favour of such methods - the fact that the L-curve does not have a sharp elbow provides an a-priori indication that the results of the minimisation should be treated with caution.

Empirical investigation suggest that the results for the generalised sum kernel can be improved by tinkering with the regularisation procedure. If, instead of, Eq. (8), we perform the minimisation

$$\min_{\mathbf{k}} \left\{ \|S\mathbf{k} - \mathbf{g}\|_2^2 + \lambda\, w(\mathbf{k}) \right\} \tag{10}$$

where

$$w(\mathbf{k}) = \sum_i \log(|k_i| + 1) \tag{11}$$

we found that the results were much better (see the triangles in the left panel of Fig. 4) although this regularisation does not seem to produce an intelligible L-curve which would help to determine λ a-priori. This form for w was found by experimentation. At present, this statement is at the level of empirical observation and requires further investigations.

5. Discussion and outlook

We conclude, as several previous authors have done, that the inverse Smoluchowski problem is technically feasible and could potentially be developed into a useful tool for the study of droplet size distributions in clouds and other applications where the underlying microphysics is still incompletely understood. The results presented here indicate that the L-curve provides a useful complementary tool to the methods developed by Wright and Ramkrishna [6] to allow the regularisation parameter to be selected a-priori in situations where the collision kernel is not known from the outset. It is worth mentioning that the results presented in Figs. (2)-(4) do not give a good indication to the casual reader of the degree of numerical sensitivity required in tackling these problems. It became clear to us during these investigations that the ill-posedness of the inverse problem represented by Eq. (7) requires that great care would need to be taken in interpreting the outputs of these methods if the kernel were not known in advance. Further validation by substituting the reconstructed kernel back into the forward problem and comparing the output against the original data would be a necessary additional step to build confidence in the results. It is also clear that further research is required, even in the simple case of self-similar time evolution, in order to make the method more robust. Our results suggest that we should consider more general functional forms for the collision kernel than Eq. (9) in order to improve this robustness. This will probably not pose much difficulty since the analytic simplification obtained by the use of Laguerre polynomials in [6] is probably less important nowadays owing to the increased computational power which can be brought to bear on the computation of matrix elements by quadrature when closed analytic forms are not available.

In the long run, however, it is clear that it is necessary to free ourselves of the assumptions of homogeneity of the kernel and self-similarity of the cluster size distribution. Some strong progress in this direction has already been made recently by Onishi et al. [7] who have been able to infer the relative importance of the turbulent and gravitational coagulation as a function of Reynolds number from direct numerical simulations of droplet-laden turbulence using inverse methods. Our approach differs slightly from this work in the sense that we would like, as far as possible, to learn the functional shape of the kernel from the data by allowing considerable freedom in the class of possible kernel functions. This approach would be more appropriate in situations when the underlying micro-physics is unknown or controversial.

Acknowledgements

Support from COST Action MP0806 is gratefully acknowledged.

References

[1] Stephens G L 2005 *J. Climate* **18** 237–273

[2] Bodenschatz E, Malinowski S P, Shaw R A and Stratmann F 2010 *Science* **327** 970–971
[3] Siebert H, Lehmann K, Wendisch M, Franke H, Maser R, Schell D, Wei Saw E and Shaw R A 2006 *Bull. Am. Meteorol. Soc.* **87** 1727–1738
[4] Reade W C and Collins L R 2000 *J. Fluid Mech.* **415** 45–64
[5] Wang L P, Ayala O, Rosa B and Grabowski G W 2008 *New J. Phys.* **10** 075013
[6] Wright H and Ramkrishna D 1992 *Comput. Chem. Eng.* **16** 1019–1038
[7] Onishi R, Matsuda K, Takahashi K, Ryoichi K and Komori S 2011 *Int. J. Multiphas. Flow* **37** 125–135
[8] Smoluchowski M V 1917 *Z. Phys. Chem.* **92** 129–168
[9] Leyvraz F 2003 *Phys. Reports* **383** 95–212
[10] Ernst M 1986 Kinetics of clustering in irreversible aggregation *Fractals in Physics* ed Pietronero L and Tosatti E (Amsterdam: North Holland) p 289
[11] Lee M 2000 *Icarus* **143** 74–86
[12] Falkovich G, Fouxon A and Stepanov M G 2002 *Nature* **419** 151–154
[13] van Dongen P 1987 *J. Phys. A: Math. Gen.* **20** 1889–1904
[14] Ball R C, Connaughton C, Stein T H M and Zaboronski O 2011 *Phys. Rev. E* **84** 011111
[15] Hansen P C 2001 The L-curve and its use in the numerical treatment of inverse problems *Computational Inverse Problems in Electrocardiology* ed P Johnston (WIT Press) pp 119–142
[16] Bhattacharjee S and Seno F 2001 *J. Phys. A–Math. Gen.* **34** 6375–6380

DNS experiments on the settling of heavy particles in homogeneous turbulence: two-way coupling and Reynolds number effects

Anne Dejoan

Modeling and Simulation Group, Centro de Investigaciones de Energéticas Medioambientales y Tecnológicas (CIEMAT) , Avenida Complutense, 22, 28040, Madrid, Spain

E-mail: anne.dejoan@ciemat.es

Abstract.
The present study addresses the effects of two-way fluid-particle interaction ("two-way coupling") on the settling of heavy particles in homogeneous turbulence. The modification of turbulence by the particles and the resulting effects on the settling velocity are analyzed by taking into account Reynolds number effects. The turbulent carrier fluid phase is resolved by direct numerical simulation (DNS) and the particle phase by the Lagrangian point-particle approximation. Results are presented for two Taylor microscale Reynolds numbers $R_\lambda = 40$ and 130. While for $R_\lambda = 40$ the Kolmogorov time scale, t_η, is found almost unaltered by the two-way coupling, it is found significantly decreased for $R_\lambda = 130$. The consequent modification of the particle Stokes number $St = \tau_p/t_\eta$, τ_p being the particle response time, shows that comparisons between one-way coupling numerical data and two-way coupling numerical or experimental data may be inconsistent. This is particularly relevant for the particle settling in turbulence which has been previously shown to be strongly influenced by the dynamical interaction of the small flow scales with the particles.

1. Introduction

The knowledge of the settling of heavy particles in turbulence is relevant to many environmental and engineering problems such as the settling of aerosol particles in the atmosphere or combustion devices.

Direct numerical simulations that only accounted for the effects of the carrier flow on the particles ("one-way coupling" DNS) have shown that the average settling velocity of heavy particles is increased by turbulence (see Wang & Maxey [1], Yang & Lei [2]). The faster settling of the particles is explained by the coupling of two mechanisms (Wang & Maxey [1]): the tendency of inertial particles to accumulate on the peripheries of vortical structures (referred to as "preferential concentration") and the tendency of particles to move along the downward fluid motion (referred to as "preferential sweeping"). These numerical studies also showed that the maximum increase of the settling velocity is obtained for particles with response time $\tau_p \sim t_\eta$ (Wang & Maxey [1]), t_η being the Kolmogorov time scale, and for terminal velocity in the still fluid $v_t \sim 0.5u'$ (Yang & Lei [2]), u' defining the turbulence intensity. Similar qualitative findings were also observed in the experiments of Aliseda *et al.* [3] and of Yang & Shy [4]. Note that the settling velocity of heavy particles in turbulent flow is a quantity very difficult to measure and that these two experiments are the main ones reported for homogeneous turbulence in the literature.

An important open issue, on which both above experiments do not agree one with each other, is the modification of turbulence by the particles ("two-way coupling") and the resulting effects on the settling velocity. On one hand, Aliseda *et al.* [3] showed that the larger is the volume fraction the larger is the increase of the settling velocity. They explained this result by concentration effects and *a-priori* discarded two-way coupling effects because of the low volume fraction considered, $\Phi_v \sim 10^{-5}$. On the contrary, for similar volume fraction, Yang & Shy [4] found a significant increase of the turbulence energy by the particles but obtained an increase of the settling velocity close to the one obtained by one-way coupling simulations. This in turn suggests that two-way coupling may not alter the settling velocity. Nonetheless, the DNS of Bosse *et al.* [5] do show an additional enhancement of the settling velocity by two-way coupling effects. However, their two-way coupling DNS results poorly compare with the experiments. It is noteworthy to mention that the most complete experimental data set provided by Yang & Shy [4] on both the modification of turbulence by the particles and on the settling velocity, corresponds to a Taylor microscale Reynolds number $R_\lambda = 120$, much higher than the one considered by Aliseda *et al.* [3] ($R_\lambda = 75$) and those used in the DNS, limited to low or moderate Reynolds numbers ($R_\lambda = 40 - 75$). Reynolds number effects are thus not excluded from the numerical and experimental observations.

The present study proposes to further investigate the two-way fluid-particle interaction in the settling of heavy particles by also addressing Reynolds number effects. For this, two-way coupling DNS is used to simulate a particle-laden homogeneous turbulence at a low and a high Reynolds numbers, $R_\lambda = 40$ and 130 (both values allow comparisons with previous numerical and experimental data). Stationary but as well decaying homogeneous turbulence are considered to assess how far the use of a forcing scheme in the simulation of the stationary turbulence may influence the results, especially when two-way coupling are accounted for and at low Reynolds numbers, when the separation between the large scales and the small scales of turbulence is not complete.

2. Methodology and flow parameters

The homogeneous and isotropic turbulence is described in the Eulerian frame by the incompressible Navier-Stokes equations

$$\frac{\partial u_i}{\partial t} + u_j \frac{\partial u_i}{\partial x_j} = -\frac{1}{\rho}\frac{\partial p}{\partial x_i} + \nu \frac{\partial^2 u_i}{\partial x_j \partial x_j} + f_i - f_i^{(p)} \tag{1}$$

where $x_i (i = 1, 2, 3)$ are the spatial coordinates, $u_i (i = 1, 2, 3)$ the velocity components, ρ the fluid density, p the pressure, ν the kinematic viscosity and f_i the external energy

Table 1. Numerical and flow parameters. Microscale Reynolds number R_λ, number of computational nodes N^3, viscosity ν, box side length L_{box}, integral length scale L, large-eddy turn-over time T, Kolmogorov length and time scales η and t_η, maximum wavenumber $k_{\max} = \sqrt{2}N/3$, forcing mode k_p and statistical integration time T_{stat} (forced flow). Values are based on the unladen flow quantities.

R_λ	N^3	ν	L/L_{box}	L/η	T/t_η	$k_{max}\eta$	k_p	T_{stat}/T
40	64^3	0.0178	0.211	31.56	15.85	1.32	3	~ 50
130	512^3	0.00059	0.114	114.24	32.96	1.61	3.5	~ 11

source term to achieve a statistically steady turbulence ($f_i = 0$ for decaying turbulence). The term $-f_i^{(p)}$ represents the net force per unit mass exerted by a number of n_p particles within the integration control fluid volume and is computed from

$$f_i^{(p)} = \frac{1}{m}\sum_{1}^{n_p} f_{p_i} \tag{2}$$

where m is the mass of fluid within the integration control volume and f_{p_i} is the drag force acting on a particle p in the x_i-direction.

The particles, with density ρ_p much larger then the fluid density ρ, are described in the Lagrangian frame by the equation of motion of Maxey and Riley [6]

$$m_p \frac{dv_i}{dt} = \underbrace{m_p \frac{(u_{x_{p_i}} - v_i)}{\tau_p}}_{f_{p_i}} + (m_p - m)g_i \tag{3}$$

where $v_i (i = 1, 2, 3)$ are the particle velocity components, $u_{x_{p_i}}$ the instantaneous fluid velocity at the particle location x_{p_i} and m_p the particle mass. The response time of the particles, τ_p, is given by $\tau_p = d^2\rho_p/(18\nu\rho)$ with d the particle diameter; the gravitational acceleration g_i is such that $g_1 = g_2 = 0$ and $g_3 = -|g|$ where $|g|$ is defined from $v_t = -\tau_p|g|(1 - \rho/\rho_p)$, v_t being the terminal velocity of the particles in the still fluid.

The Navier-Stokes equations are solved on a cubic fluid box of side length $L_{\text{box}} = 2\pi$, discretized into N^3 computational nodes, with periodic boundary conditions. A fully pseudo-spectral algorithm with a de-aliasing truncation technique (referred to as the "2/3 rule") is used with a second-order Runge-Kutta time-stepping for the non-linear terms and an analytic integrating factor for the viscous terms. The forcing is achieved by introducing a power input distributed over a narrow band of wavenumbers k that satisfy $k_p - 1 \leq k \leq k_p + 1$, where k_p defines the peak forcing mode (see Rogallo [7] and Jímenez & Wray [8] for further computational details). The average mean flow integrated over the flow domain is imposed to be zero throughout the two-way coupling simulations. This is done by applying a mean pressure gradient that balances the net weight of the particle phase in the periodic domain. This avoids any further acceleration

Table 2. Particle Stokes number $St = \tau_p/t_\eta$ and normalized diameter; all particles have a density ratio $\rho_p/\rho = 5000$. Values are based on the unladen flow quantities.

St	0.36	1	2
d/η	0.0359	0.0597	0.0847

of the particles in the gravity direction due to a non-zero net volume flux (see Maxey & Patel [9] and Bosse *et al.* [5]).

In a Lagrangian-Eulerian approach, the computation of the drag force f_{p_i} requires the fluid velocity to be interpolated from the Eulerian frame to the particle location (forward interpolation). In addition, when two-way coupling is accounted for, the force exerted by the particles on the fluid, $f_i^{(p)}$, has to be assessed from the particle data to the Eulerian frame (backward interpolation). Various forward and backward interpolation schemes are proposed in the literature and recently discussed in Garg *et al.* [10]. In the present study, a fourth-order Lagrangian polynomial interpolation is used for the forward interpolation while a linear projection of the particle force over the eight grid nodes surrounding the particle is used for the backward interpolation. These two schemes were shown to be reasonably accurate for a flow with a high spectral content in the fluid velocity field. For the time discretization, the same Runge-Kutta scheme as used for the carrier fluid is used for the particle equation.

The main numerical and flow parameters are provided in table 1 and the particle properties in table 2. Note that the particle diameter is well below the Kolmogorov length scale so that the corresponding particle Reynolds number, Re_p, is very small, which justifies the use of the linear Stokes drag law in equation (3). The condition $Re_p \ll 1$ was also checked to be satisfied during the simulation runs. Finally, the ratio of the number of computational particles to the number of real particles is set to one in all the low Reynolds number simulations and does not exceed 3 in the high Reynolds number simulations (up to 2.56×10^8 particles were tracked).

3. Results

Results on the effects of two-way coupling on the settling velocity rate are presented for the Stokes numbers $St = 0.36, 1$ and 2 in decaying homogeneous turbulence at the Reynolds numbers $R_\lambda = 40$ and 130. For this flow configuration identical initial conditions for the fluid (issued from a velocity field of the forced turbulence simulation) and for the particles (random distribution) is used for the three Stokes numbers considered. In the stationary turbulence, the statistics are reported for the Stokes number $St = 1$. They were performed once the average settling velocity exhibited a stationary behavior. The statistical time integrations are given in table 1.

3.1. Reynolds number effects: overview

The average increase of the settling velocity normalized by the terminal velocity, $< \Delta v_s > /v_t =< (v_s - v_t) > /v_t$ (where v_s is the particle settling velocity), is given as a

Table 3. Forced turbulence, $R_\lambda = 40$ and 130: average settling rate $< \Delta v_s > /v_t$ and particle Stokes number St; $\Phi_v = 3 \times 10^{-5}$, $v_t = v_\eta$.

	$R_\lambda = 40$		$R_\lambda = 130$	
	St	$< \Delta v_s > /v_t$	St	$< \Delta v_s > /v_t$
1-way	1	0.479	1	0.630
2-way	1.064	0.712	1.7	0.574

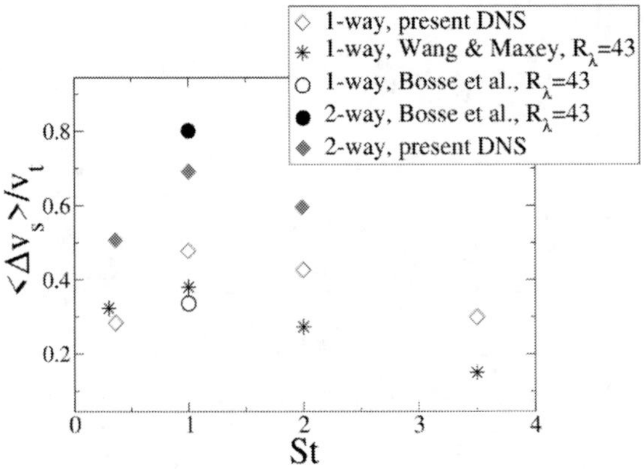

Figure 1. Forced turbulence, $R_\lambda = 40$, $v_t = v_\eta (= \eta/t_\eta)$: comparison with previous DNS of the normalized average settling rate $< \Delta v_s > /v_t$ as a function of the Stokes number St. For both two-way coupling DNS $\Phi_v = 3 \times 10^{-5}$.

function of the Stokes number St in figure 1 for the one-way and two-way coupling DNS performed for the forced turbulence at $R_\lambda = 40$. In agreement with Wang & Maxey [1] and Bosse *et al.* [5], the present results reproduce the maximum of the settling rate for $St \sim 1$ and also show an additional increase effect by the two-way coupling which is significant for the three Stokes numbers considered. The observed quantitative discrepancies can be mainly attributed to the smaller forcing radius used in the present simulations compared to the one used in the DNS of Wang & Maxey [1] and of Bosse *et al.* [5]. Similar forcing effects were reported by Wang & Maxey [1] for a lower Reynolds number (see figure 7 in [1]). The significant enhancement of the settling velocity by the two-way coupling is also confirmed in the decaying turbulence for identical flow and particle parameters as shown in figure 2(a).

By contrast, the DNS results obtained for $R_\lambda = 130$ show that clear conclusions are difficult to extract regarding the effects of the two-way coupling on the settling velocity. Figure 2(c) exhibits a two-way coupling settling rate well above the corresponding one-way coupling settling rate for $St = 0.36$. However, for the particles with higher inertia, the two-way coupling DNS provide results close to the one-way coupling DNS except at large times where the settling velocity is found larger in the two-way coupling DNS. Regarding the stationary turbulence, the settling rate is found smaller in the two-way coupling simulation (see table 3). Nevertheless, it has to be stressed out that for $R_\lambda = 130$ a direct comparison between two-way and one-way coupling DNS

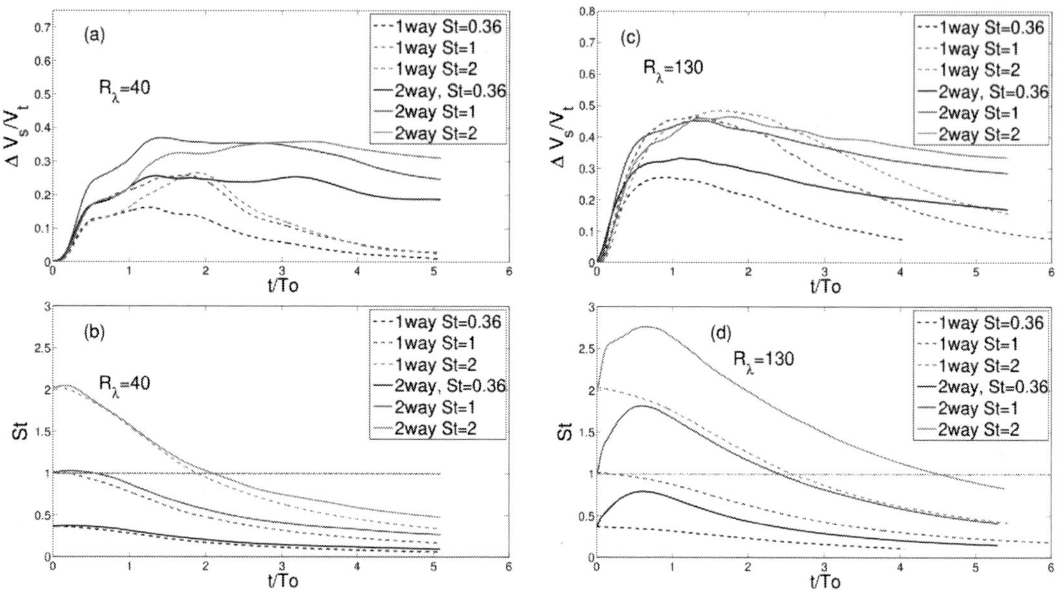

Figure 2. Decaying turbulence, $R_\lambda = 40$ and 130: time evolution of the settling rate, $\Delta v_s/v_t$, and of the particle Stokes number, St; $\Phi_v = 3 \times 10^{-5}$, $v_t = v_\eta$.

is not consistent because of the large increase of the Stokes number resulting from the interaction of the particles with the fluid. This is shown in figure 2(d) for the decaying turbulence and in table 3 for the stationary turbulence. The enhancement of the particle inertia reflects a significant decrease of the Kolmogorov time scale by the presence of the particles. By comparison, in the low Reynolds number case, figure 2(b) and table 3 show that the Kolmogorov time scale is only slightly reduced by the two-way coupling. Reduction of length flow scales was also observed experimentally by Poelma *et al.* [11] in particle-laden turbulence submitted to gravity effects for different flow and particle parameters from those considered herein. Note that $k_{\max}\eta$ based on the Kolmogorov length scale computed from the two-way coupling simulations takes the minimum instantaneous values 1.04, 1.12 and 1.23 for the respective Stokes numbers St=0.36, 1 and 2 in the decaying turbulence case while it takes the value 1.15 for $St = 1$ in the forced turbulence case. These values ensure an adequate numerical resolution.

3.2. Turbulence energy redistribution

Figure 3 compares the instantaneous unladen and laden turbulent flow energy spectra obtained in the decaying turbulence for a given time at which the particle Stokes numbers differ slightly from their initial value. For the low and high Reynolds numbers and for all the Stokes numbers considered, a considerable energy transfer from the particles to the fluid is observed in the large wavenumber range while the low wavenumbers are not significantly modified. In agreement with Yang & Shy [4], the maximum turbulence augmentation at the dissipative scales (very large wavenumbers) is obtained for $St \sim 1$,

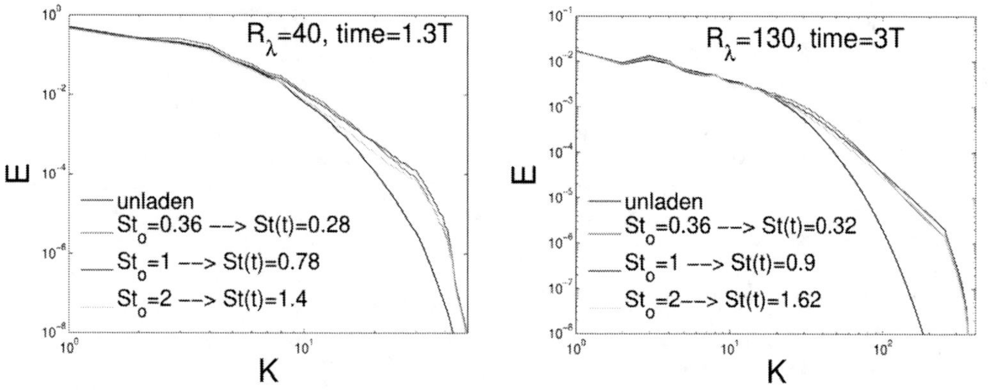

Figure 3. Decaying turbulence, $R_\lambda = 40$ and 130: instantaneous three-dimensional energy spectra; St_o refers to the initial Stokes number value and $S(t)$ to the Stokes number value at the considered time. $\Phi_v = 3 \times 10^{-5}$, $v_t = v_\eta$.

at which the preferential concentration is commonly reported to be the most significant. At the intermediate wavenumbers, the turbulence energy spectra present similar features as those found in absence of gravity acceleration by Ferrante & Elghobashi [12].

The redistribution of the turbulence energy is visualized by the energy spectra of the horizontal, E_{11} and E_{22}, and vertical, E_{33}, components given in figure 4. The spectra show that, under gravity, the turbulence becomes anisotropic with a redistribution of energy from the horizontal to the vertical components. This corroborates the DNS findings of Elghobashi & Truesdel [13], Ferrante & Elghobashi [12] and also the experimental observations of Poelma *et al.* [11]. For the high Reynolds number, the redistribution of energy occurs up to very low wavenumbers while at the low Reynolds number it is damped by viscous dissipation at relatively larger wavenumbers. This may partly explain the significant reduction of the Kolmogorov scales by the particles observed for the high Reynolds number. At the low Reynolds number the energy added by the particles at the small flow scales and redistributed to larger scales is dissipated over the full range of the flow scales as a consequence of the not-complete separation between the energetic and dissipative scales. On the other hand, at the high Reynolds number, dissipation mainly occurs at the small turbulence scales so that the Kolmogorov scales further stretch (and thus decrease) to dissipate part of the additional energy.

The energy spectra obtained in the forced turbulence for $R_\lambda = 40$ and 130 and $St = 1$ (not shown here) exhibit similar features as those reported above for the decaying turbulence.

3.3. Particle force field

An illustration of the preferential concentration effect is supplied by figure 5 which represents the instantaneous particle positions drawn in a transversal plane of the fluid box in the case of the decaying turbulence for the two Reynolds numbers and for $St = 1$. Compared to the low Reynolds number, the particle clusters tend to concentrate in small

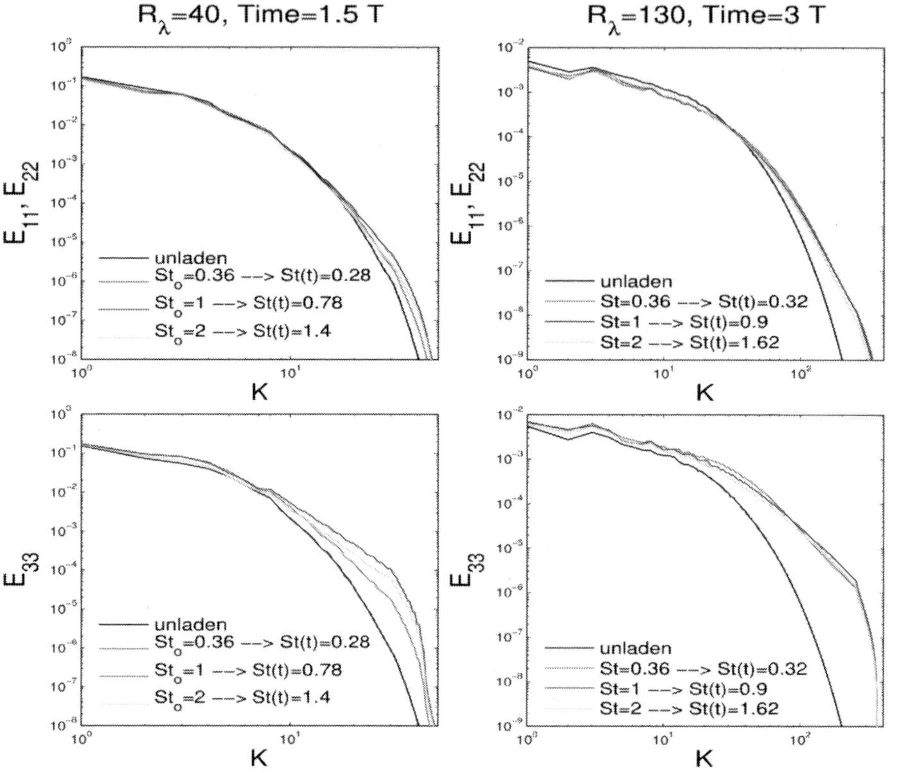

Figure 4. Decaying turbulence, $R_\lambda = 40$ and 130: instantaneous vertical and horizontal flow velocity components energy spectra; see legend in figure 3.

flow regions in the high Reynolds number turbulence (see inset for $R_\lambda = 130$). This indicates that particles mainly accumulate at the small turbulence scales, in agreement with the observations of Aliseda *et al.* [3] and Yang & Shy [4]. The force exerted by the particles on the fluid is analyzed in figure 6, by showing the probability density function (PDF) of the particle force density. The large probability for zero force is associated with regions of fluid devoid of particles (see figure 5) and reflects a highly non-homogeneous particle force field due to concentration effects. As shown in figure 6, concentration effects are very small for $St = 0.36$. The particles are shown to mainly exert a negative force on the fluid or equivalently a downward force which in turns accelerates the fluid in the gravity direction. Also, the PDF of large values of the downward force is found to increase when the preferential accumulation of the particles becomes more significant. This is well shown by comparing the PDF obtained for the less inertial particles ($St = 0.36$) with the more inertial ones ($St = 1$ and 2). Note that the downward force exerted by the particles on the fluid must result in an enhancement of the settling velocity. This was clearly observed for $R_\lambda = 40$ but not for $R_\lambda = 130$. The preferential concentration of the particles highly depends on their Stokes number. Thus, for $R_\lambda = 130$, the large

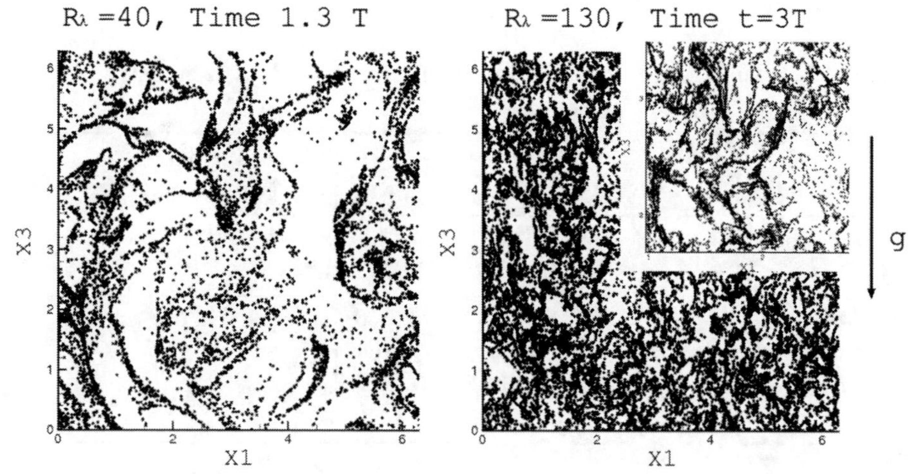

Figure 5. Decaying turbulence, $R_\lambda = 40$ and 130: instantaneous particle positions drawn in the plane $X_2 = 2\pi/2$; $\Phi_v = 3 \times 10^{-5}$, $v_t = v_\eta$. The inset for $R_\lambda = 130$ is a inbox zoom of the particle positions field.

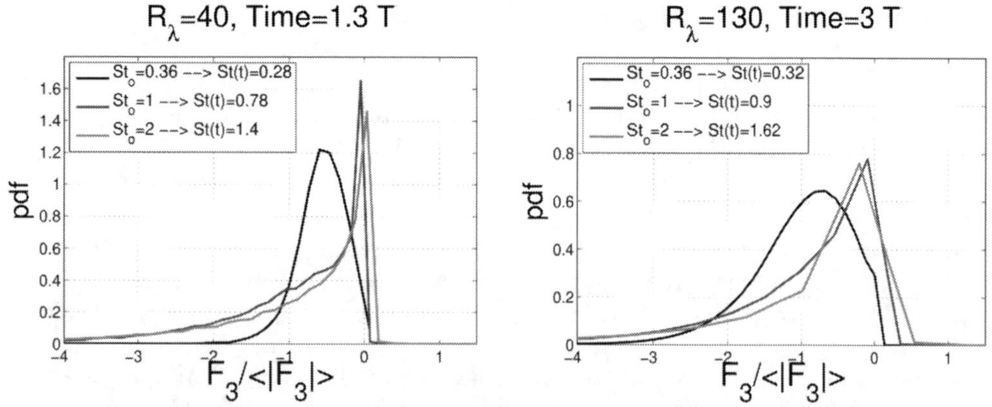

Figure 6. Decaying turbulence, $R_\lambda = 40$ and 130: PDF of the force density exerted by the particles on the fluid along the gravity direction; $\Phi_v = 3 \times 10^{-5}$, $v_t = v_\eta$.

increase of St by two-way coupling significantly alters the concentration effects. This in turn implies that the resulting effect of the downward force on the settling velocity is strongly dependent on the modification of the small turbulence scales by the two-way coupling (see comment at the end of next section).

Journal of Physics: Conference Series **333** (2011) 012006 doi:10.1088/1742-6596/333/1/012006

Table 4. Forced turbulence, $R_\lambda = 130$: comparison with experiment. $v_t = 0.5u'$ ($v_t = 3.46v_\eta$).

	DNS, $R_\lambda = 130$ 1-way	2-way	2-way	Experiment, $R_\lambda = 120$ Yang & Shy [4]
$\Phi_v \times 10^{-5}$	-	1.5	3	5
Φ_m	-	0.075	0.15	~ 0.05
$< \Delta v_s > /v_t$	0.216	0.326	0.366	0.24
$< \Delta v_s > /u'$	0.106	0.153	0.165	0.13
u'/u'_o	1	1.029	1.08	1.2
St	1	1.52	2.23	-

3.4. Comparison with experiments

Table 4 compares the present particle and flow DNS data with experimental data given by Yang & Shy [4]. In both the numerical and experimental studies the carrier turbulence is stationary homogeneous and isotropic, the gravity acceleration is such that $v_t = 0.5u'$ and the Stokes number is $St = 1$ (value based on the unladen-flow Kolmogorov time scale). The Reynolds number, volume fraction and mass loading ($\Phi_m = \rho_p \Phi_v / \rho$) are somewhat different in the DNS but still very close to the experimental values. For $\Phi_v = 3 \times 10^{-5}$ the value of $k_{\max}\eta$ is 1.1 so that no higher volume fraction was considered for the sake of numerical resolution.

The average settling rate computed from the one-way and two-way coupling DNS compare reasonably well with the experiment, a better agreement being obtained with the two-way coupling DNS and when the mass loading is closer to the one reported in the experiment. Note that as the volume fraction increases, the settling rate increases as well. This behavior corroborates the experimental observations of Aliseda *et al.* [3]. Table 4 shows that the two-way coupling DNS reproduce the enhancement of the turbulence energy by the particles as observed in the experiments but the ratio of the turbulence intensity of the laden flow to the one of the unladen flow, u'/u'_o, is found smaller in the simulations. More important, as previously observed in section 3.1, the DNS show that the particle Stokes number computed *a-posteriori* in the two-way coupling DNS is significantly above its value based on the unladen flow quantities (one-way coupling St). This again implies that the Kolmogorov scales have been significantly decreased by the interaction of the particles with the turbulence. This effect can not be experimentally measured as there is no data supplied regarding the modification of the dissipative scales by the particles. The significant alteration of the small flow scales by the presence of the particles found in the present DNS shows that comparisons between one-way coupling simulations and experiments might be not-consistent and that further studies on the modification of the structure of the small turbulence scales by the particles are required.

Note that in section 3.1 (for $v_s = v_\eta$), the two-way coupling effects are shown to result in a net increase of the particle settling velocity for $R_\lambda = 40$ (for all values of St considered) while for $R_\lambda = 130$ they are more ambiguous to interpret. Section 3.3

shows that the resulting force exerted by the particles on the fluid must have an increase effect on the particle settling whatever the Reynolds number considered. Keeping in mind that particle concentration effects are strongly dependent on St (commonly found to be optimum for $St \sim 1$), the significant decrease of the small scales of turbulence (or equivalently the significant increase of St) observed in the two-way coupling simulations for the high Reynolds number will in turn alter the particle concentration compared to the one-way coupling simulations. Thus, the enhancement of the particle settling through the downward force may be reduced or increased by a less or more efficient concentration mechanism of the particles depending on the change in St. By considering the DNS results reported in table 4, the increase of St (becoming larger than unity) with Φ_v may suggest a less effective particle concentration as the volume fraction is enhanced and explain that the settling velocity obtained from the one-way and two-way coupling simulations are not drastically different. By contrast, for the low Reynolds number simulations, because of the slight alteration of the dissipative scales by the two-way coupling effects it can be argued that particle concentration effects are similar in the one-way and two-way coupling simulations and that the downward force exerted by the particles on the fluid has a net increase effect on the settling velocity (see table 3). The Reynolds number dependence of two-way coupling effects on the Kolmogorov scales may in part explain the apparent contradiction between the experimental results of Aliseda *et al.* [3] and Yang & Shy [4].

4. Conclusions

The present study reported preliminary results from one-way and two-way coupling DNS of heavy particles settling in a low and high Reynolds number homogeneous turbulence. Decaying and stationary turbulence were considered as well as particle inertia effects.

The main results of this study are found similar for both Reynolds numbers in the decaying and forced turbulence simulations. Thus, it can be concluded that the use of a forcing scheme in the simulation of the stationary turbulence does not alter fundamentally the results.

For the low Reynolds number and all the particle Stokes numbers considered, the force exerted by the particles on the fluid has an additional increase effect on the settling velocity when compared to the one-way coupling DNS. This is explained by the alignment of the force with the gravity direction which in turns accelerates the fluid and thus the particles. It is found that the more significant is the preferential concentration of the particles, the larger is the probability of finding large values of this downward force.

For the high Reynolds number, it is shown that the interaction of the particles with the fluid causes a significant decrease of the Kolmogorov time scale. The consequent modification of the Stokes number brings out an important consequence: direct comparison between one-way and two-way coupling DNS is not consistent. The two-way fluid-particle interaction is obviously inherent in experiments so that the consistency of comparisons between one-way coupling DNS and experiments is also questioned. The two-way coupling DNS results are found in fair agreement with the experiment of Yang & Shy [4].

The particle inertia and the small flow scale dynamics play a fundamental role in the settling of particles in turbulence. In this sense, the present results evidence that the interaction of the particles with the small turbulent vortical structures under gravity must be further investigated in a two-way coupling approach.

Acknowledgments

This research was supported by the Ramón y Cajal research program of the Spanish Ministerio de Educación y Ciencia

References

[1] Wang L P and Maxey M R 1993 *J. Fluid Mech.* **256** 27–68
[2] Yang C Y and Lei U 1998 *J. Fluid Mech.* **371** 179–205
[3] Aliseda A, Cartellier A, Hainaux F and Lasheras J C 2002 *J. Fluid Mech.* **468** 77–105
[4] Yang T S and Shy S S 2005 *J. Fluid Mech.* **526** 171–216
[5] Bosse T, Kleiser L and Meiburg E 2006 *Phys. Fluids* **18** 027102
[6] Maxey M and Riley J 1983 *Phys. Fluids* **26** 883–889
[7] Rogallo R 1981 NASA Techn. Report.
[8] Jímenez J and Wray A A 1998 *J. Fluid Mech.* **373** 255–282
[9] Maxey M R and Patel B K 2001 *Int. J. Multiphase Flow* **27** 1603–1626
[10] Garg R, Narayanan C, Lakehal D and Subramaniam S 2007 *Int. J. Multiphase Flow* **33** 1337–1364
[11] Poelma C, Westerweel J and Ooms G 2007 *J. Fluid Mech.* **589** 315–351
[12] Ferrante A and Elghobashi S 2003 *Phys. Fluids* **15** 315–329
[13] Elghobashi S and Truesdell G C 1993 *Phys. Fluids* **A 5 (7)** 1790–1801

Clustering and turbulence modulation in particle laden shear flows

P. Gualtieri, F. Picano, G. Sardina & C.M. Casciola

Dipartimento di Ingegneria Meccanica e Aerospaziale, Università di Roma *La Sapienza*, Via Eudossiana 18, 00184 Roma *Italy*

E-mail: paolo.gualtieri@uniroma1.it

Abstract. Turbulent fluctuations induce the commonplace phenomenology on the transport of small inertial particles known as clustering. Particles spread disuniformly and form aggregates where their local concentration is much higher than it is in nearby rarefaction regions, the voids, where in extreme cases not even a single particle can be found. The underlying physics has been exhaustively analyzed in statistically homogeneous and isotropic flows under the so called one-way coupling regime, i.e. in conditions where the momentum exchange between the carrier fluid and the disperse phase is negligible. Recently it has been shown that the addition of a mean flow might have dramatic effects on the disperse phase, i.e. the mean flow, through its large scale anisotropy, induces a preferential orientation of the clusters. Due to inertial effects, their directionality can even increase in the smallest scales, contrary to the expectation based on the isotropy recovery behavior of velocity fluctuations. This finding opens new issues in presence of large mass loads, when the momentum exchange between the two phases becomes significant and the back-reaction of the particles on the carrier flow cannot be neglected. These aspects are discussed here by addressing direct numerical simulation data of particle laden homogeneous shear flow in the two-way coupling regime. Consistently with previous findings we observe an overall depletion of turbulent fluctuations. In particular, particles with order Kolmogorov scale relaxation time induce the energy depletion of the classical inertial scales and the amplitude increase of the smallest ones where the particle back-reaction pumps energy into the turbulent eddies increasing their energy content. We find that increased mass loads result in the substantial broadening of the energy co-spectrum thereby extending the range of scales driven by anisotropic production mechanisms. This is due to the clusters which form the spatial support of the back-reaction field and give rise to a highly anisotropic forcing active down to the smallest scales.

1. Introduction

Transport of inertial particles is involved in several fields of science such as droplets growth and collisions in clouds [1, 2], the plankton accumulation in the oceans [3, 4] or the plume formation in the atmosphere [5]. At the same time multiphase flows are the base of several technological applications. The inertial particles dynamics is crucial in designing injection systems of internal combustion engines [6], to prevent sediment accumulation in pipelines [7, 8] or for the appropriate dimensioning of several industrial devices [9].

The relevant physical aspect in particles dynamics consists in their finite inertia which prevents them from following the fluid trajectories. Most evident is the "preferential accumulation" which, in inhomogeneous flows such as wall bounded flows, occurs in the form of the so called "turbophoresis", i.e. preferential localization of particles in the near wall region

[10, 11]. An exhaustive review of the subject can be found e.g. in [12], see also [13] and the recent paper [14] for a physical explanation in terms of statistical properties of velocity fluctuations in the near wall region.

When the idealized conditions of isotropic turbulence are addressed, preferential accumulation manifests itself in the form of small scale clustering. The disperse phase forms small scale aggregates where most particles concentrate, separated by void regions of small particle density, see e.g. [15] and references therein.

So far the effect of turbulent transport on particle dynamics has been studied extensively in many flow configurations. Much less is known about the effect the disperse phase may have on the carrier flow demanding for a renewed effort in this direction see e.g. [16, 17]. It is expected that, under proper coupling conditions, the momentum exchange between the two phases might become relevant in driving the turbulent fluctuations away from their universal equilibrium state predicted by Kolmogorov in the early forties. Clearly, in contrast to the one-way coupling regime, addressing these effects calls into play the more realistic two-way coupling mechanism, where the disperse phase provides an active modulation of velocity fluctuations.

In this context, many authors first studied the simplest flow configuration, i.e. decaying homogeneous isotropic turbulence [18, 19, 20, 21, 22], by addressing the decay rate of turbulence intensity and the modification of the energy spectra under two-way coupling regime. Other numerical investigations analyzed the statistically steady state of isotropic turbulence [23] showing a non uniform distortion of the energy spectra, i.e. a damping of turbulent fluctuations at large scales and a relative enhancement at the smallest ones. In some cases, the overall effect of back reaction on the carrier phase is reported as an attenuation of turbulence fluctuations controlled by the mass load ratio defined as the ratio of total disperse phase mass to fluid mass, see e.g. [24, 25]. Similar effects have also been observed for particles larger than the Kolmogorov scale showing that the crossover wavenumber characterizing the transition between dumped and enhanced energy modes is related to the particles finite size [26]. Other important issues such as the increase of the particles settling velocity under gravity have been addressed both numerically [27] and experimentally [28]. By breaking isotropy gravity leads to a preferential augmentation of turbulence intensities along the vertical direction while in the transverse directions augmentation is only observed below the Taylor scale. Other interesting effects of the momentum exchange between the carrier fluid and the disperse phase is observed in the context of grid generated spatially decaying turbulence. In fact, starting from an isotropic state the feedback of the particles seems to lead to an anisotropic flow as turbulent fluctuations develop downstream [29, 30]. An extensive review of the back-reaction effects on isotropic turbulence might be found in Ref. [31]. Concerning wall bounded flows many authors considered the classical geometry of the channel [32, 33, 34] or of the pipe flow [35, 36]. Due to inhomogeneity and anisotropy new features emerge such as the preferential suppression of turbulence intensities in the wall normal direction [34], the enhancement of large scale anisotropy as measured by velocity variances [33] or even the occurrence of drag reduction [32]. A substantial modification of turbulent kinetic energy production has been reported in Ref. [37] in the context of a time evolving homogeneous shear flow.

In any case, as isotropy is broken by the presence of gravity or by the mean streamwise advection, the back-reaction of the disperse phase seems to immediately originate strong anisotropies in the carrier phase. Motivated by recent findings in the context of anisotropic clustering [38, 39], we consider here the modulation of turbulence by transported particles addressing the particle laden homogeneous shear flow in the two-way coupling regime. Such flow can be considered as a sort of bridge between the idealized conditions of isotropic turbulence and the more realistic geometries of wall bounded flows, since it preserves spatial homogeneity and retains the anisotropy of shear flows.

It is worth recalling that, under shear, turbulent fluctuations are strongly anisotropic at the largest scales due to production of turbulent kinetic energy via interaction of the mean velocity gradient and turbulent fluctuations. At smaller scales, below the so-called shear scale $L_S = \sqrt{\epsilon/S^3}$–where ϵ is the energy dissipation rate and S the mean shear–inertial energy transfer usually prevails. In such conditions re-isotropization of turbulent fluctuation takes place following a route described in [40, 41, 42]. However, turbulent fluctuations in the one-way coupling regime are found to induce an anisotropic clustering of the disperse phase which persists down to the smallest scales [38]. Actually, in contrast to the small scale behavior of velocity fluctuations, particles aggregates do not lose their directionality. Their anisotropy even increases down to the viscous scales where clusters still keep memory of the spatial orientation of the large scale coherent motions.

Here we consider the same flow under the two-way coupling scenario. In this more complex case the physical result we achieve consists in a view of multi-phase turbulent flows consistent with the multiscale forcing addressed in [43]. Small scales anisotropic clusters acts as a source/sink of momentum for the turbulent motions distributed all along the range of scale. They deplete the energy from the largest inertial scales which is in part retrieved in the low-inertial/dissipative range. Here the clusters keep the velocity fluctuations to a higher éxcitation state than expected on the basis of the standard Kolmogorov theory. As we will see such back energy scatter is highly anisotropic in shear flow, hence the small scales fluid motions are prevented from recovering isotropy and eventually increment their anisotropy.

2. Methodology

Concerning the carrier fluid, the velocity field $\mathbf{v}$ is decomposed into a mean flow $\mathbf{U} = Sx_2\,\mathbf{e}_1$ and a fluctuation $\mathbf{u}$ where $\mathbf{e}_1$ is the unit vector in the streamwise direction, x_2 denote the coordinate in the direction of the mean shear S and x_3 is in the spanwise direction. Rogallo's technique [44] is employed to rewrite the Navier-Stokes equations for velocity fluctuations in a deforming coordinate system convected by the mean flow according to the transformation of variables $\xi_1 = x_1 - Stx_2$; $\xi_2 = x_2$; $\xi_3 = x_3$; $\tau = t$. The resulting system

$$\nabla \cdot \mathbf{u} = 0\,; \qquad \frac{\partial \mathbf{u}}{\partial \tau} = (\mathbf{u} \times \boldsymbol{\zeta}) - \nabla\pi + \nu\nabla^2\mathbf{u} - Su_2\mathbf{e}_1 + \mathbf{F}\,, \tag{1}$$

is numerically integrated by a pseudo-spectral method combined with a fourth order Runge-Kutta scheme for temporal evolution. In equations (1) $\boldsymbol{\zeta}$ is the curl of $\mathbf{u}$, π is the modified pressure which includes the fluctuating kinetic energy $|\mathbf{u}|^2/2$, ν is the kinematic viscosity and $\mathbf{F}$ denote the back-reaction due to the disperse phase. The latter consists of diluted particles with mass density ρ_p much larger than the carrier fluid ρ_f. The approximation of point particles can be adopted whenever the particle diameter d_p is much smaller than the typical turbulence scales. It follows that the only relevant force is the Stokes drag [45]. Accordingly, the equations for particles position $x_i^p(t)$ and velocity $v_i^p(t)$ read

$$\frac{dx_i^p}{dt} = v_i^p\,; \qquad \frac{dv_i^p}{dt} = \frac{1}{\tau_p}\left[v_i(x^p,t) - v_i^p(t)\right]\,, \tag{2}$$

where $v_i(x^p,t)$ is the instantaneous fluid velocity evaluated at $x_i^p(t)$ and $\tau_p = \rho_p d_p^2/(18\nu\rho_f)$ is the Stokes relaxation time. Particles' velocities are decomposed as $v_i^p = U_i[x_k^p(t)] + u_i^p$ where u_i^p denotes the particle velocity deviation with respect to the local mean flow of the carrier fluid. Finally by using Rogallo's transformation eqs. (2) can be written as

$$\frac{d\xi_i^p}{d\tau} = u_i^p - S\tau u_2^p\delta_{i1}\,; \qquad \frac{du_i^p}{d\tau} = f_i^p. \tag{3}$$

Equations (1) and (3) are integrated in computational space. In fact, the computational box gets distorted in physical space by the mean flow advection. The re-meshing procedure is periodically applied to allow long time integrations. By exploiting periodicity in the ξ_1 direction, the computational domain is transformed back into a non skewed domain every $\Delta t_r = 2/S$, preventing the unbounded growth of the wavenumbers, see [46] for further details on the numerics adopted. In equation (3) $f_i^p = \tau_p^{-1}\left[u_i(\xi^p,\tau) - u_i^p(\tau)\right] - Su_2^p\delta_{i1}$ is the expression of the Stokes drag acting on the pth particle. In the so-called two way coupling regime an equal and opposite force acts on the carrier fluid accounting for the momentum exchange between the two phases.

Modeling the back reaction in numerical simulations is an issue [47]. The local distortion of turbulence due to the disperse phase can be captured only resolving the boundary of each particle on the computational grid. In the so-called resolved particle simulations several approaches have been proposed ranging from finite volume schemes [48] to Lattice Boltzman Methods [49]. However other approaches are possible once it has been recognized that the flow close to a small particle can be locally approximated as a Stokes Flow. In Ref. [50] the Stokes solution is used to provide appropriate boundary conditions to the Navier-Stokes equations close to each particle. Clearly these approaches, even though they provide the exact momentum coupling between the carrier and the disperse phase, are feasible only for a relatively small number of particles. When particles are much smaller than the turbulent scales they can be considered as material points i.e. as point source/sinks of momentum for the carrier fluid. Within this approximation several other methodologies are available. For instance, in the force coupling method [51] the disturbance flow produced by the particles is modeled as a regularized steady Stokes solution corresponding to a force monopole and quadrupole while the singular steady Stokes solution is employed in Ref. [34]. In our case we adopt the particle in cell method introduced in Ref. [52] where the particle is regarded as a point source of momentum for the carrier fluid. This is the simplest approach to model the two way coupling regime widely adopted in the literature even if the methodology might suffer of lack of numerical convergence in estimating the interphase momentum transfer [53]. Accordingly, the resulting force on the fluid is computed as

$$\mathbf{F} = -\frac{N_c}{N_p}\Phi\sum_{p}^{n(\xi)}\mathbf{f}^p \tag{4}$$

where the sum is extended to all the $n(\xi)$ particles belonging to the computational cell centered at point ξ. In eq. (4) N_c denote the number of Eulerian cells, N_p is the total number of particles and Φ denote the mass load ratio i.e. the ratio between the mass of the disperse phase $M_p = N_p\pi\rho_p d_p^3/6$ and the carrier fluid $M_f = \rho_f V_f$ where V_f is the volume of the computational box. Fluid properties are known in an Eulerian frame while particles evolve along their own Lagrangian trajectories requiring a first interpolation when the fluid velocity is computed at the particle position, namely $u_i(\xi^p,\tau)$. An other interpolation is required when the back-reaction on the fluid is computed since f_i^p is known at the particle position. In fact, the force acting on the pth particle is re-distributed via inverse interpolation to the nearest Eulerian grid points where the fluid velocity is defined. Equations (3) are integrated by the same fourth order Runge-Kutta scheme used for the Navier-Stokes equations and the interpolation adopt a tri-linear scheme.

The data we analyze come from a statistically steady Direct Numerical Simulation (DNS) of a homogeneous shear flow. The mean shear induces velocity fluctuations which are strongly anisotropic at the larger scales driven by production while, at smaller separations, the classical energy transfer mechanisms become effective in inducing re-isotropization. The position of the shear scale L_S was found crucial to determine whether small scale isotropy recovery eventually occurs. For these reasons L_S plays an important physical role in shear turbulence where it enters the two basic control parameters [54, 55]. The first one is the Corrsin parameter, $S_c^* = S(\nu/\epsilon)^{1/2} = (\eta/L_S)^{2/3}$, where $\eta = (\nu^3/\epsilon)^{1/4}$ is the Kolmogorov scale. The Corrsin parameter can be recast in terms of the Taylor-Reynolds number $1/S_c^* \propto \mathrm{Re}_\lambda = \sqrt{5/(\nu\epsilon)}\langle u_\alpha u_\alpha\rangle$, where

u_α is the $\alpha^{\rm th}$ Cartesian component of the velocity fluctuation and the angular brackets denote ensemble averaging. The second parameter is the shear strength $S^* = S\langle u_\alpha u_\alpha\rangle/\epsilon = (L_0/L_S)^{2/3}$, with $L_0 = \langle u_\alpha u_\alpha\rangle^{3/2}/\epsilon$ the integral scale of the flow. The former determines the extension of the range of scales below the shear scale and above the Kolmogorov length η. When this range is sufficiently extended, i.e. the Taylor-Reynolds number is large enough, small scale isotropy recovery is likely to occur in the fluid velocity field. The latter, instead, fixes the range of scales directly affected by the geometry of the forcing. This is the anisotropic range between integral and shear scale.

The Navier-Stokes equations are integrated in a $4\pi \times 2\pi \times 2\pi$ periodic box see table 1 for a full description of the dataset. The Kolmogorov scale is $\eta = 0.07$ which corresponds to $K_{max}\eta$ ranging from 5.8 to 15.5. Note the very accurate resolution of the smallest scales required to have smooth fields in view of accurate interpolations and re-distribution of the back-reaction on the carrier fluid. Accurate resolution of the smallest scales is also crucial when addressing the statistical properties of the viscous dissipation field [56, 57].

Concerning the disperse phase, the dynamics is controlled by the ratio of the particles relaxation time τ_p to a characteristic flow time scale, typically the Kolmogorov time scale $\tau_\eta = (\nu/\epsilon)^{1/2}$, i.e. the relevant control parameter is the Stokes number $St_\eta = \tau_p/\tau_\eta$. When the two-way coupling regime is considered other non dimensional parameters are required to describe the momentum exchange between the two phases, namely the density ratio ρ_p/ρ_f – assumed to be much larger than unity– and the mass load fraction $\Phi = M_p/M_f$. In fact, the mass load can be expressed as $\Phi = (\rho_p/\rho_f)\,\Phi_v$ where $\Phi_v = V_p/V_f$ is the volume fraction. When $\rho_p/\rho_f \gg 1$, the mass load Φ can be order one when the volume fraction is still very small, order 10^{-3} in the present case. In such conditions inter-particles collisions might be safely neglected see e.g. [16] while the momentum coupling between the two phases must be retained. For the simulations summarized in table 1 particles are injected in an already fully developed turbulent flow. Their position is initialized at random homogeneous points with initial velocity matching the fluid velocity at particle position. The total number of particles is changed to achieve different values of the mass load parameter while the Stokes number is kept constant. To this purpose, after discarding an initial transient, 150 statistically independent snapshots are used to compute the relevant statistical observables. The snapshots are collected at each remesh time

run	$N_x \times N_y \times N_z$	Φ	N_p
a	$192 \times 96 \times 96$	–	–
b	$384 \times 192 \times 192$	0.2	327500
c	$384 \times 192 \times 192$	0.4	655000
d	$384 \times 192 \times 192$	0.8	1310000
d_l	$192 \times 96 \times 96$	0.8	1310000
d_h	$512 \times 256 \times 256$	0.8	1310000

Table 1. Direct Numerical Simulation dataset. Navier-Stokes equations are integrated in a $4\pi \times 2\pi \times 2\pi$ periodic box with a resolution of $N_x \times N_y \times N_z$ Fourier modes. The 3/2 dealiasing rule is adopted to compute the non linear terms. The Taylor Reynolds number is $Re_\lambda = 50$ and the shear parameter is $S^* = 7$. N_p is the number of particles whose Stokes number based on the Kolmogorov time is $St_\eta \simeq 1$ with $\tau_p = 0.4$. The mass load parameter (see text for definition) is denoted by Φ, $\rho_p/\rho_f = 2000$ indicate the density ratio between the disperse and carrier phase. In our case $d_p/\eta \simeq 0.1$. The first four entries of the table provide for the effect of increasing the mass load ratio for a given grid resolution, runs $a - d$. Comparison of run d_l and d_h with the reference case d provide the sensitivity to grid resolution for a given mass load ratio.

when the computational box comes back to its unskewed configuration.

3. Particles Clustering

A visual impression of instantaneous particles' configurations is provided in figure 1 where slices of the domain in selected coordinate planes are displayed for different values of the mass load parameter Φ at $St_\eta = 1.05$. Hereafter the Stokes number is defined with respect the Kolmogorov time of the simulation operated in the one-way coupling regime. In fact, as we will see in §4, the back-reaction leads to an overall depletion of the viscous energy dissipation rate and consequently to an alteration of the Kolmogorov time-scale $\tau_\eta = (\nu/\epsilon)^{1/2}$. In all the four cases

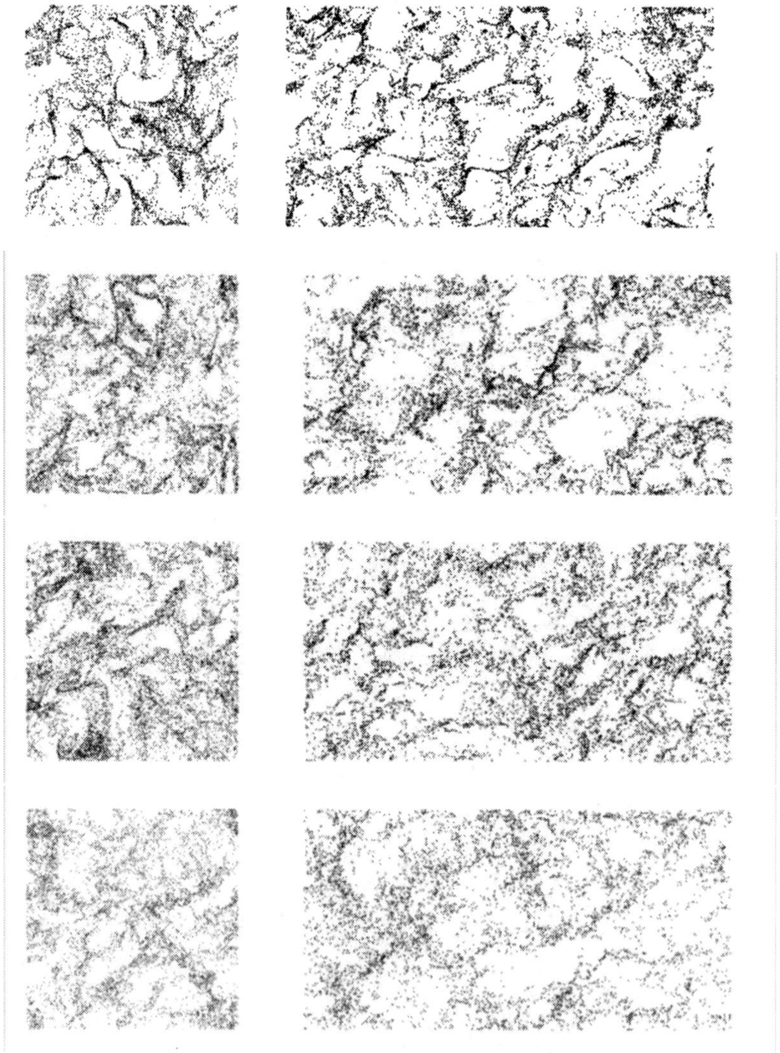

Figure 1. Snapshots of particle positions for increasing values of the mass load parameter $\Phi = 0,\ 0.2,\ 0.4,\ 0.8$ from top to bottom. For all the datasets $St_\eta = 1$. Left column thin slice in the $y-z$ plane; right column slice in the $x-y$ plane. The slice thickness is of the order of a few Kolmogorov scales.

reported in figure 1, the disperse phase is characterized by a multi-scale distribution of particles concentration and voids. From the figure the shear induced orientation of the clusters is apparent even though the concentration of the particles in the accumulation regions seems to be reduced as the coupling effects become relevant. At a first sight this effect might be explained in terms of an alteration of the Kolmogorov time scale resulting in a smaller effective Stokes number i.e. for a given relaxation time–$\tau_p = 0.4$ in our case–particles might be expected to behave more like tracers in the two-way coupling regime. However, a more quantitative analysis shows that this is not the case. The modification of the Kolmogorov time-scale, even though the energy dissipation rate is significantly changed at high mass loads, is small and corresponds to an alteration of the Stokes number which can not explain the observed depletion of the clustering intensity in the two-way coupling regime. For instance, the Stokes number ranges from $St_\eta = 1.05$ when turbulence is not modified by the particles to $St_\eta = 0.90$ in the most severe coupling conditions at $\Phi = 0.8$. In conclusion, the observed depletion of the clustering intensity calls into play non trivial physical effects which are more likely related to an overall alteration turbulent fluctuations rather than an oversimplified explanation in terms of an alteration of the Stokes number.

In this paper we limit ourselves to the description of anisotropic clustering under the two-way coupling conditions by considering the Angular Distribution Functions (ADF) introduced in [38]. Essentially the ADF measures the number of particles' pairs at separation r in the direction $\hat{\mathbf{r}}$ and is defined as

$$g(r, \hat{\mathbf{r}}) = \frac{1}{r^2} \frac{d\nu_r}{dr} \frac{1}{n_0}, \tag{5}$$

where $n_0 = 0.5 N_p(N_p - 1)/V_f$ is the volume density of particles' pairs and $\nu_r(r, \hat{\mathbf{r}}) d\Omega$ is the numbers of particles' pairs contained in the spherical cone of radius r with axis along $\hat{\mathbf{r}}$ and amplitude $d\Omega$. The spherical average of the ADF $g_{00}(r) = 1/(4\pi) \int_\Omega g(r, \hat{\mathbf{r}}) d\Omega$ is called the Radial Distribution Function (RDF) and has been already used to characterize particles clustering in isotropic conditions. The ADF extend the tool to anisotropic conditions retaining information on the directionality of the clusters. The behavior of the RDF near the origin $g_{00}(r) \propto r^{-\alpha}$

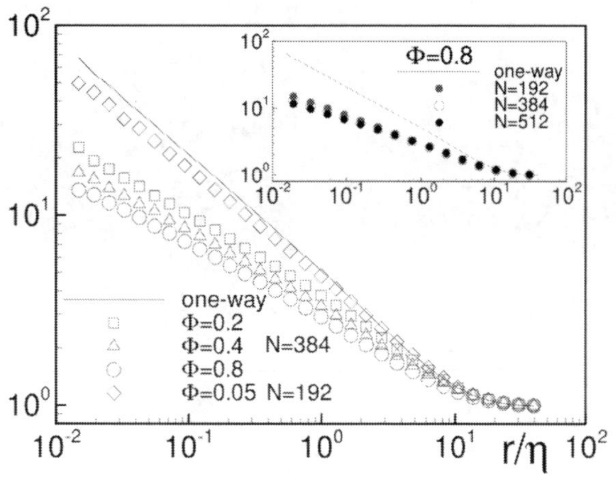

Figure 2. Projection of the ADF in the isotropic sector g_{00} as a function of separation. The main panel address the effect of an increasing mass load for a given resolution of the carrier phase. The inset provides sensitivity of the data to the grid resolution for the highest mass load $\Phi = 0.8$. Error-bars (not shown) are of the order of the symbols size in the entire range of scales.

can be shown to be related to important geometrical features of the clusters. In particular $\mathcal{D}_2 = 3 - \alpha$ is the so-called correlation dimension [58] of the multi-fractal measure associated with the particle density. A positive α indicates the occurrence of small scale clustering.

The RDF is shown in figure 2 for the different values of the mass load parameter Φ. The data show the quantitative alteration of the clustering intensity. In fact, the back-reaction progressively attenuates the particles concentration at small scales leading to a smaller value of the scaling exponent α. The modification is substantial. As the mass load is progressively increased the number of particles pairs at small scales is reduced about seven times as inferred by comparing the small scale values of g_{00} obtained in the one-way coupling simulation (solid line) against the two-way coupling data at $\Phi = 0.8$ (open circles) in figure 2. The plot shows the smooth convergence of the two-way coupled simulation to the uncoupled case for very small values of the mass load, see for instance the data at $\Phi = 0.05$ in comparison with those at $\Phi = 0$. Concerning statistical accuracy, in the case of the isotropic projection of the ADF shown in figure 2 the error-bars (not shown) are smaller than the symbol size. In conclusion, the behavior confirms the qualitative impressions gained by simple inspection of the instantaneous particles patterns reported in figure 1 where the clusters are definitely less defined in the case $\Phi = 0.8$ than at $\Phi = 0$, compare bottom and top panels of the figure, respectively. In any case clustering, though partially attenuated, is a persistent feature of the two-way coupling regime as measured by the exponent α. In order to exclude any bias in our results due to the numerics, in the inset of figure 2 we address the RDF in the most severe coupling conditions. For the highest mass load $\Phi = 0.8$ we have performed two additional simulations with a coarser and finer resolution with respect to the reference case of run d, namely run d_l and d_h of table 1 respectively. The data confirm that the attenuation of the clustering is a genuine physical effect resulting from the momentum exchange between the carrier and disperse phase.

In view of turbulence modification, the anisotropy of particles' clusters will play a crucial role to be discussed in the next section where the analysis of the back-reaction effects on the carrier

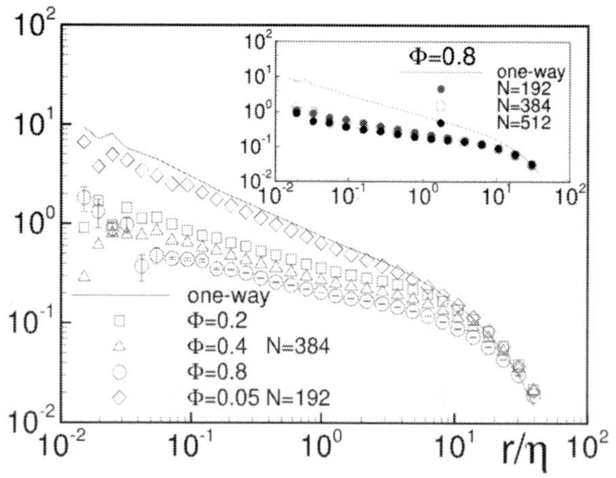

Figure 3. Projection of the ADF in the most energetic anisotropic sector $g_{2,-2}$ as a function of separation. The main panel addresses the effect of the mass load for a given resolution of the carrier phase. The inset provides the sensitivity to the grid resolution for the highest mass load $\Phi = 0.8$. Error-bars for the data at $\Phi = 0.05, 0.2, 0.4$ (not shown) are of the same order of those reported for the most severe coupling case at $\Phi = 0.8$.

fluid is addressed. For this reason, we complete the analysis of clustering under two-way coupling regime by characterizing the anisotropy of the aggregates. This can be done by exploiting the directionality properties of the ADF. Its angular dependence $g(r, \hat{\mathbf{r}})$ can be resolved in terms of spherical harmonics

$$g(r, \hat{\mathbf{r}}) = \sum_{j=0}^{\infty} \sum_{m=-j}^{j} g_{jm}(r) \, Y_{jm}(\hat{\mathbf{r}}) \,, \tag{6}$$

achieving a systematic description both in terms of separation r, accounted for by the coefficients $g_{jm}(r)$, and in terms of directions, described by the shape of the basis functions $Y_{jm}(\hat{\mathbf{r}})$. Each successive subspace, here labeled j, accounts for increasing levels of anisotropy consistently with the geometrical meaning of the spherical harmonics. This tool, introduced by [59], allowed for the systematic study of small scale isotropy recovery, see for instance [60, 61, 62, 40] and the exhaustive review in [63].

In figure 3 we show the most energetic anisotropic projection of the ADF namely g_{2-2}. As in the one-way coupling regime, by reducing the scale separation the intensity of the anisotropic projection increases indicating directionality in the particles aggregates. As for the RDF, the small scale behavior of g_{2-2} is controlled by the mass load ratio while the numerical resolution seems to play a negligible role, see the right panel of figure 3 where g_{2-2} is shown for $\Phi = 0.8$ and different resolutions. The statistical convergence of this observable might be an issue since it samples only few preferential directions. The error-bars on the data for the worse case at $\Phi = 0.8$ are always smaller than the symbol size up to $r/\eta = 0.1$ below which the error becomes appreciable on the scale of the plot. To correctly infer about the persistency of directionality at small scales, in figure 4 we show the most energetic anisotropic projection normalized by the RDF, namely the ratio g_{2-2}/g_{00}. This indicator quantifies the average level of anisotropy of particles' clusters at that particular scale. We observe that the back-reaction on the carrier fluid partially reduces the small scale anisotropy of the disperse phase when compared with

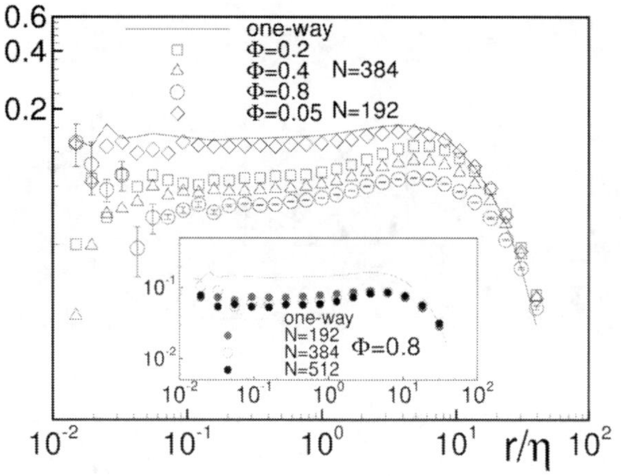

Figure 4. Normalized anisotropic projection of the ADF g_{2-2}/g_{00} as a function of separation. The left panel addresses the effect of the mass load for a given resolution of the carrier phase. The right panel show the sensitivity to the grid resolution for the highest mass load $\Phi = 0.8$ The error-bars for the data at $\Phi = 0.05, 0.2, 0.4$ (not shown) are of the same order of those reported for the most severe coupling case at $\Phi = 0.8$.

the one-way coupling regime. In contrast to the saturation observed under one-way coupling conditions (solid line), the ratio g_{2-2}/g_{00} seems to slightly decrease to eventually saturate to a lower value when the smallest scales are approached. In any case, going down the scale range, the anisotropy of the clusters is still present also in the two-way coupling regime. The overall behavior is not substantially altered from what we already described in previous papers [38] on the one-way coupling regime denoting a persistency of the clusters' directionality up to the smallest viscous scales.

4. Turbulence Modulation

In presence of two-way coupling, the support of the particle reaction field on the fluid is mainly given by the sets where most of the disperse phase concentrate. Actually clusters are organized in multi-scale sets whose orientation is controlled by the large scales anisotropy [38]. The effect is strong and the particles never achieve small scale isotropy, despite the Richardson cascade, in absence of back reaction, drives the fluid velocity fluctuations towards an isotropic state. We infer that the fluid is excited by a highly anisotropic, spectrally non-compact forcing, quite an unusual circumstance in turbulence. In these conditions, the back-reaction of the disperse phase is expected to deeply alter the structure of turbulence. Energy extraction/injection is now strongly anisotropic and active from the largest down to the smallest scales of the flow leaving no opportunity for a classical inertial range.

We start our description of turbulence modulation observing that, in the homogeneous shear flow, the mean velocity profile is imposed both in the one-way and the two-way coupling regimes. This excludes the modification of the mean flow itself as it could happen instead in channels or jet flows. Any alteration of turbulence can be directly ascribed to an intrinsic distortion of velocity fluctuations due to the inter-phase momentum exchange. The simplest quantities which

Figure 5. Left: effect of the mass load Φ on the turbulent fluctuations. Data normalized with the corresponding values of the un-coupled case. Open symbols refers to simulations $a-d$ in table 1. Filled symbols refers to cases d_l (grey) and d_h (black). $\langle u_1^2 \rangle$ (square), $\langle u_2^2 \rangle$ (delta), $\langle u_3^2 \rangle$ (diamond), $1/2\langle u_i\, u_i \rangle$ (gradient), $-\langle u_1\, u_2 \rangle$ (circle). The vertical tick of amplitude 0.04 provides the statistical error of the data. Right: turbulent kinetic energy budget. Data normalized with the production rate of the un-coupled case $\mathcal{P}_0$. Production $\mathcal{P}$ (circles), viscous energy dissipation ϵ (squares), energy flux intercepted by particles ϵ_p (delta). The algebraic sum $\mathcal{P} - \epsilon - \epsilon_p$ is also reported (diamonds) to check the quality of the budget. Filled symbols refers to cases d_l (grey) and d_h (black). Error-bars are of the order of the symbol size.

characterize the response of turbulence are single point statistics such as the turbulent kinetic energy, the energy dissipation rate and, as always for shear flows, the Reynolds shear stress. The turbulent kinetic energy is addressed in figure 5 as a function of the mass load ratio. In addition, in the same figure, data at fixed mass load $\Phi = 0.8$ and different grid resolutions, namely run d_l and d_h, are reported to check the grid sensitivity. Consistently with previous findings in the context of isotropic turbulence and wall bounded shear flows, by increasing the mass load ratio the turbulent fluctuations are found to be progressively attenuated.

The directionality of the mean flow leads to a significant difference in the three normalized velocity variances. The streamwise component $\langle u_1^2 \rangle$–squares in figure 5–is the least affected by the back-reaction. Nonetheless the trend is toward its slight increase in comparison with the un-coupled case. The other two components, $\langle u_2^2 \rangle$ (delta) and $\langle u_3^2 \rangle$ (diamonds), are progressively attenuated as the coupling becomes more relevant. The depletion–up to 50% at $\Phi = 0.8$– overwhelms the slightly increasing trend of the streamwise variance leading to an overall reduction of turbulent kinetic energy. The resolution check for case $\Phi = 0.8$ (gray and black filled squares in the figure) confirms the quality of the results. The impression is that the back-reaction enhances the large scale anisotropy of the flow. The concurrent depletion of the Reynolds shear stress is noteworthy, since it drags along the reduction of turbulent kinetic energy production.

The steady state turbulent kinetic energy balance in the two-way coupling regime, $\mathcal{P} = \epsilon + \epsilon_p$, involves a new term, the energy exchanged with the particles through the Stokes drag $\epsilon_p = \langle \mathbf{F} \cdot \mathbf{u} \rangle$, whereas the production term $\mathcal{P}_0 = -S \langle u_1 \, u_2 \rangle_0$ simply balances the viscous dissipation ϵ_0 in the uncoupled case (the subscript stands for $\Phi = 0$, no back-reaction). Actually at steady state the average particle kinetic energy is constant implying $\epsilon_p = \langle \mathbf{F} \cdot \mathbf{u} \rangle > 0$ since $\langle \mathbf{F} \cdot \mathbf{v_p} \rangle = 0$. Hence the energy provided by the Reynolds stress work is partially intercepted by the back-reaction field which represents an alternative dissipation channel, see [64, 65, 66] for a similar mechanism in the context of polymer laden flows and [67, 68] for more recent work on the

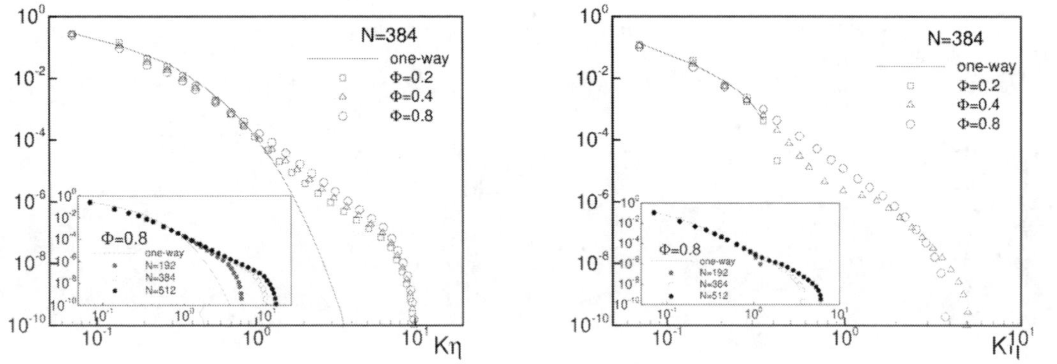

Figure 6. Effects of the mass load on the energy spectrum (left) and cospectrum (right) for a fixed spectral resolution of the carrier phase. Data of the one-way coupled simulation run a (solid line) are compared with two-way coupled simulations operated for incresing mass loads $\Phi = 0.2, 0.4, 0.8$ (symbols) which correspond to run b (square), c (delta) and d (circle) respectively, see table 1. In the inset effect of the grid resolution at the fixed mass load ratio $\Phi = 0.8$. Data of the one-way coupled simulation (solid line, run a in table 1) are compared against two-way coupled data at $\Phi = 0.8$ obtained with different resolutions of the carrier phase. Symbols correspond to runs d_l, d and d_h of table 1.

subject. The energy budget is addressed in graphical form in figure 5. The striking effect of the coupling is the considerable reduction of the production rate $\mathcal{P}$–20% at $\Phi = 0.8$–which entails the reduction of the viscous dissipation ϵ. Although typically smaller than ϵ, intriguing is also the behavior of the Stokes dissipation ϵ_p (deltas in figure 5) which achieves its maximum–order 15%–at intermediate mass loads.

A more detailed analysis of turbulence modulation calls into play the spectral distribution of turbulent kinetic energy. In the spectra shown in figure 6, as Φ increases, turbulent fluctuations are first attenuated in an intermediate range of scales to be eventually enhanced at small scales before being shut down by viscosity. In other words, the Stokes drag intercepts energy from the classical cascade at intermediate scales where velocity fluctuations are damped. The intercepted energy is partly pumped back to the fluid at small scales increasing the energy content by orders of magnitude, compare solid line (one-way coupling) and open symbols (two-way coupling) in figure 6. The overall effect is a reduction of turbulent kinetic energy, essentially given by the low wave number part of the spectrum. The emerging picture of particle-laden turbulence amounts to a standard transfer across the inertial range, a partial removal of energy from the cascade by the Stokes drag and its partial re-injection in the small scales. Under certain respects, the conceptual framework is similar to that operating in polymeric solutions.

The cospectrum, figure 6, shows a definite alteration of the spectral distribution of the turbulent shear stress and allows to identify the scales affected by the shear stress forcing. Though hardly appreciable in log scale, the cospectrum slightly decreases at large scales as the mass load ratio increases, consistently with the overall reduction of the shear stress. Meanwhile the range affected by production is progressively enlarged. These are the back reaction effects of the highly directional multi-scale clusters which force anisotropic motions in a progressively wider range of scales, see figure 6. Results at $\Phi = 0.2$ present only small deviations from case $\Phi = 0$. Increasing the mass load, $\Phi = 0.4 \div 0.8$, the back-reaction excites the Reynolds shear stress in almost the entire range of scales. The cospectrum vanishes much more slowly at high wave numbers resulting in unexpected energy injection at small scales. This keeps the turbulent fluctuation excited at scales much smaller than expected on the sole basis of the overall energy

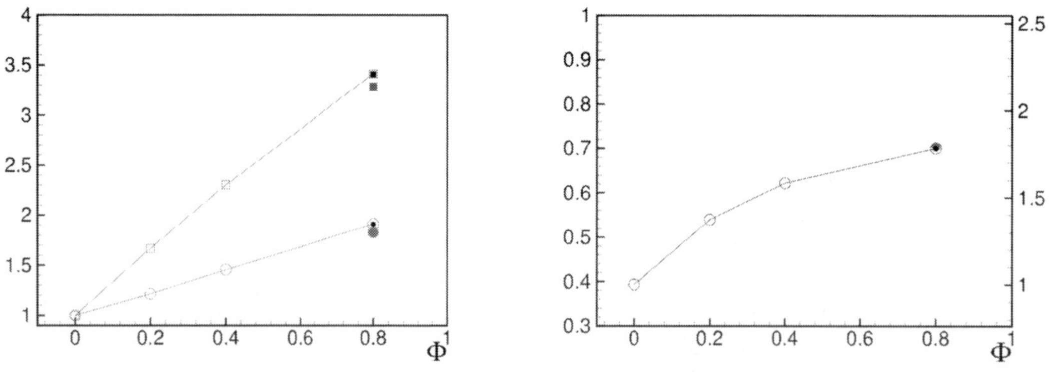

Figure 7. Left: normalized norm of the deviatoric component of the Reynolds stresses tensor $||b||/||b||_0$ (open circles) and of the pseudo dissipation tensor $||a||/||a||_0$ (open squares) as a function of the mass load ratio Φ. For comparison data at $\Phi = 0.8$ namely runs d_l and d_h (grey and black symbols) have been reported. Right: ratio $r = ||a||/||b||$ (circles left axis) and r/r_0 (circles right axis) as a function of the mass load ratio. Filled symbols as in the left panel.

production which, as already recalled, diminishes. The insets in figure 6 illustrate the results of grid refinement tests.

The present results apparently question the small scale isotropy assumption for particle-laden flows under considerable mass loads, which were classically motivated by the orientational randomization expected of the Richardson cascade.

Concerning the large scales, their anisotropy is quantified by the deviatoric component of the Reynolds stresses tensor $b_{ij} = \langle u_i\, u_j \rangle / \langle u_k\, u_k \rangle - 1/3\delta_{ij}$. Since anisotropy is already present at large scales in the Newtonian case (no back-reaction), to isolate the effect of the back-reaction we address, shown by open circles in the left panel of figure 7, the ratio $\|b\|/\|b\|_0$, where the subscript stands for $\Phi = 0$ and the norm is $\|b\| = \sqrt{b_{ij}\, b_{ij}}$. As anticipated in the discussion of figure 5, the back-reaction enhances the anisotropy. This mainly occurs by the attenuation of the transversal velocity variances $\langle u_2^2 \rangle$ and $\langle u_3^2 \rangle$ associated with a slight augmentation of the longitudinal fluctuation $\langle u_1^2 \rangle$. Concerning the dissipative scales, their anisotropy, shown by open squares in the left panel of figure 7, is analogously quantified by the deviatoric component of the pseudo-dissipation tensor $\epsilon_{ij} = 2\,\nu \langle \partial_k u_i\, \partial_k u_j \rangle$, i.e. by computing $a_{ij} = \epsilon_{ij}/\epsilon_{kk} - 1/3\delta_{ij}$ and its norm $\|a\| = \sqrt{a_{ij}\, a_{ij}}$. The data show a dramatic growth of the gradient anisotropy which under the most severe coupling conditions increases by 350% with respect to the reference Newtonian case. Two features emerge. First, the disperse phase is able to alter the anisotropy of the largest scale. Second, the small scale anisotropy level normalized with the Newtonian value increases even more steeply with the mass load, due to the spectrally non compact anisotropic forcing operated by the disperse phase on the carrier fluid. Actually the ratio $r = \|a\|/\|b\|$ of small to large scale anisotropy, right panel of figure 7, almost doubles going from $r = 0.4$ for $\Phi = 0$ to $r = 0.7$ for $\Phi = 0.8$, scale on the left axis of the plot.

5. Final Remarks

Clustering is a well known feature of inertial particle configurations in turbulent flows. The classical explanation of this small scale behavior is the centrifugal effect due to the vortical structures that populate a turbulent flow, which tend to expel heavy inertial particles from the vortex cores, to localize in the interstices between the vortices. The resulting segregation may be extremely intense for particles at Stokes number order one. A peculiar feature of these systems is their response to the anisotropy of the flow. In recent investigations of a particle-laden homogeneous shear flow it was found, [38], that particle clusters maintain the anisotropy down to their smallest scales, in ranges where, typically, the carrier velocity field already recovered isotropy under the action of the Richardson cascade.

When the mass load ratio is increased, the back-reaction of the suspended phase on the carrier fluid can be no longer neglected, opening the issue of the ultimate fate of the clusters in the strong two-way coupling regime, and poses the question as to how the structure of the turbulence is altered in the carrier phase. This is the regime we have explored in the present paper. As main results, we find that the coupling of the carrier fluid with the suspended phase leads to less pronounced segregation effects, with the clusters that tend to become progressively smeared and less neatly defined with respect to the Newtonian simulation with advected passive inertial particles. At the same time, the anisotropy of the clusters is also reduced. Overall the trend seems to proceed towards a reduction of the peculiar segregation and anisotropy amplification effects recently described for vanishingly small mass loads.

The intriguing aspect is here the response of the carrier fluid to the particle back-reaction. The oriented clusters, though fatter and more isotropic than in the one-way coupling regime, constitute the geometrical support of the force field exerted back on the fluid. This forcing is distributed on the entire range of scales spanned by the clusters and present significant anisotropic features at all active scales. The result is an increased large scale anisotropy of the carrier fluid turbulence, manifested by the Reynolds stress anisotropy indicator. The effect consists in the alteration of the energy carrying large scales of the fluid, which is accompanied by a reduction of turbulent kinetic energy. The Reynolds shear stress also diminishes, though less than its isotropic component, resulting in the increase of the anisotropy indicator mentioned above. It follows the reduction in the turbulent kinetic energy production rate, which, at equilibrium, entails a lowering of turbulent energy dissipation rate.

The observation that the back-reaction forcing occurs throughout the range of scales spanned by the clusters, implies that a non-negligible amount of forcing takes place also in the smallest scales of the turbulence, which are usually passively fed only by the energy cascade from larger scales. Consistently, the energy spectrum is broadened, with the excitation of scales significantly smaller than the Kolmogorov scale expected from the energy content of the system. As we have shown, the clusters retain their anisotropy down to the smallest scales. Consistently, the Reynolds shear stress forcing is also much less compact in spectral space than found in classical shear turbulence. This leads to a substantial increase of anisotropy at the level of the instantaneous velocity gradients, as measured by the anisotropy indicator of the pseudo-dissipation tensor.

In conclusion, the structure of the system of particles and fluid is significantly altered by the coupling, an effect which clearly increases with the amount of mass loading of the suspension. Our findings could have a certain impact on turbulence modeling of multiphase flows which, both in the context of Reynolds averaged (RANS) and filtered equations (LES), heavily rely upon the concepts of inertial energy cascade and of a presumed universal small scale statistics. Actually, we have shown that the cornerstone Kolmogorov theory no longer safely applies, since the classical energy cascade is overwhelmed by the anisotropy-enhancing back-reaction of the particles.

6. Acknowledgments

Authors thanks the COST Action MP0806 "Particles in Turbulence" for supporting the present work and the CASPUR consortium where the simulations were performed.

[1] Falkovich G, Fouxon A and Stepanov M 2002 *Nature* **419** 151
[2] Wang L, Ayala O, Rosa B and Grabowski W 2008 *New J. of Phys.* **10** 1–40
[3] Károlyi G, Péntek A, Scheuring I, Tél T and Toroczkai Z 2002 *Proc. Natl. Acad. Sci.* **97** 13661
[4] Lewis D and Pedley T 2000 *J. Thoretical Biology* **205** 377–408
[5] Woods W A 2010 *Ann. Rev. Fluid Mech* **42** 391–412
[6] Post S and Abraham J 2002 *Int. Journal Mult. Flows* **28** 997–1019
[7] Rouson D and Eaton J 2001 *J. Fluid Mech.* **428** 149
[8] Portela L, Cota P and Oliemans R 2002 *Powder Technology* **125** 149–157
[9] Derksen J, Sundaresan S and Akker H V D 2006 *Powd. Tech.* **163** 59–68
[10] Reeks M 1983 *J. Aerosol Sci.* **14** 729–739
[11] Brooke J, Kontomaris K, Hanratty T and McLaughlin J 1992 *Phys. Fluids* **A 6** 825–834
[12] Soldati A and Marchioli C 2009 *Int. J. Multiphase flow* **35(9)** 827–839
[13] Marchioli C and Soldati A 2002 *J. Fluid. Mech.* **468** 283
[14] Picano F, Sardina G and Casciola C 2009 *Phys. Fluids* **21** 093305
[15] Bec J, Biferale L, Cencini M, Lanotte A, Musacchio S and Toschi F 2007 *Phys. Rev. Lett.* **98(8)** 084502
[16] Balachandar S and Eaton J 2010 *Ann. Rev. Fluid Mech* **42** 111–133
[17] Eaton J 2009 *Int. J. Multiphase flow* **35(9)** 792–800
[18] Elghobash S and Truesdell G 1993 *Physics of Fluids A: Fluid Dynamics* **5(7)** 1790–1801
[19] Druzhinin O and Elghobashi S 1999 *Physics of Fluids* **11(3)** 602–610
[20] Druzhinin O 2001 *Physics of Fluids* **13(12)** 3738–3755
[21] Sundaram S and Collins L 1999 *J. Fluid Mech.* **379** 105–143
[22] Ferrante A and Elghobashi S 2003 *Physics of Fluids* **15(2)** 315–329
[23] Boivin M, Simonin O and Squires K 1998 *J. Fluid Mech.* **375** 235–263
[24] Hwang W and Eaton J 2006 *Int. J. Multiphase flow* **32** 1386–1396
[25] Hwang W and Eaton J 2006 *J. Fluid Mech.* **564** 361–393

[26] Yeo K, Dong S, Climent E and Maxey M 2010 *Int. J. Mult. Flow* **36(3)** 221–233
[27] Bosse T, Kleiser L and Meiburg E 2006 *Phys. Fluids* **18** 0271021
[28] Yang T and Shy S 2005 *J. Fluid Mech.* **526** 171–216
[29] Geiss S, Dreizler A, Stojanovic Z, Chrigui M, Sadiki A and Janicka J 2004 *Exp Fluids* **36** 344–354
[30] Poelma C, Westerweel J and Ooms G 2007 *J. Fluid. Mech.* **589** 315
[31] Poelma C and Ooms G 2006 *Appl. Mech. Rev.* **50** 78–90
[32] Zhao L, Andersson H and Gillisen J 2010 *Phys. Fluids* **22** 0817021
[33] Yiming L, McLaughlin J, Kontomaris K and Portela L 2001 *Phys. Fluids* **13(10)** 2957–2967
[34] Pan Y and Banerjee S 2001 *Phys. Fluids* **8(10)** 2733–2755
[35] Rani S, Winkler C and Vanka S 2004 *Powder Tech.* **141** 80–99
[36] Ljus C, Johansson B and Almsedt A 2002 *Int. J. Multiphase flow* **28** 1075–1090
[37] Ahmed A and Elghobashi S 2000 *Phys Fluids* **12(11)** 2906–2930
[38] Gualtieri P, Picano F and Casciola C 2009 *J. Fluid Mech.* **629** 25–39
[39] Shotorban B and Balachandar S 2006 *Phys. Fluids.* **18** 065105
[40] Casciola C, Gualtieri P, Jacob B and Piva R 2007 *Phys. Fluids* **19** 101704
[41] Casciola C, Gualtieri P, Jacob B and Piva R 2005 *Phys. Rev. Lett.* **95** 024503
[42] Jacob B, Casciola C, Talamelli A and Alfredsson P 2008 *Phys. Fluids* **20(4)** 045101
[43] Stresing R, Peinke J, Seoud R and Vassilicos J 2010 *Physical Review Letters* **104** 194501 ISSN 1079-7114
[44] Rogallo R 1981 *Nasa T-M* **81315**
[45] Maxey M and Riley J 1983 *Phys. Fluids* **26** 2437
[46] Gualtieri P, Casciola C, Benzi R, Amati G and Piva R 2002 *Phys. Fluids* **14(2)** 583–596
[47] der Hoef M V, Annald M V S, Deen N and Kuipers J 2008 *Ann. Rev. Fluid Mech* **40** 47–70
[48] Burton T and Eaton J 2005 *J. Fluid Mech.* **545** 67–111
[49] Poesio P, Ooms G, Cate A T and Hunt J 2006 *J. Fluid Mech.* **555** 113–130
[50] Zhang Z and Prosperetti A 2005 *J. Comp. Phys.* **210** 292–324
[51] Lomholt S and Maxey M 2003 *J. Comp. Phys.* **184** 381–405
[52] Crowe C, Sharma M and Stock D 1977 *J. Fluid Eng.* **99** 325
[53] Garg R, Narayanan C and Subramaniam S 2009 *Int. J. Mult. Flow* **35** 376–388
[54] Gualtieri P, Casciola C M, Benzi R, Amati G and Piva R 2002 *Physics of Fluids* **14** 583–596
[55] Shumacher J 2004 *Phys. Fluids* **16(8)** 3094–3102
[56] Shumacher J, Eckhardt B and Doering C 2010 *Phys. Lett. A* **374** 871
[57] Boeck T, Krasnov D and Schumacher J 2010 *Physca D* **239** 1258
[58] Grassberger P and Procaccia I 1983 *Phys. Rev. Lett.* **50** 346
[59] Arad I, L'vov V and Procaccia I 1999 *Phys. Rev. E* **59** 6753
[60] Biferale L and Toschi F 2001 *Phys. Rev. Lett.* **86** 4831
[61] Biferale L, Boffetta G, Celani A, Lanotte A, Toschi F and Vergassola M 2003 *Phys. Fluids* **15** 2105–2112
[62] Chen S, Dhruva B, Kurien S, Sreenivasan K and Taylor M 2005 *J. Fluid Mech.* **533** 183
[63] Biferale L and Procaccia I 2005 *Phys. Rep.* **414** 43
[64] De Angelis E, Casciola C, Benzi R and Piva R 2005 *J. Fluid Mech.* **531** 1–10
[65] Casciola C and De Angelis E 2007 *J. Fluid Mech.* **581** 419–436
[66] Vaithianathan T, Robert A, Brasseur J and Collins L 2006 *J. Non-Newtonian Fluid Mech.* **140** 3
[67] Cai W, Li F and Zhang H 2010 *J. Fluid Mech.* **665** 334
[68] Perlekar P, Mitra D and Pandit R 2010 *Phys. Rev. E* **82** 066313

Volcanic ash in the free atmosphere: A dynamical systems approach

T Haszpra and T Tél

Institute for Theoretical Physics, Eötvös Loránd University, Pázmány Péter sétány 1/A, H-1117 Budapest, Hungary

E-mail: hatimi@caesar.elte.hu, tel@general.elte.hu

Abstract. The motion of volcanic ash particles is described on large scales by an ordinary differential equation, and the dynamics is chaotic. Deposition with a terminal velocity is found to be influenced by all three components of the wind velocity. The particles sooner or later leave the free atmosphere and enter the planetary boundary layer. Chaos is thus transient, and we point out that quantities like escape rate, average lifetime and topological entropy can efficiently be applied to characterize the volcanic ash dynamics in the free atmosphere. These quantities are evaluated for particular eruption examples of volcano Eyjafjallajökull, as well as for volcanic averages taken at different geographical locations, and for global averages over the Earth.

1. Introduction

The spreading of different forms of air pollutants in the atmosphere is an intensively studied subject (see e.g. [1]). The eruptions of the Iceland's Eyjafjallajökull mid April–mid May 2010, which were relatively weak compared to other volcanic eruptions, called the attention to the particular case of volcanic ash as they caused enormous disruption of air traffic across Europe. In order to characterize the advection and study the dispersion (see e.g., [2, 3]), we apply here a single-particle-approach. Our main aim is to understand the spreading of volcanic ash in the free atmosphere from a dynamical systems point of view (in a different context, see [4]).

It will be demonstrated that the large-scale dynamics can be described by the equations of passive scalar advection in the horizontal direction, but in the vertical direction a well-defined terminal velocity should be taken into account as a term added to the vertical wind velocity. The background wind field data are taken from the ERA-Interim data bank of the European Centre for Medium-Range Weather Forecasts [5].

The simulations show that the advection of volcanic ash is chaotic, but since particles spend a finite period of time in the free atmosphere (before entering the planetary boundary layer, where they become subject to a qualitatively different, noisy dynamics), chaos is transient. We point out that quantities well known from the theory of transient chaos [6] can efficiently be applied to characterize the dynamics of volcanic ash in the free atmosphere. They characterize the deposition as a decay process, the average lifetime outside of the planetary boundary layer, and the stretching rate of material line segments of volcanic ash. We evaluate these measures particular cases of Eyjafjallajökull eruptions, as well as, for global averages taken over several model volcano eruptions distributed over the globe.

The paper is organized as follows. Section 2 gives an overview of the basic equation, of the ECMWF reanalysis wind field and the numerical methods used to determine the trajectories. In Section 3, we present three basic quantities: escape rate, average lifetime, and topological entropy, to be used to characterize the spreading of volcanic ash. The results for local and global averages are presented in Sections 4 and 5, respectively. A summary is given in Section 6.

2. Equations, data and methods

The dynamics of inertial particles in turbulence has attracted recent attention (for a review see [7, 8]). The equations of motion for small, inertial, spherical particles of radius r in a viscous fluid advected by a flow deterministically are given by the Maxey-Riley equations [9, 10]. For heavy particles of density ρ_p much larger than that of the ambient medium ρ, the dimensionless equations for the particles trajectory $\mathbf{r}_p(t)$ read as

$$\ddot{\mathbf{r}}_p = \frac{1}{St}\left(\mathbf{v}(\mathbf{r},t) - \dot{\mathbf{r}}_p - W_{terminal}\mathbf{n}\right), \tag{1}$$

where $\mathbf{v}(\mathbf{r},t)$ is the flow field, $W_{terminal}$ is the dimensionless terminal velocity in still fluid, and $\mathbf{n}$ is a unit vector pointing upwards. Velocity and distance are measured in units of a characteristic velocity U and L, respectively. The Stokes number St appearing here is the dimensionless relaxation time of inertial particles subject to Stokes drag. In a fluid of kinematic viscosity ν, it can be expressed as

$$St = \frac{2r^2\rho_p U}{9\rho\nu L}. \tag{2}$$

Since in the free atmosphere diffusion is negligible on large scales (scales larger than a few 10 kms) [11], Eq. (1) applies to volcanic ash particles if wet deposition and other chemical reactions do not play a role.

It has been observed [12] that the size of typical aerosol particles ranges between $r = 1$ and $r = 12$ μm. Their density is approximately twice as large as that of water. With $\rho_p/\rho = 2000$, $\nu = 10^{-5}$ m^2/s as the kinematic viscosity of air, and a typical large scale atmospheric horizontal distance $L = 1000$ km, the Stokes number is calculated for two typical U values in Table 1, and is found to be very small.

Table 1. Typical smallest and largest Stokes number in the problem.

r [μm]	U [m/s]	St
1	1	$4.44 \cdot 10^{-11}$
12	10	$6.40 \cdot 10^{-08}$

Since the limit of $St \to 0$ in (1) implies finite acceleration only if the paranthesis on the right hand side vanishes, the large-scale equation of motion for aerosol particles becomes even simpler than (1). We conclude that inertial effects are negligible, but in the vertical direction deposition has to be taken into account with a terminal velocity.

The dimensional equation of motion for volcanic ash particles thus follows from (1) as

$$\mathbf{v}_p \equiv \dot{\mathbf{r}}_p = \mathbf{v}(\mathbf{r}_p(t),t) - w_{terminal}\mathbf{n}. \tag{3}$$

The expression for the terminal velocity, from the Stokes law [13], is:

$$w_{terminal} = \frac{2}{9}r^2\frac{\rho_p}{\rho\nu}g. \tag{4}$$

Equation (3) is similar to that used in FLEXPART [14] and NAME [3] but without any stochastic component, and we apply it to different volcanic ash particles, individually. This equation is known to have chaotic solutions for typical velocity fields.

In principle, precise atmospheric trajectories can be determined directly from high resolution wind observations by interpolating between the measuring locations and time instants. In practice, however, trajectory calculations are mostly based on gridded output of numerical models (weather forecasts or reanalyses) such as the ones provided by the European Centre for Medium-Range Weather Forecasts (ECMWF).

In this work, the 3rd generation ECMWF reanalysis ERA-Interim data bank is exploited, which is almost up to date from 01/01/1989 [5]. Zonal (u), meridional (v) and vertical (w) wind velocity components of global geographic coverage are evaluated for the year of 2010. Four values are available each day for 00, 06, 12 and 18 h UTC (Universal Time Coordinated) at each geographic location with a spatial resolution of $1.5° \times 1.5°$ (lat/long). Note that the gridded wind fields are rather smooth, subgrid scale turbulence is not resolved. Although the wind velocity at a given site and time is intended to represent an instantaneous value, direct comparison with high resolution wind tower measurements indicates that the velocity should be regarded as a 6-hour mean value [15].

Global ERA-Interim wind fields are available on 37 pressure levels in the range of 1000–1 hPa [16]. Since we are interested in free atmospheric flow, which takes place above the planetary boundary layer (PBL) (in average above 850 hPa), wind fields were downloaded at pressure levels in the range of 850–100 hPa.

When solving (3), the velocities were interpolated at the particle location (linear in time and in the vertical direction and using bicubic spline interpolation in the horizontal direction). Then the differential equation was solved by using Euler's method with a time step of 45 minutes. We choose $dt = 45$ min, because $dt = 6$ h would lead to unrealistically large displacements of a time step. A decrease of dt from 6 h leads to a change in trajectories and dispersion, but for values smaller than $dt = 45$ min the dispersion of particles does not differ significantly.

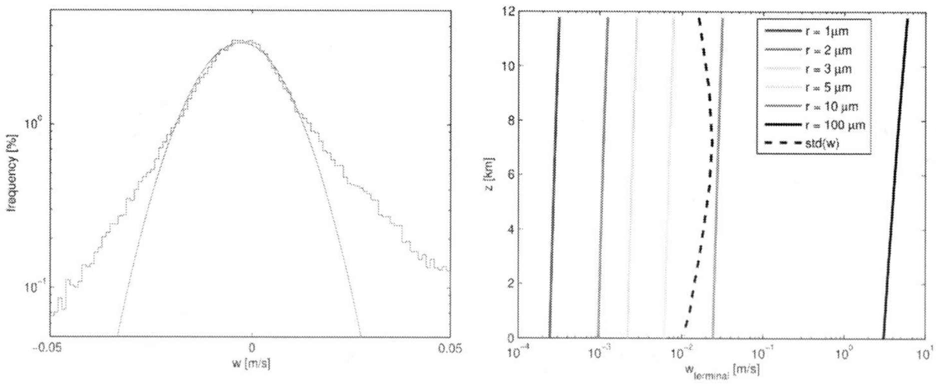

Figure 1. Left panel: Logarithm of the probability distribution of the vertical wind velocity w at 500 hPa on a given day (01.01.2010 at 00, 06, 12, 18 UTC) over the globe. A clear deviation from a Gaussian form (parabola) can be seen with fat tails. Right panel: Terminal velocity $2r^2\rho_p g/(9\rho(z)\nu(z))$ in ICAO standard atmosphere for different particle sizes as a function of height.

Before turning to the results it is worth comparing the histogram of the vertical wind velocity, for example, on the level of 500 hPa (which corresponds approximately to 5.5 km), with the terminal velocity of particles of different sizes in different heights in the ICAO (International Civil Aviation Organization) standard atmosphere [17], we can notice two important features

(Fig. 1). First, the terminal velocity is hardly height-dependent: the decrease in air density appears to be compensated by the increase of the kinematic viscosity due to lower temperatures. Therefore, in (3) we use a constant terminal velocity $\bar{w}_{terminal}$ obtained as an average over height. Second, the magnitude of the standard deviation of the wind velocity fluctuations (black dashed line in the right panel) is larger or is of the same order as the terminal velocity for particles of 1 to 10 μm. Thus, the terminal velocity cannot be assumed to be attained instanteously, and Eq. (3) describes how the vertical wind component influences the deposition dynamics.

Since the downloaded wind data are given in latitude and longitude coordinates on pressure levels, we determined trajectories in the pressure coordinates. The third variable besides the horizontal wind velocity components u and v is the time derivative ω of the pressure coordinate (symbol ω should not be confused with w, that of the vertical wind velocity). The dimension of ω is Pa/s, so the terminal velocity should also be transformed in this variable. For $w_{terminal} > 0$, we expect $\omega_{terminal}$ to be negative. The third component of (3) is

$$\dot{z}_p \equiv w_p = w(\mathbf{r}_p(t), t) - w_{terminal}. \quad (5)$$

Similarly to (5), the vertical equation of motion for a particle in pressure coordinate system is

$$\frac{dp(\mathbf{r}_p(t), t)}{dt} \equiv \omega_p = \omega(\mathbf{r}_p(t), t) - \omega_{terminal}. \quad (6)$$

From the total derivatives we have

$$\omega = \frac{\partial p}{\partial t} + u\frac{\partial p}{\partial x} + v\frac{\partial p}{\partial y} + w\frac{\partial p}{\partial z}, \text{ and } \omega_p = \frac{\partial p}{\partial t} + u_p\frac{\partial p}{\partial x} + v_p\frac{\partial p}{\partial y} + w_p\frac{\partial p}{\partial z}. \quad (7)$$

Since from (3) $u_p = u$ and $v_p = v$, we find that

$$\omega_{terminal} = (w - w_p)\frac{\partial p}{\partial z} = w_{terminal}\frac{\partial p}{\partial z}. \quad (8)$$

If the hydrostatic approximation is valid, i.e., $\partial p/\partial z = -\rho g$, then

$$\omega_{terminal} \approx -\frac{2}{9}r^2\frac{\rho_p}{\nu}g^2. \quad (9)$$

We have checked that the deviation $(\partial p/\partial z - \rho g)/(\rho g)$ is at most one percent over the full ERA-Interim data bank, which implies that approximation (9) is satisfactory.

As an appropriate choice with the structure of downloaded data, we use spherical coordinates and solve the following equations

$$\frac{d\lambda_p}{dt} = \frac{u(\lambda_p, \varphi_p, p_p, t)}{R_E \cos\varphi_p}, \frac{d\varphi_p}{dt} = \frac{v(\lambda_p, \varphi_p, p_p, t)}{R_E}, \frac{dp_p}{dt} = \omega(\lambda_p, \varphi_p, p_p, t) - \omega_{terminal}, \quad (10)$$

where λ_p and φ_p are the longitude and latitude coordinates of a particle, $p_p(t) \equiv p(\mathbf{r}_p(t), t)$ is the pressure coordinate of a particle, and R_E is the radius of the Earth.

3. Dynamical systems characteristics

Since deposition occurs, volcanic ash particles ultimately reach the planetary boundary layer. Here, however, the character of their equation of motion basically changes due the strong turbulent fluctuations. Within the framework of individual particle simulations turbulence can be taken into account by converting (3) into a stochastic differential equation by adding a noise term of strength proportional to the turbulent eddy diffusivity. The equation is then no longer determinisic, and the dynamics changes drastically. Here we concentrate on the behaviour in the free atmosphere. Particles are regarded to have escaped from the free atmosphere when entering the boundary layer. Chaotic dynamics is then necessarily of transient character.

A natural quantity characterizing the decaying process is the survival probability, which is proportional to the number $N_p(t)$ of particles staying in the free atmosphere up to time t at least. We consider the lower boundary of the free atmoshpere to be the 850 hPa pressure level. Based on the theory of transient chaos [6], we expect the proportion $N = N_p/N_0$ of non-escaped particles out of an ensemble of N_0 particles to decay exponentially, i.e.,

$$N(t) \sim \exp(-\kappa t) \tag{11}$$

which is valid after some time. The rate of the decay, called the escape rate, is denoted by κ. It represents the basic escape statistics independent of the initial distribution of the particles. The average lifetime τ of chaos is often estimated as $\tau \sim 1/\kappa$, but this estimate is known to be rather rough. In the knowledge of the particular initial distribution, a direct computation of the average lifetime τ is always useful.

An initially short material line becomes strongly stretched within a short time. The length $L(t)$ of the filament grows rapidly in time and this growth is found to be exponential after some time with an exponent called the topological entropy, h [18]:

$$L(t) \sim \exp(ht). \tag{12}$$

The topological entropy is similar in spirit to, but different in value from the average Lyapunov exponent λ: inequality $h \geq \lambda - \kappa$ should always hold [6]. Our experience shows that the topological entropy is technically much easier to measure than the Lyapunov exponent. When calculating the largest Lyapunov exponent, one selects a reference trajectory and initiates a test particle close to a point of it. When the distance of the pair becomes large enough, one takes another test particle nearby the reference trajectory with an initial distance in the same direction, etc. Since the initial distance cannot be shorter than the grid size, the distance reaches thousands of kilometers within a few days. From the local deviations finite time Lyapunov exponents are obtained which belong, however, to different geographical locations and different times. Averaging them over months provides the value of the average Lyapunov exponent over the entire globe for which we have obtained 0.3 day^{-1} (passive tracers, $r = 0$ μm). We prefer the use of the topological entropy to that of the Lyapunov exponent since even the geographical distribution of the former can be easily obtained as illustrated by Fig. 8.

As characteristic numbers from dynamical system theory, we propose the use of κ, τ and h to measure the dynamics of volcanic ash particles in the free atmosphere. Besides these characteristic numbers, different pictorial representations of the particle cloud are also useful.

4. Local results

We simulated the dispersion of the ash due to the volcanic eruption in Iceland with some simplifications. A single initial column of 1° × 1° × 140 hPa size ($\sim$ 100 km × 100 km × 4 km) was started at 6 UTC on the 14th of April on the 300 hPa level with $N_0 = 27000$ particles.

The group of charts in Fig. 2 shows the spreading of particles of 10 μm. Due to the relatively large terminal velocity, they are continuously sinking from the initial height of 9 km, and in two days time they reach the level of approximately 600 hPa (the altitude of which is 4.2 km on average). It is also interesting to see in simulations where particles leave the free atmosphere. The last panel shows the location of particles when they reach the upper boundary (850 hPa) of the boundary layer (for the last particles this happens before April 22). The distribution is rather inhomogeneous with strong separation of connected patches, stretching across the width of Siberia. As mentioned simulations were carried out in the free atmosphere, where the effect of

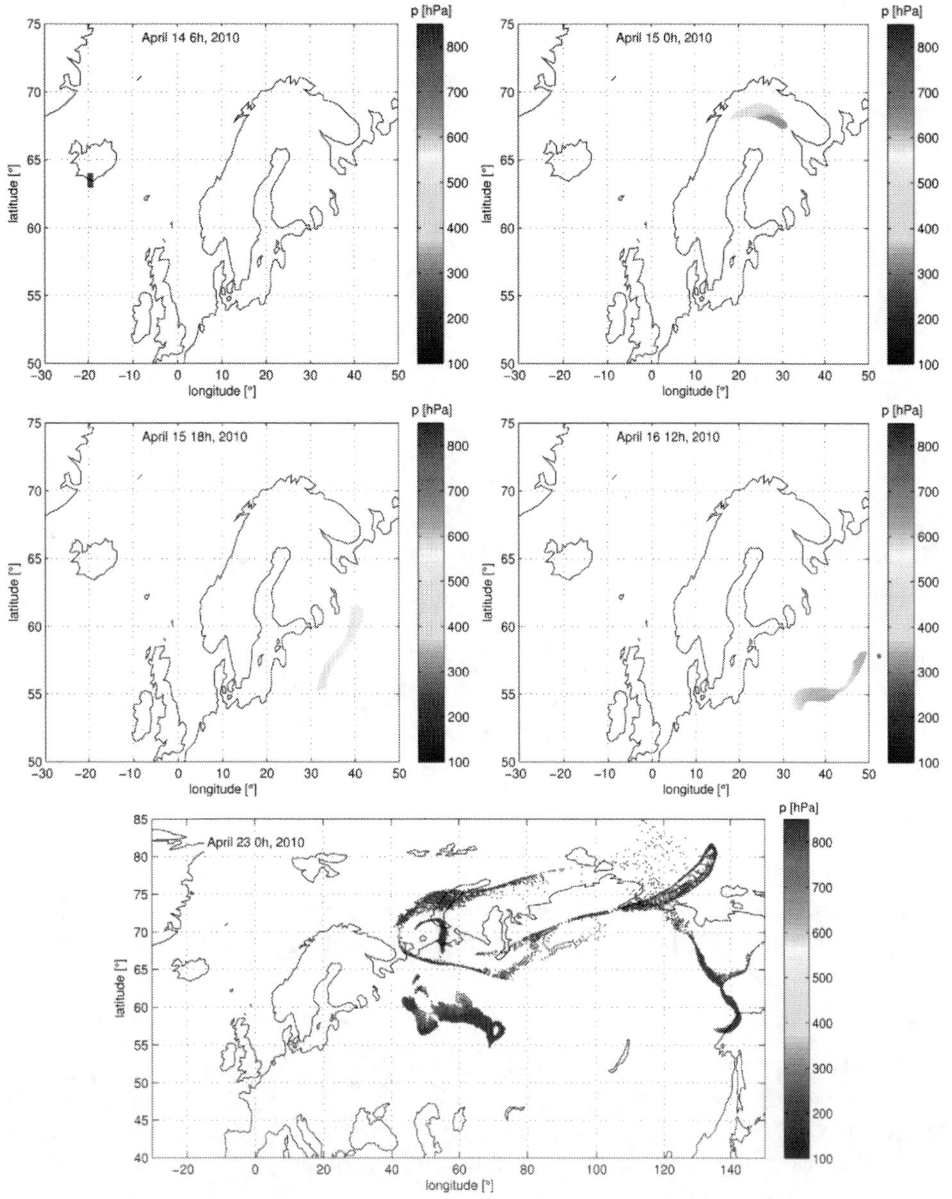

Figure 2. Simulation of the spreading of the volcanic ash ($r = 10\ \mu$m) from Eyjafjallajökull. A single initial column of $1^\circ \times 1^\circ \times 140$ hPa size (~ 100 km $\times$ 100 km $\times$ 4 km) was started at 06 UTC on 14.04.2010 on the 300 hPa level with $N_0 = 27000$ particles. The particular time instants are indicated on the panels.

the turbulent diffusion is known to be negligible: The characteristic scale over which the effect of diffusion can spread in a chaotic flow of average Lyapunov exponent λ is $d = \sqrt{D/\lambda}$ [11]. On scales larger than d, diffusion does not play any role. With $\lambda = 0.3$ day^{-1}, even with a rather large turbulent diffusion coefficient of $D = 10$ m^2/s we obtain $d \sim 2$ km only. We thus conclude that the pattern seen in the free atmosphere (see e.g., Fig. 2) is not influenced at all by diffusion (at most fractality disappears on scales smaller than 2 km, but this horizontal scale is not resolved in any of our large scale distributions). After sinking through the planetary

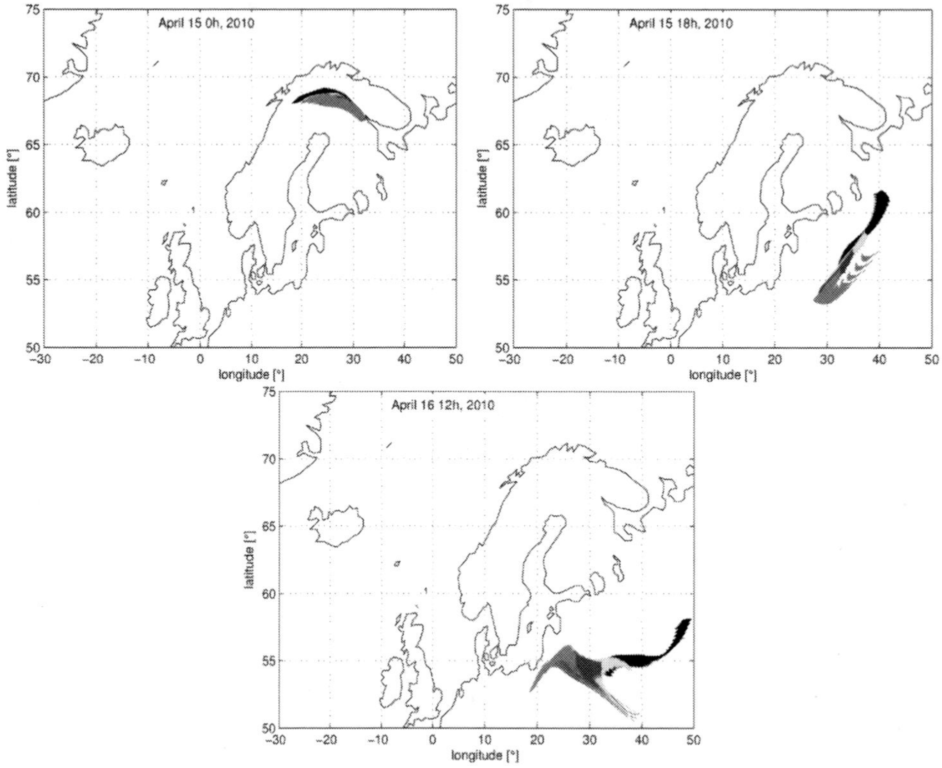

Figure 3. Dispersion of the simulated volcanic ash of different sizes from Eyjafjallajökull (red: $r = 1$ μm, blue: $r = 2$ μm, green: $r = 5$ μm, black: $r = 10$ μm). A single initial column of $1° \times 1° \times 140$ hPa size ($\sim$ 100 km $\times$ 100 km $\times$ 4 km) was started at 06 UTC on 14.04.2010 on the 300 hPa level with $N_0 = 27000$ particles. The particular time instants are indicated on the panels.

boundary layer, the true deposition pattern is expected to be qualitatively similar to that seen on the level of 850 hPa.

The group of charts (Fig. 3) illustrates the spreading of particles of different size. A separation can be seen: different types of particles reach different regions being subject to different winds because of their different terminal velocities.

Fig. 4 demonstrates that the particle distribution is wider if another eruption, taking place somewhat later, is also taken into account. The observed total distribution of the volcanic ash on the 16th of April can be seen in the left panel. The accumulation of volcanic ash south of Denmark is apparent in both pictures, but numerical simulation indicates particles in regions further away. Difference might be due to meteorological conditions disturbing satellite observations and to the fact that we only investigated three artificial sudden eruptions, instead of the continuous eruption over the days. In addition, wet deposition was not taken into account.

As Fig. 5 shows, after about 4 days a strong decrease of the survivors takes place. This can be roughly approximated locally by a linear segment in the log-linear representation of the left panel, corresponding to an exponential decay. The approximate value of the escape is found from the range 4.5–8.75 days to be $\kappa = 0.82$ day^{-1}. The average lifetime over the full process is found to be $\tau = 4.86$ day. The a shape of an initially short and straight line of volcanic ash particles becomes quickly stretched and folded as the inset in the right panel shows. The growth rate of the length of this material line is, after about 3 days, clearly exponential (right panel of

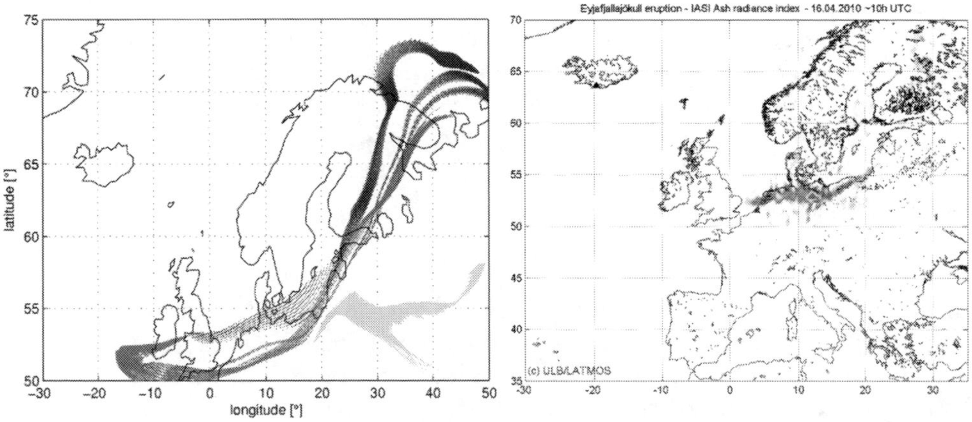

Figure 4. Left panel: Volcanic ash distribution of three model eruptions of Eyjafjallajökull ($r = 1, 2, 5, 10$ μm) at 12 UTC on 16.04.2010. A single initial column of 1° × 1° × 140 hPa size (∼ 100 km × 100 km × 4 km) was started at 06 (green), 12 (blue) and 18 (red) UTC on 14.04.2010 on the 300 hPa level with $N_0 = 27000$ particles. Right panel: observed volcanic ash distribution at 10 UTC on 16.04.2010 from the homepage of the European Organisation for the Exploitation of Meteorological Satellites (EUMETSAT [19], snapshot taken from a video).

Fig. 5) and a value of $h = 0.68$ day^{-1} follows for this eruption.

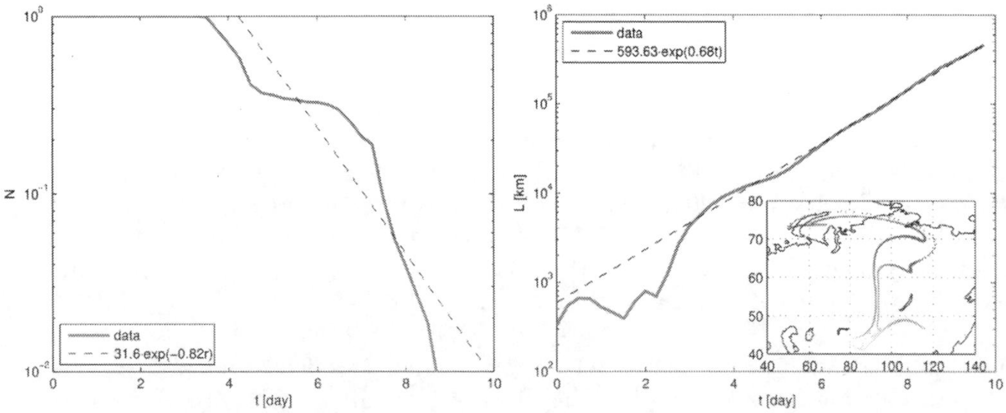

Figure 5. Left panel: The proportion of non-escaped particles as a function of time ($N_0 = 27000$, $r = 10$ μm, $p_0 = 300$ hPa, initial conditions as in Fig. 2). Right panel: the length of a material line of volcanic ash ($L_0 = 3° \approx 330$ km, $N_0 = 10000$, $r = 1$ μm, $p_0 = 300$ hPa) as a function of time. The inset shows the shape of this line at 00 UTC on 20.04.2010.

5. Global results

In order to gain global characteristics of volcanic ash dynamics not related to specific eruptions, we consider several instantaneous model volcanic eruptions distributed over the Globe. Alternatively, we also study an ensemble of a large number of particles distributed uniformly over the Earth at a certain pressure level. In both cases much more precise averages can be obtained (see Fig. 6) then for the particular eruptions on Iceland.

The left panel shows that the number of survivors in the free atmosphere exhibits a clear exponential decay valid over a week or so. By investigating the dependence of the escape rate on the particle radius and the pressure level on which the particles are initially uniformly distributed, we find that $\kappa(r)$ shows a very strong increase with the size (right panel). The escape rate increases by a factor of about 100 between radius 1 and 12 μm. The fit shows that a roughly exponential dependence is found with the size:

$$\kappa(r) \sim \exp(kr), \tag{13}$$

with coefficient $k \approx 0.37$ for any level. Such a strong dependence of the escape rate on a system parameter is found in so-called supertransients occuring in spatio-temporal chaotic models, and in particular in pipe turbulence where this strong dependence on a parameter (the Reynolds number) was also observed in experiments [20].

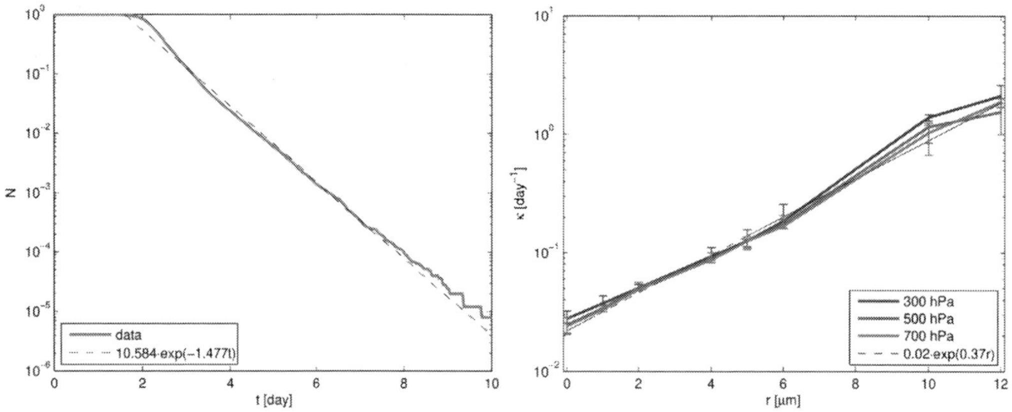

Figure 6. Left panel: The decay of the number of non-escaped particles of radius $r = 10$ μm started from 300 hPa, at 01.01.2010 0 UTC. Right panel: the dependence of the escape rate on size and pressure level. $N_0 = 250000$ particles were distributed uniformly over the whole globe on the level of 300, 500, and 700 hPa. Note that $r = 0$ values are also shown which correspond to the dynamics of air parcels. The non-zero values of the escape rate for $r = 0$ indicate that air parcels also leave the free atmosphere within a finite time.

Concerning the outfall, we estimated the average lifetime (residence time) of particles in two ways. First, as a naive estimate, by using the average terminal velocity of particles between the initial height and the planetary layer's upper boundary in the ICAO standard atmosphere we obtain:

$$\tau_w = \frac{z_{initial} - z_{850hPa}}{\bar{w}_{terminal}}, \tag{14}$$

where $z_{initial} - z_{850hPa} \approx 1560$ m, 4120 m and 7700 m corresponding to the initial level of 700 hPa, 500 hPa and 300 hPa. The results are illustrated by the dotted lines on the different initial levels in Fig. 7. Alternatively, we determined the average lifetime τ from simulations, where 250000 particles were distributed uniformly over the whole globe. For smaller particles enhanced deposition can be observed compared with the naive estimate. The difference is due to the fact that (14) is valid in a continuum at rest, while τ is obtained from solving (3). This shows that the presence of vertical winds is essential for the deposition of volcanic ash, and its role is dominant for small size particles, as also follows from Fig. 1.

The size-dependence is found to be exponential again:

$$\tau(r) \sim \exp(-k'r), \tag{15}$$

with $k' \approx 0.30$. The difference between $\tau(r)$ and $1/\kappa(r)$ is due to the fact that with the chosen initial condition all particles spend a long period of time in the free atmosphere, and the decay in the number of survivors starts after a few days only.

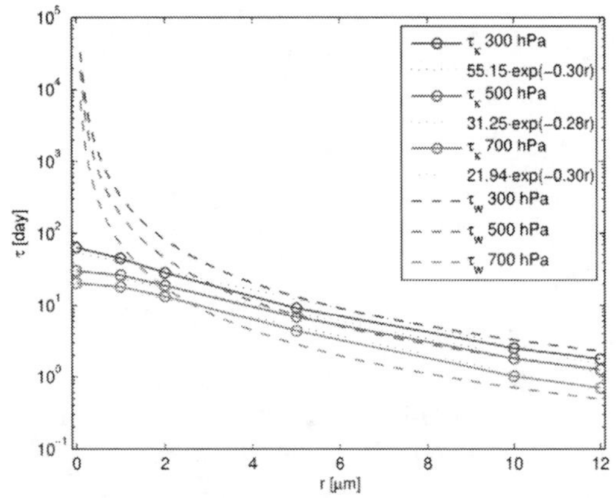

Figure 7. Average residence time τ and the naive estimate τ_w as a function of size and initial pressure level. $N_0 = 250000$ particles were distributed uniformly over the whole globe on the level of 300, 500, and 700 hPa at 01.01.2010 00 UTC.

A detailed investigation of the topological entropy has also been carried out by using several identical short line segments distributed on a grid over the globe initially. Naturally, the topological entropy depends on the actual location implying different meteorological conditions. The topological entropy is found to depend on the initial geographic location, and has mainly zonal distribution (Fig. 8).

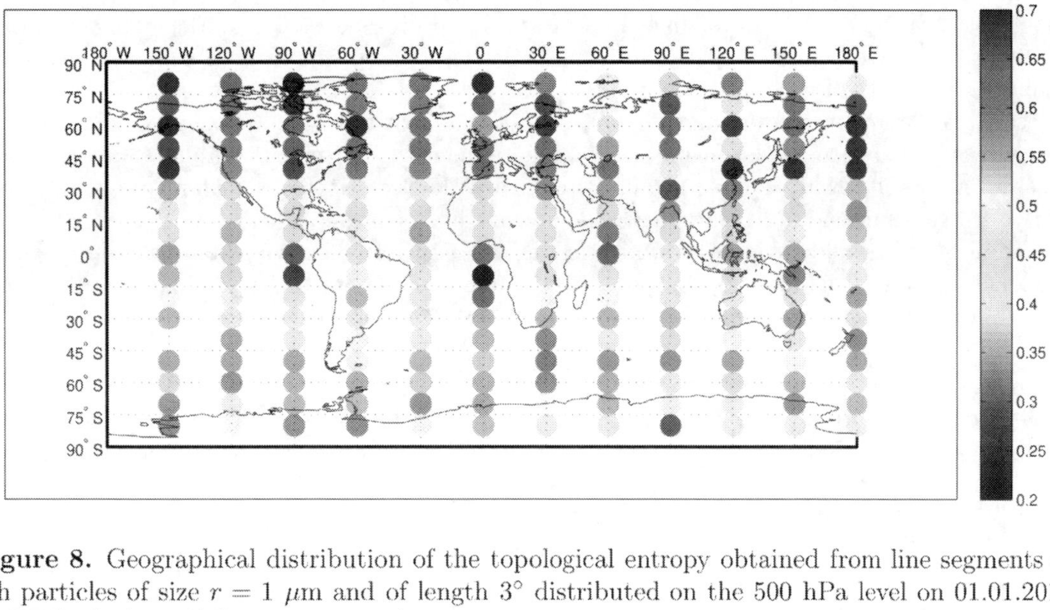

Figure 8. Geographical distribution of the topological entropy obtained from line segments of ash particles of size $r = 1$ μm and of length 3° distributed on the 500 hPa level on 01.01.2010 00 UTC which are followed over 10 days.

There are strong differences between the values close to the Equator and in extratropics. The largest values appear in the mid-latitudes (Fig. 9), especially in winter, due to the strong mixing

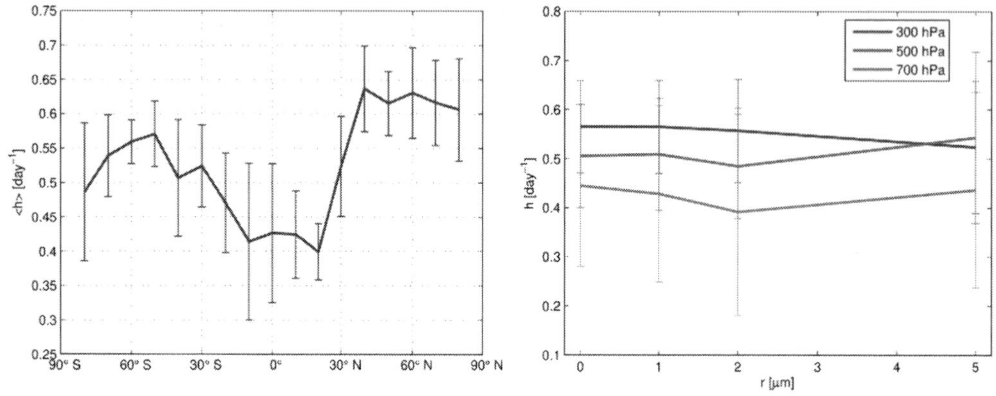

Figure 9. Left panel: zonal average of the topological entropies of Fig. 8 plotted as a function of latitude. Right panel: the dependence of the average topological entropy on particle size and the initial pressure level.

and shearing effects of cyclones on particles, while the smallest values (less than 0.5 1/day) are located in the tropics. Average topological entropy hardly depends on the particle size (Fig. 9). For lines initially started higher in the atmosphere (300 and 500 hPa) the topological entropy is larger than on lower levels.

6. Summary and discussion

In summary, we can say that the advection of volcanic ash is chaotic and particles spend a finite period of time in the free atmosphere. Obviously, the higher the initial height and the smaller the particle, the larger the average life time. An initially localized pollutant cloud becomes strongly stretched in a few days, fine filamentary structures develop and the filaments grow rapidly in time. We demonstrated that transient chaos related quantites like κ, τ, and h are useful characteristics of the statistics of volcanic ash in the free atmosphere.

Because of stretching and folding, the tracers cover almost the whole hemisphere after about a few weeks. Therefore in the future we intend to calculate the encompassing time for small particles, in a similar way as done for air parcels in [21]. Besides this, we plan to estimate the effective fractal dimension of pollutant clouds during the spreading.

In addition, we would also like to study the dispersion in the planetary boundary layer. Here turbulent diffusion plays an important role, and the anisotropy of diffusion, i.e. the difference between horizontal and vertical diffusivities should be taken into account.

Acknowledgments

Useful discussions with T. Bódai, A. Horányi, I.M. Jánosi, I. Lagzi, T. Práger, P. Tasnádi and M. Vincze are acknowledged. The project is supported by the European Union and co-financed by the European Social Fund (grant agreement no. TAMOP 4.2.1./B-09/1/KMR-2010-0003), as well as by OTKA (grant No. NK72037) and the COST Action "Particles in Turbulence".

References

[1] Stohl A, Huntrieser H, Richter A, Beirle S, Cooper O, Eckhardt S, Forster C, James P, Spichtinger N, Wenig M, Wagner T, Burrows J and Platt U 2003 *Atmos. Chem. and Phys. Disc.* **3** 2101–2141

[2] Leadbetter S J and Hort M C 2011 *J. Volcanol. Geotherm Res.* **199** 2030

[3] Dacre H F, Grant A L M, Hogan R J, Belcher S E, Thomson D J, Devenish B J, Marenco F, Hort M C, Haywood J M, Ansmann A, Mattis I and Clarisse L 2011 *J. Geophys. Res.* (*Preprint* DOI:10.1029/)

[4] Várai A, Vincze M, Lichtenberger J and Jánosi I M 2011 *this volume*

[5] http://www.ecmwf.int/research/era/do/get/era-interim
[6] Lai Y C and Tél T 2011 *Transient Chaos: Complex Dynamics on Finite-Time Scales* (Springer)
[7] Toschi F and Bodenschatz E 2009 *Annu. Rev. Fluid. Mech.* **41** 375
[8] Cartwright J H E, Feudel U, Károlyi G, de Moura A, Piro O and Tél T 2010 *Nonlinear Dynamics and Chaos: Advances and Perspectives* ed Thiel M, Kurths J, Romano M, Károlyi G and de Moura A (Berlin: Springer)
[9] Maxey M R and Riley J J 1983 *Phys. Fluids* **26** 883
[10] Auton T R, Hunt J and Prud'homme M 1988 *J. Fluid. Mech.* 241
[11] Tél T *et al.* 2000 *Chaos* **10** 89
[12] http://cimss.ssec.wisc.edu/goes/blog/wp-content/uploads/2010/04/
[13] Kundu P K and Cohen I M 2008 *Fluid Mechanics* (Academic Press)
[14] Stohl A, Forster C, Frank A, Seibert P and Wotawa G 2005 *Atmos. Chem. Phys.* **5** 2461
[15] Kiss P, Varga L and Jánosi I M 2009 *J. Renewable Sustainable Energy* **1** 033105
[16] http://data-portal.ecmwf.int/data/d/interim_daily/
[17] http://www.atmosculator.com/The Standard Atmosphere.html
[18] Newhouse S E and Pignataro T 1993 *J. Stat. Phys.* **72** 1331
[19] http://www.eumetsat.int/Home/Main/News/Features/717626
[20] Tél T and Lai Y C 2008 *Phys. Rep.* **460** 245
[21] Haszpra T, Kiss P, Tél T, and Jánosi I M 2011 Advection of passive tracers in the atmosphere: Batchelor scaling Submitted to Int. J. Bif. and Chaos

New experimental set-up to analyse cryogenic flows by visualisation

Marco LA MANTIA, Miloš ROTTER and Ladislav SKRBEK

Department of Low-Temperature Physics, Faculty of Mathematics and Physics, Charles University in Prague, V Holešovičkách 2, 180 00 Praha 8, Czech Republic

lamantia@nbox.troja.mff.cuni.cz

Abstract. The first laboratory in Europe for visualisation of liquid helium flows is currently being established at the Charles University in Prague. The use of such a valuable experimental approach for the analysis of cryogenic flows of normal and superfluid ^{4}He is introduced and its specific features discussed. More importantly, it is shown that the newly implemented flow visualisation equipment is potentially capable of obtaining novel results, i.e. further understanding of the involved physics.

1. Introduction

Flow visualisation techniques have been recently applied at very low temperatures for the investigation of various liquid helium flows [1, 2]. Quantitative techniques, such as PIV (Particle Image Velocimetry) and PTV (Particle Tracking Velocimetry), have been indeed proven very fruitful in many scientific and industrial areas of research during the last decades [3]. However, such a promising experimental approach is still in its infancy in the analysis of cryogenic flows and the ways to optimise it are yet to be fully investigated due to a number of technical and fundamental difficulties, e.g. the optical access to the helium bath and choice of suitable tracer particles.

Gaseous and liquid ^{4}He are generally used as working fluids in cryogenic fluid mechanics. At the pressure of 1 bar ^{4}He is gaseous, if the temperature is larger than 4.2 K. At lower temperatures it becomes liquid and, if the temperature is larger than 2.17 K, liquid ^{4}He is called normal helium or He I. This fluid is characterised by extremely low values of the kinematic viscosity (of the order of 10^{-4} cm^2/s), compared to those of air (of the order of 10^{-1} cm^2/s) and water (of the order of 10^{-2} cm^2/s) [4]. If the temperature decreases further, liquid ^{4}He is called He II. Its viscosity can be considered null at 0 K, i.e. the fluid is assumed inviscid in the zero-temperature limit: this is why He II is also called superfluid.

The unique properties of He II are described by the two-fluid model, e.g. see [5, 6]. It is assumed that He II is made of two fluids, normal and superfluid helium. The former is viscous and carries entropy while the latter is inviscid and does not carry entropy. The total density ρ of He II is defined as the sum of the densities of its normal and superfluid components, ρ_n and ρ_s, respectively, and depends weakly on temperature. ρ_n and ρ_s have instead a much stronger dependence on temperature. The ratio between ρ_s and ρ increases indeed steeply as the temperature decreases and, for example, is equal to 0.986 at 1.1 K [7]. In other words, He II can be considered entirely superfluid at temperatures below 1 K. To highlight the uniqueness of He II an example can be made on the basis of the just sketched two-fluid model. If a volume of He II is heated, the normal component flows away from the heater while the superfluid component moves towards the heater in order to have a null flow rate, i.e. to conserve the mass of He II. This phenomenon is called thermal counterflow and has not any equivalent in classical fluid mechanics.

Besides, the superfluid component of He II has been described as a quantum fluid, that is, by a macroscopic wave function, see again [5, 6] for further details. This leads to the result that superfluid flow is irrotational, i.e. ω = curl $v_s = 0$, where ω is the flow vorticity and the superfluid velocity v_s is proportional to the gradient of the wave function phase. It follows that for a simply connected fluid region the circulation of the superfluid velocity is null. If the region is instead multiply connected, the circulation is not null and equal to an integer multiple of the quantum of circulation k, which is the ratio of the Planck constant and ^{4}He atomic mass, i.e. $k = 9.98\ 10^{-4}$ cm^2/s [7]. This result can be seen as a quantum restriction to the superfluid motion. In other words, just quantised vortices – line singularities where the superfluid density is null – can exist in superfluid helium. These vortices usually arrange themselves in a tangle and such a tangle of vortices constitutes what is generally called quantum turbulence.

The description of superfluid helium (He II) as a quantum fluid is specifically relevant for the implementation of flow visualisation techniques at low temperatures. For example, the complex interactions between tracer particles, quantised vortices and macroscopic eddies in cryogenic flows are far from being completely understood and novel experiments are required to verify the current theoretical understanding of these coupled phenomena [8, 9]. Moreover, the interesting and puzzling results recently obtained in overseas laboratories [1, 2] are posing more questions than giving answers, e.g. the mechanisms of particles' trapping into the quantised vortices' cores and the vortical structures observed around cylinders in thermal counterflow deserve further attention and study. These outcomes show consequently the need of more detailed experimental analyses by flow visualisation, which is being proven as a very valuable technique to study cryogenic flows.

2. Experimental apparatus

In order to fulfil such a need of novel experimental data we are currently establishing the first laboratory in Europe for flow visualisation at low temperatures. Experimental investigation of selected cryogenic flows over wide ranges of governing dynamical parameters, spanning from laminar to developed turbulent regime, using all forms of cryogenic ^{4}He as working fluids are being planned. All these classical and quantum flows will be mainly probed by using quantitative flow visualisation techniques, i.e. PIV and PTV. The cryogenic flows will be obtained by various means. Flows around bodies oscillating in stationary fluid will be studied, due to their relative simplicity of implementation at low temperatures. A bellows system, connected to a precise linear motor [10], will be used to generate flows of various velocities around bluff bodies and heaters will be employed to obtain thermal counterflow, the aim of these experiments being also the comparisons of dynamically similar flows in normal and superfluid ^{4}He.

The flow visualisation equipment, which already is available in our laboratory, consists of the following parts. A custom-made low-loss cryostat equipped with five sets of 25 mm diameter windows that minimise the heat input into the helium bath, enabling horizontal as well as vertical optical access, was designed and manufactured, see Figure 1. A seeding system with a fast computer-controlled valve to supply the helium bath with the desired amount of hydrogen and deuterium micron-sized solid tracers was built and is being tested. These low-temperature related parts are to be used with an off-the-shelf 5 W continuous wave solid state laser, a very fast digital camera (up to 6273 fps at full resolution, i.e. 1 MP) and relevant hardware and software to implement the PIV and PTV techniques for cryogenic flows, purchased from Dantec Dynamics.

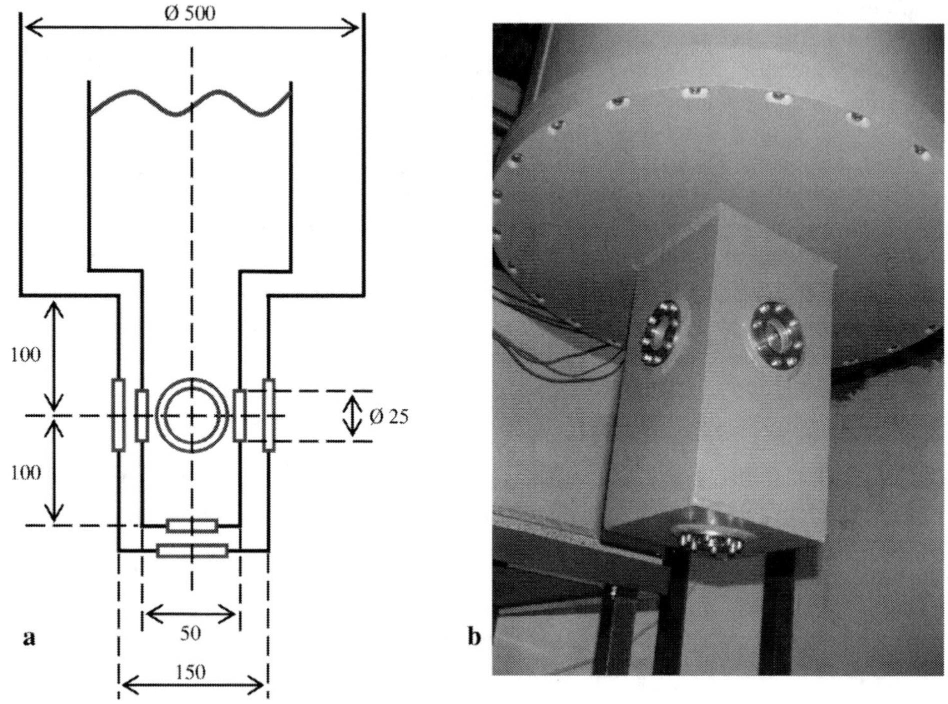

Figure 1. a: Sketch of the low-loss optical cryostat for flow visualisation at low temperatures (not to scale, dimensions in mm); b: Cryostat optical tail.

Our facility is being designed to perform novel experiments, that is, not to be just a copy of existing systems but to be potentially capable of obtaining new results. For example, the much faster camera will most likely allow a more detailed analysis of cryogenic flows' dynamics, compared to [1, 2]. As already mentioned, the complex interactions between tracer particles, quantised vortices and macroscopic eddies could be studied experimentally in unprecedented detail and, for example, the theoretical models that predict the conditions for particles' trapping into the quantised vortices' cores verified [8, 9]. This may also have an impact on the study of turbulent multiphase flows, which is a very active field of research in classical fluid mechanics [11, 12]. Besides, the PTV technique, especially chosen to study quantum flows, has so far been rarely used for such applications and the use of PIV also appears controversial for these peculiar flows. In other words, particularly for turbulent flows, it is not completely clear if and when the PIV computed velocities are those of the normal component of He II or those of the superfluid portion of it or a mixture of the two. This is particularly evident in the case of thermal counterflow that, as mentioned above, is characterised by the fact that the normal and superfluid component of He II are flowing in opposite directions. A plausible option would then be to study the same flow by using both techniques and consequently assess which technique is the most suitable at low temperatures. However, a number of technical difficulties have to be tackled. For example, even though the set-up was carefully designed, the seeding system needs to be tuned in order to obtain the most suitable tracers. The PIV and PTV equipment is also currently being tested on various experiments in water, such as well-known flows past bluff bodies and around oscillating objects, see next Section. The first results at low temperatures are expected in the coming months. More precisely, thermal counterflow experiments are being planned, similar to those discussed in [2], as well as experiments on cryogenic flows around oscillating objects, such as cylinders and spheres, which to our best knowledge have not been yet analysed by visualisation.

3. Visualisation of water flows

Figure 2 shows the mean velocity field for a slightly bent water jet, directed downwards, as calculated by the Dynamic Studio software (200 images were taken at the maximum frame rate and a 2D PIV analysis with micron-sized buoyant particles was performed). It can be noted that the camera field of view is quite small, ca. 4 cm^2. Besides, the Reynolds number is equal to ca. 18,000, i.e. the flow is turbulent, as the jet diameter ≈ 12 mm and its mean velocity ≈ 1.5 m/s. In other words, such a flow field could not have been probed with a much slower camera. In Figure 3 the corresponding mean vorticity field is plotted and the clock-wise region of vorticity can be explained by considering that the jet was slightly bent to the left of the field of view, see Figure 2. This might appear clearer if it is assumed that the jet is made by two smaller jets, close to each other. The first one, on the right of the field of view, is directed vertically downwards while the second one, less energetic than the first one and similarly directed downwards, is also bent to the left of the field of view. The interaction between these two close jets can then justify the presence of two local minima with negative vorticity.

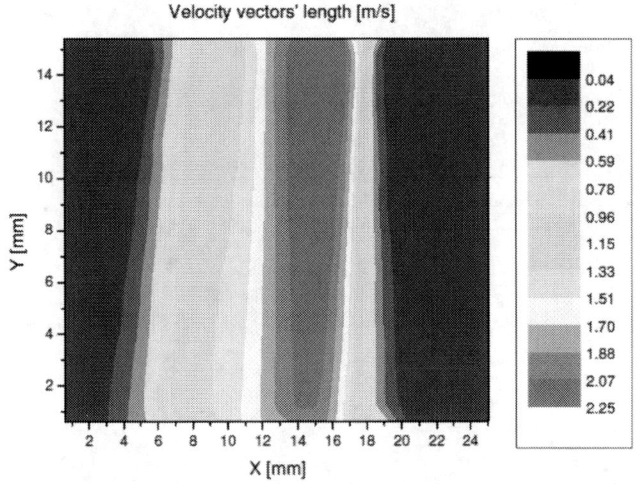

Figure 2: Velocity field for a slightly bent water jet (Re ≈ 18,000).

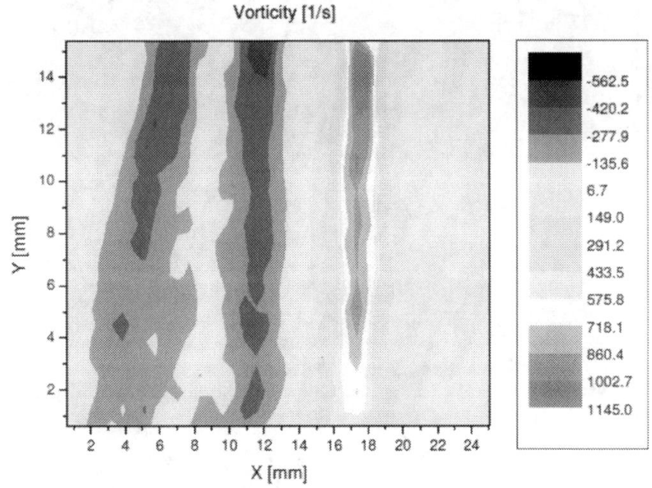

Figure 3: Vorticity field for a slightly bent water jet (Re ≈ 18,000).

Figure 4 displays the mean vorticity field of the flow past a circular cylinder, as calculated by the Dynamic Studio software (100 images were taken at the maximum frame rate). The cylinder, placed at the left of the field of view, is in a slightly bent water jet, directed horizontally from the left to the right of the field of view The Reynolds number is in this case equal to ca. 1,000, as the cylinder diameter = 5 mm and jet mean velocity ≈ 0.2 m/s, i.e. the flow is still turbulent.

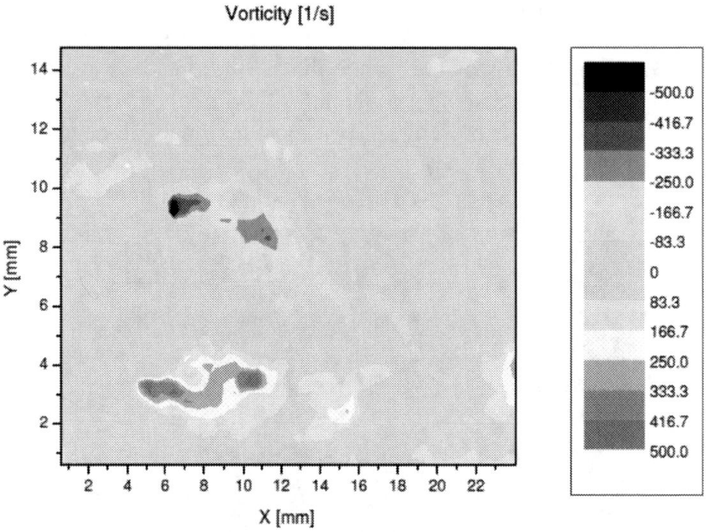

Figure 4: Vorticity field for the flow past a cylinder in a slightly bent water jet (Re ≈ 1,000).

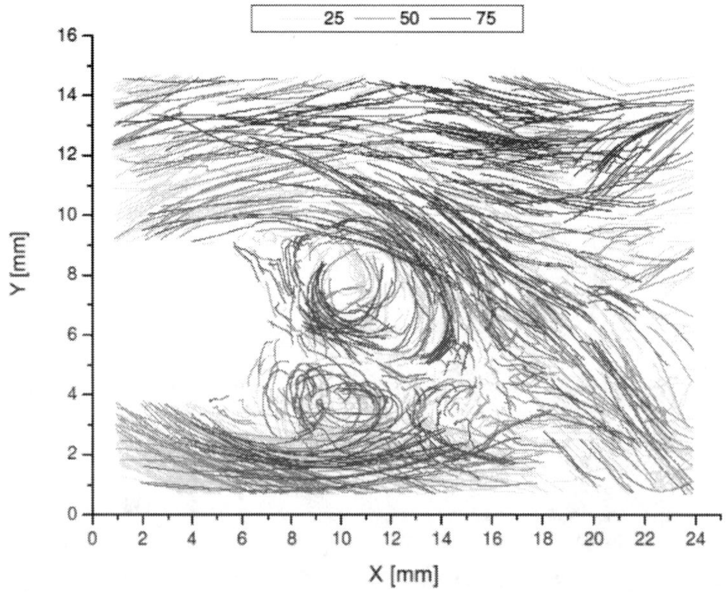

Figure 5: Particles' tracks for the flow past a cylinder in a slightly bent water jet (Re ≈ 1,000). The legend indicates the minimum number of points for the tracks shown in that colour.

In Figure 5 relevant particles tracks are plotted, as computed by processing the results obtained by the Dynamic Studio software. The colour code for the tracks indicates the minimum number of points for the tracks shown in that colour. For example, there are ca. 1,400 tracks with at least 25 points and ca. 300 with at least 75 points. It can be seen that the recirculation zone behind the cylinder is well captured by both visualisation techniques.

Figure 6 shows the instantaneous vorticity field generated by a cylinder oscillating harmonically in water, as calculated by the Dynamic Studio software (images taken at 1000 fps). The cylinder was oscillating vertically (the lowest position of the harmonic cycle was just above the bottom of the field of view) and the vorticity field was computed just after the cylinder left the field of view moving upwards. The Reynolds number Re $\approx$ 100, as the cylinder diameter = 5 mm and its maximum velocity $\approx$ 0.02 m/s. The oscillation frequency was set to ca. 0.5 Hz and its amplitude to 0.02 m. This corresponds to a viscous penetration depth $\delta = (2\nu/\omega)^{0.5} \approx 0.8$ mm, where ν is the water kinematic viscosity, 10^{-6} m^2/s, and ω the cylinder oscillation frequency in rad/s. Vortices rotating in opposite direction were consecutively shed by the cylinder, i.e. a Von Karman vortex street is clearly visible. Besides, the weak clock-wise vortex at the bottom left of the figure was shed at the end of the previous stroke, when the cylinder reached its lowest position, and pushed away while the cylinder moved upwards.

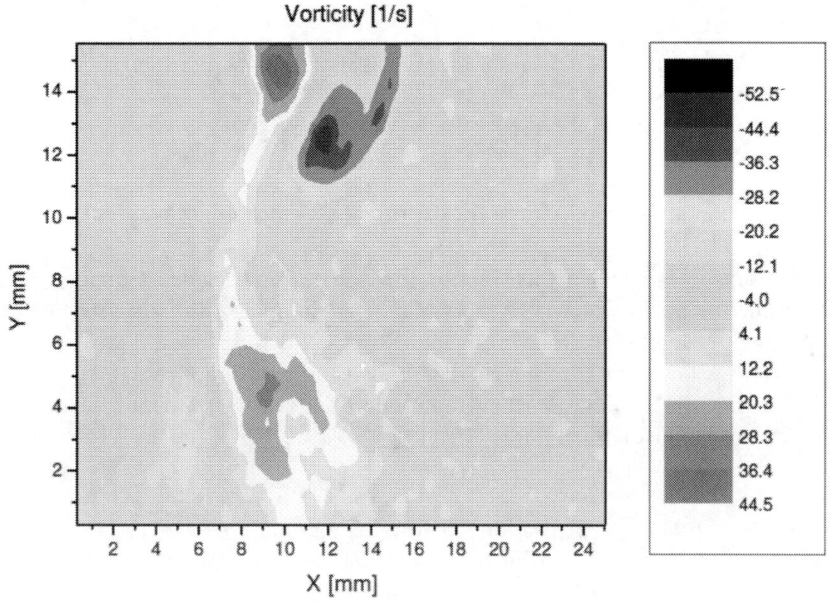

Figure 6: Instantaneous vorticity field for the flow past an oscillating cylinder (Re $\approx$ 100).

In Figure 7 the vorticity field at the same point of the harmonic cycle, i.e. the cylinder just left the field of view moving upwards, is shown for Re $\approx$ 20, as the cylinder maximum velocity $\approx$ 0.004 m/s and its oscillation frequency $\approx$ 0.1 Hz (the cylinder diameter and oscillation amplitude did not change), the viscous penetration depth being in this case equal to ca. 1.8 mm. Images were taken at 100 fps and processed by the Dynamic Studio software, as above. Single vortices were not shed and a region of vorticity can be seen behind the cylinder, i.e. the boundary layer. It has also to be noted that the vorticity absolute value is about one order of magnitude smaller than that shown in Figure 6.

Further investigations, at different oscillation velocities and by using different cylinder section geometries, e.g. square instead of circular, are being currently performed and planned, their aim being also a systematic study of the characteristics of the wake shed by oscillating bodies [13-15].

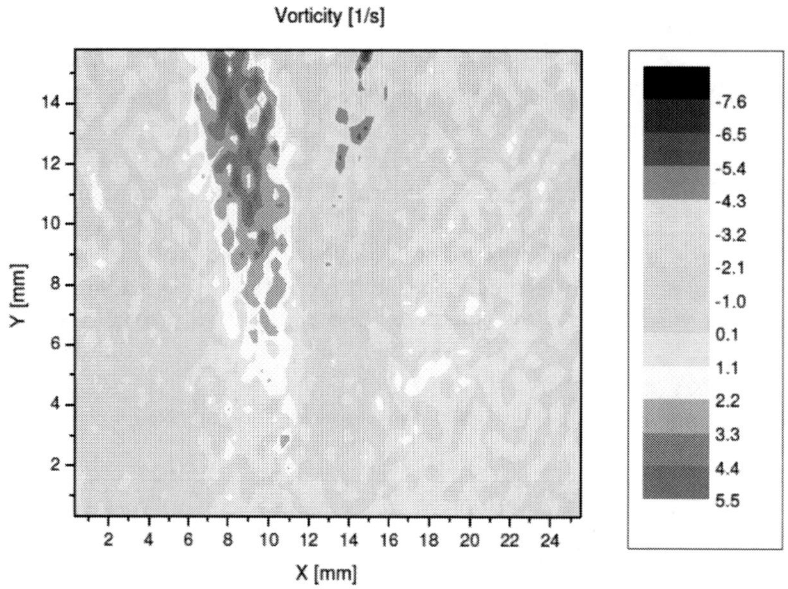

Figure 7: Instantaneous vorticity field for the flow past an oscillating cylinder (Re $\approx$ 20).

However, the results presented in this Section are well known, at least qualitatively. The main purpose of these experiments in water is to explore the capabilities of the visualisation system that is soon going to be used to analyse various cryogenic flows, as detailed in the previous Section. More precisely, it was briefly shown that the chosen visualisation apparatus appears to be well suited for the task of analysing cryogenic flows, at least in the range of investigated parameters, as the latter are similar to those expected at low temperatures [16].

4. Conclusion

In summary, the use of flow visualisation at low temperature seems to be a very promising experimental tool that is certainly capable of improving our general knowledge of classical and quantum flows. Flow instabilities, the fine details of the transition to turbulence and the properties of developed and decaying classical and quantum turbulent flows could be studied systematically over wide ranges of dynamical parameters by using cryogenic helium, a remarkable working fluid with well-known, unique and easily tuneable properties.

Acknowledgments

We would like to thank Simone Babuin, Gregory Bewley, Tymofiy Chagovets, Veronika Pilcová, David Schmoranzer, František Soukup, Mathias Stammeier, Josef Šebek, Václav Uruba and Bohumil Vejr for fruitful discussions and valuable help. Besides, we acknowledge the support of COST Action MP0806, MS 0021620834 and GAČR P203/11/0442.

References

[1] GP Bewley, DP Lathrop and KR Sreenivasan, 2006: *Visualisation of quantized vortices*. Nature 441, 588

[2] T Zhang and SW Van Sciver, 2005: *Large-scale turbulent flow around a cylinder in counterflow superfluid ^{4}He (He II)*. Nature Phys 1, 36-38

[3] M Raffel, CE Willert, ST Werely and J Kompenhans, 2007: *Particle image velocimetry – a practical guide*. Springer

[4] JJ Niemela and KR Sreenivasan, 2006: *The use of cryogenic helium for classical turbulence: promises and hurdles*. J Low Temp Phys 143, 163-212

[5] DR Tilley and J Tilley,1990: *Superfluidity and superconductivity*. Institute of Physics Publishing

[6] M La Mantia, M Rotter and L Skrbek, 2010: *Feasibility of low-temperature flow visualisation*. Proceedings of the International Conference Experimental Fluid Mechanics 2010 (Edited by T Vit and P Dančová), Technical University of Liberec, 330-337

[7] WF Vinen and JJ Niemela, 2002: *Quantum turbulence*. J Low Temp Phys 128, 167-231

[8] YA Sergeev and CF Barenghi, 2009: *Particles-vortex interactions and flow visualisation in ^{4}He*. J Low Temp Phys 157, 429-475

[9] SW Van Sciver and CF Barenghi, 2009: *Visualisation of quantum turbulence*. Progress in Low Temperature Physics: Quantum Turbulence (Edited by M Tsubota and WP Halperin), Springer, 247-303

[10] S Babuin, M Rotter, M Stammeier and L Skrbek, 2010: *A new flow source for the study of quantum turbulence*. Proceedings of the International Conference Experimental Fluid Mechanics 2010 (Edited by T Vit and P Dančová), Technical University of Liberec, 48-57

[11] C Poelma and G Ooms, 2006: *Particle-turbulence interaction in a homogeneous, isotropic turbulent suspension*. Appl Mech Rev 59, 78-90

[12] S Balachandar and JK Eaton, 2010: *Turbulent dispersed multiphase flow*. Annu Rev Fluid Mech 42, 111-133

[13] D Schmoranzer, M Král'ová, V Pilcová, WF Vinen and L Skrbek, 2010: *Experiments relating to the flow induced by a vibrating quartz tuning fork and similar structures in a classical fluid*. Phys Rev E 81, 066316

[14] M La Mantia and P Dabnichki, 2011: *Influence of the wake model on the thrust of oscillating foil*. Eng Anal Bound Elem 35, 404-414

[15] D Schmoranzer, M La Mantia, G Sheshin, I Gritsenko, A Zadorozhko, M Rotter and L Skrbek, 2011: *Acoustic emission by quartz tuning forks and other oscillating structures in cryogenic ^{4}He fluids*. J Low Temp Phys 163, 317-344

[16] V Pilcová, M La Mantia, M Rotter and L Skrbek, 2011: *Testing the performance of a cryogenic visualisation system on selected room-temperature water flows*. Proceedings of the 25th Symposium on Anemometry (Edited by Z Chára and L Klaboch), Institute of Hydrodynamics of the Academy of Sciences of the Czech Republic, 51-58

Particle transport in incompressible MHD Kolmogorov flow

C C Lalescu[1,2,*], D Carati[1], M Negrea[2] and I Petrisor[2]
[1]Statistical and Plasma Physics, Université Libre de Bruxelles, Campus Plaine, CP 231, B-1050 Brussels, Belgium
[2]Faculty of Physics, University of Craiova, A.I. Cuza 13, 200585 Craiova, Romania
E-mail: *clalescu@ulb.ac.be

Abstract. In previous work [ ], an investigation of Kolmogorov flow for incompressible magnetohydrodynamics (MHD) was performed. It consists of a three-dimensional periodic flow driven by a unidirectional forcing varying on a transverse direction. In practice, the forcing is chosen to be $f_x = A\sin(k_f y)$, and $f_y = f_z = 0$. It was found that vorticity structures with long lifetimes can be formed in the turbulent regime. The presence of such structures may affect the transport of particles interacting with the flow as shown in this preliminary study.

1. Theoretical framework

1.1. Introduction

The general purpose of this work is the description of particle transport in turbulent plasmas. A plasma is generally composed of several species of charged particles as well as neutrals. Its complete description may thus require the use of several coupled kinetic equations. However, in the fluid limit, the large scale features of the plasma can be described using the magneto-hydrodynamic equations. Depending on the level of description, single fluid or multi-fluid MHD approaches can be developed. Here, only the simplest fluid description of the plasma will be considered and its evolution is assumed to be driven by the incompressible MHD equations. However, an ensemble of test particles, interacting with the plasma and the electromagnetic fields, will also be considered. These particles are treated as passive. It is a common approximation for impurities in fusion plasmas [,].

A detailed discussion of MHD Kolmogorov flow is proposed in [ ]. Such a geometry is used here because it displays large two dimensional structures with long lifetimes. The presence of persistent large-scale structure is reminiscent of zonal flows observed in various fusion plasma situations. Moreover, since particle transport in two-dimensional Navier Stokes Kolmogorov flow displays anomalous transport properties [ ], the same situation is expected in three-dimensional flows.

The system of equations describing the evolution of the MHD fields is given by:

$$\partial_t u_i = -\partial_j(u_i u_j - b_i b_j) + \nu\partial_j\partial_j u_i - \partial_i p + f_i \tag{1}$$
$$\partial_t b_i = -\partial_j(b_i u_j - u_i b_j) + \eta\partial_j\partial_j b_i \tag{2}$$
$$\partial_i u_i = 0 \tag{3}$$
$$\partial_i b_i = 0 \tag{4}$$

Journal of Physics: Conference Series **333** (2011) 012010 doi:10.1088/1742-6596/333/1/012010

where the flow is assumed to be incompressible and isothermal to simplify the discussion. The only parameters appearing in these equations are the kinematic viscosity of the fluid ν and the magnetic diffusivity η. The pressure p represents the sum of an hydrodynamic and a magnetic contributions rescaled by the density. Thanks to the incompressibility condition, no state equation is needed for p which is obtained by imposing the constraint (3) into the equation (1). The flow is driven by the Kolmogorov force, given $\mathbf{f} = (A\sin(k_f y), 0, 0)$ and appears to be periodic but anisotropic. Indeed, the three directions play different "roles": the forcing is along the x direction, it varies along y, and it should be invariant to the z direction. The structures themselves are in the xy plane, so z is the only direction of homogeneity, in a statistical sense. Indeed, once turbulence develops, the flow exhibits fluctuation in all three directions. Statistical results concerning the turbulent fields are always averaged over z, for faster convergence [5].

In [4], fluid particle transport in 2D Kolmogorov flow is investigated. Here, test particles interacting with both the fluid velocity field and the electromagnetic fields generated in 3D MHD Kolmogorov flow are considered. Indeed, the MHD equations do not only provide the fields $\mathbf{u}$ and $\mathbf{b}$, but also a self-consistent electric field given by:

$$\mathbf{e} = -\mathbf{u} \times \mathbf{b} + \eta \nabla \times \mathbf{b} \tag{5}$$

All these fields depend implicitly on both the position and time. Several types of test particles can be considered. The simplest example is given by the "Velocity field lines" (VFL) which correspond to fluid particles, or simply tracers:

$$\frac{d}{dt}\mathbf{r} = \mathbf{u} \tag{6}$$

Such particles simply follow the fluid without particular interaction with the electric and magnetic fields. However, if charged particles are considered, they might experience a triple interaction consisting of a friction with the fluid, an acceleration due to the electric field and a locally helical motion due to the magnetic field. The motion can then become extremely complex and very difficult to track numerically. For these reasons, various approximations have been designed. The simplest one, reasonably valid when the magnetic field is very strong, is given by the first guiding center (GC) approximation with the parallel velocity $v_{\|}$:

$$\frac{d}{dt}\mathbf{r} = v_{\|}\frac{\mathbf{b}}{|\mathbf{b}|}, \tag{7}$$

where $v_{\|}$ is a parameter with the dimensions of a velocity, and its value has to be determined by the specific conditions of the system studied. For instance, a charged particle, whose velocity projected on the local magnetic field is $v_{\|}$, will approximately follow a spiral with an axis that is precisely this guiding center. The guiding center approximation can be improved by adding extra terms to the right hand side of (7) to account for various effects such as the possible curvature of the magnetic field or the interaction with the electric field. Such more sophisticated approximations will be treated in future studies. The approximation of a "large" charged impurity (CP for "charged particle"), for which collisions can be approximated by a drag term is given by:

$$\begin{aligned} &\frac{d}{dt}\mathbf{r} = \mathbf{v} \\ &\frac{d}{dt}\mathbf{v} = \alpha(\mathbf{e} + \mathbf{v} \times \mathbf{b}) + \chi(\mathbf{u} - \mathbf{v}) \end{aligned} \tag{8}$$

where α corresponds to the charge to mass ratio in the Lorentz force, and χ is the inverse of a friction timescale. This model is a natural extension of the usual model for particles with mass moving in turbulent Navier-Stokes fluids [6].

1.2. Characterization of the dynamic regimes

For the MHD Kolmogorov flow, the same length and velocity scales defined for the Navier-Stokes case [5] are used:

$$L = \frac{1}{k_f}, \quad U = \left(\frac{A}{k_f}\right)^{1/2} \tag{9}$$

(i.e. based on the characteristics of the forcing, $\mathbf{f} = (A\sin(k_f y), 0, 0)$). Kinetic and magnetic Reynolds numbers corresponding to these units can then easily be defined:

$$Re = \frac{UL}{\nu} \equiv \frac{A^{1/2}}{\nu k_f^{3/2}}, \quad Re_m = \frac{UL}{\eta} \equiv \frac{A^{1/2}}{\eta k_f^{3/2}}, \tag{10}$$

and the magnetic Prandtl number is defined as the ratio of the two Reynolds numbers: $Pr_m = Re_m/Re$. A timescale can also be introduced:

$$\tau_f = \frac{L}{U} \equiv \frac{1}{(Ak_f)^{1/2}} \tag{11}$$

Throughout this work, Alfven units are used for the fields ($[\mathbf{b}] = [\mathbf{u}] = ms^{-1}$ and $[\mathbf{e}] = m^2 s^{-2}$), so α (proportional to the charge to mass ratio) is the inverse of a length and χ is the inverse of a time. An alternative is then to note $\ell \equiv 1/\alpha$ and $\tau \equiv 1/\chi$ — this allows for a straightforward construction of adimensional parameters. The following three adimensional numbers can then be defined:

$$\beta = \frac{v_\parallel}{U} \equiv \frac{v_\parallel A^{1/2}}{k_f^{1/2}}, \quad \gamma = \frac{\ell}{L} \equiv \frac{k_f}{\alpha}, \quad St = \frac{\tau}{\tau_f} \equiv \frac{(Ak_f)^{1/2}}{\chi} \tag{12}$$

to characterize the behaviour of the particles. The Stokes number St is well-known in the Navier-Stokes turbulence literature; note that the timescale defined by the forcing has been used here for St rather than the dissipation timescale. Parameters might be used differently in the literature: in [7] for instance, the classical Stokes number definition is used, and a different adimensional parameter is defined using ℓ.

2. Simulations

2.1. Method

The MHD equations are solved using TURBO [8], a pseudo-spectral solver. This code has been extended to study the transport of passive particles interacting with the fields generated by the MHD evolution. Particle trajectories in turbulent flows have been studied numerically for quite some time, but there are a few particularities to charged particles. The electric field is constructed as a product of two turbulent fields summed with derivatives of another turbulent field. Its spectrum might thus be quite different from the velocity field spectrum and, in particular, it might exhibit a slower decrease in k. As a consequence, the interpolation scheme needed to estimate the electric field at the particle position must be well behaved. Furthermore, because particles can have characteristic timescales much smaller than those of the MHD flow, it can be necessary to use high order integration schemes for tracking their position, which in turn, may impose high order spline spatial interpolations for the fields [9].

The particle trajectories are integrated in time using a fourth order implicit Runge Kutta solver [10], coupled with a seventh order spline computed over six grid points ($S^{(7,6)}$ from [9]). It has been checked that improving the time integration scheme or the interpolation method does not affect the results. Also, double precision floating point arithmetic is used for all the calculations, as this can have a noticeable effect on particle trajectory integration, [11].

2.2. Parameters

A moderately turbulent regime of $Re = Re_m = 100$ has been chosen. The fluid equations are solved with the pseudo-spectral solver, in a periodic box of dimensions $6\pi L \times 2\pi L \times 2\pi L$, using a uniform, rectangular grid with $784 \times 128 \times 128$ nodes. Vorticity structures with very long (possibly infinite) lifetimes, [1, 5] are then observed in the flow. It is worth mentioning that, when these vorticity structures are present, very strong vertical streams are also formed, i.e. in the y direction. They are formed in pairs, so the average fluid flow on the vertical direction is still null. This geometry is reminiscent of stratified turbulence, with alternating vertical sheets of movement in opposite directions.

Once the quasi-stationary regime is reached, particles are placed uniformly in the field, on a $15 \times 5 \times 5$ grid. The initial velocity for CP is random and of the same order as the fluid velocity $|\mathbf{v}| \sim U$. Because of the drag term, it is expected that the particle's velocities will be close to the local fluid velocity after a relaxation time. Two values of β (1 and 10) are investigated, and four pairs (γ, St): $(1/2, 1), (1/2, 1/10)$, $(1/4, 1)$ and $(1/4, 1/10)$.

2.3. Mean squared displacements

Preliminary results show that particles are very much influenced by the large scale structures in the Kolmogorov flow. Figure 1 shows that, after the ballistic regime, VFL along the y direction seem to be super-diffusive. The number of particles appears to be small, as there are significant irregularities in the curves; it must however be recognised that in evolving turbulence bursts in the fluid flow can affect the short time behavior of particles as well. The black dash-dot line corresponding to normal diffusion is shown for comparisons.

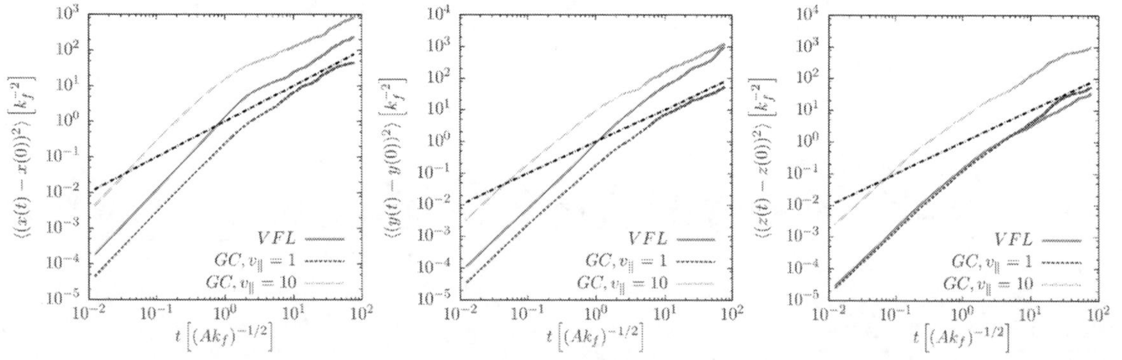

Figure 1. Mean squared displacements for both the velocity field lines and the guiding center approximation.

Charged particles (see Figure 2) also exhibit anisotropic diffusion properties. Again, the mean-square displacement is very close to classical diffusion in the forcing (x) and transverse (z) directions. However, due to the large-scale structures that correspond to strong flows, a super-diffusive regime is observed in the y direction.

2.4. Velocity PDF

The velocity distribution functions for VFL and CP are shown in figure 3. It seems that, for these values of the parameters, the drag term controls CP behaviour (because CP results are very close to VFL results). It is noted that the distribution function for v_y is apparently the sum of two Gaussian distributions. This further confirms the presence of two "streams" along the y axis. Superdiffusive transport for VFL is explained by these two maxima — particles may be

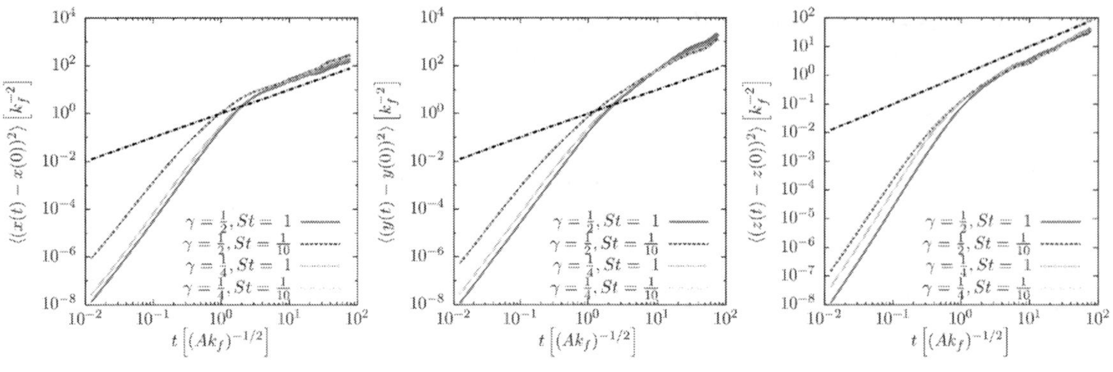

Figure 2. Mean squared displacements for charged particles submitted to both the Stokes and the Lorentz forces.

caught in one vertical stream or the other, and they can experience long flights. Superdiffusive transport was observed for stratified turbulence in [12, 13], but it is not obvious if the generating mechanism is similar (for instance the forcing is different here).

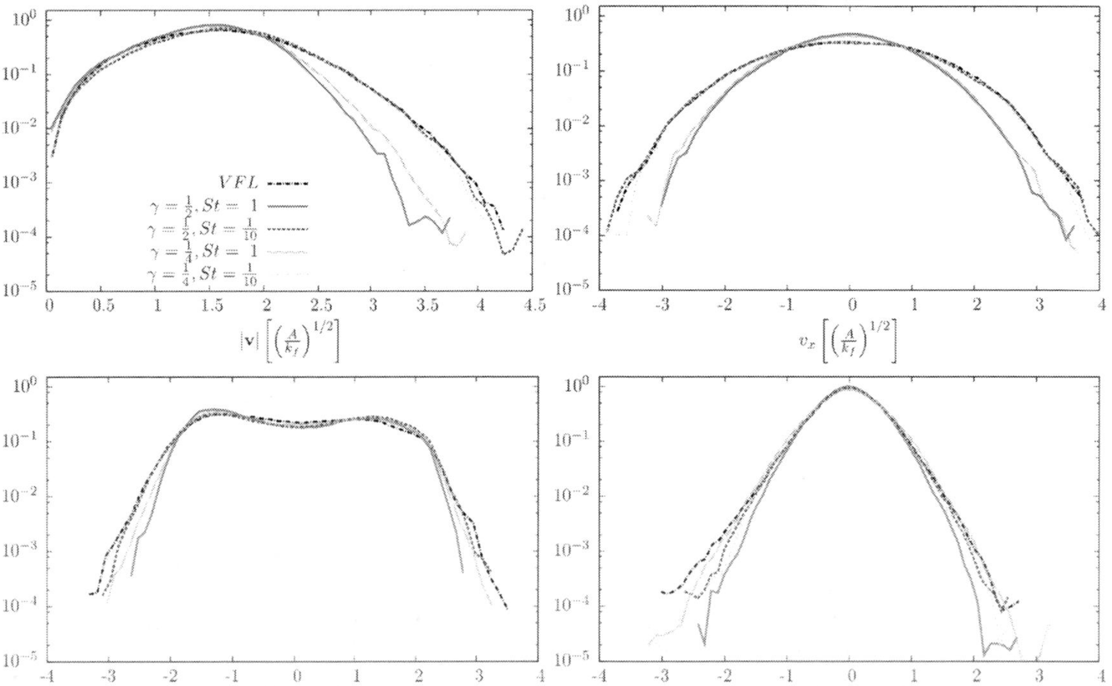

Figure 3. Velocity PDF for VFL and CP

Another interesting result is obtained when looking at the GC approximation. Since these particles are moving at a constant velocity $v_\parallel$ (in amplitude) along the local magnetic field, a Dirac delta distribution has to be observed for the norm of their velocity $|\mathbf{v}|$. However, the distributions for individual coordinates do show interesting properties, as seen in figure 4.

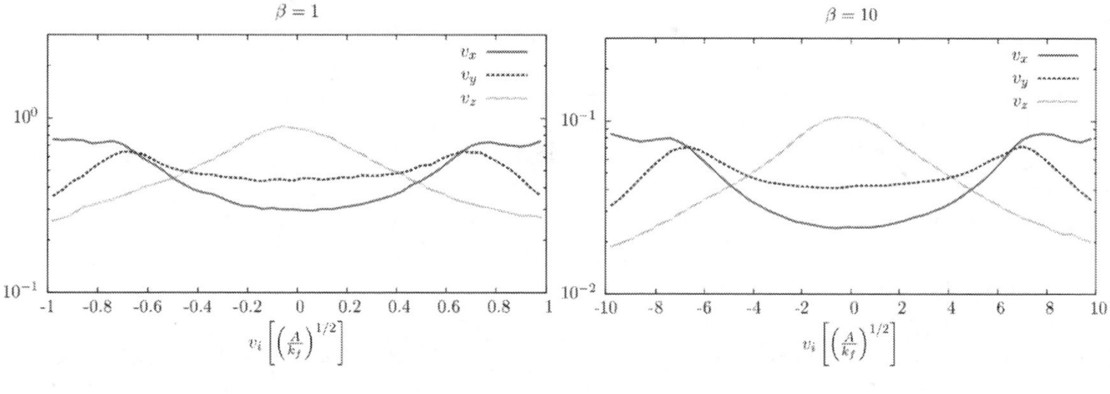

Figure 4. Velocity PDF for GC.

Because these particles have a fixed velocity amplitude, all three PDFs have finite support. Some correlation between the magnetic field and the velocity field is to be expected, but it is interesting that both the x and the y components of the velocity have distributions with two maxima.

2.5. Acceleration PDF

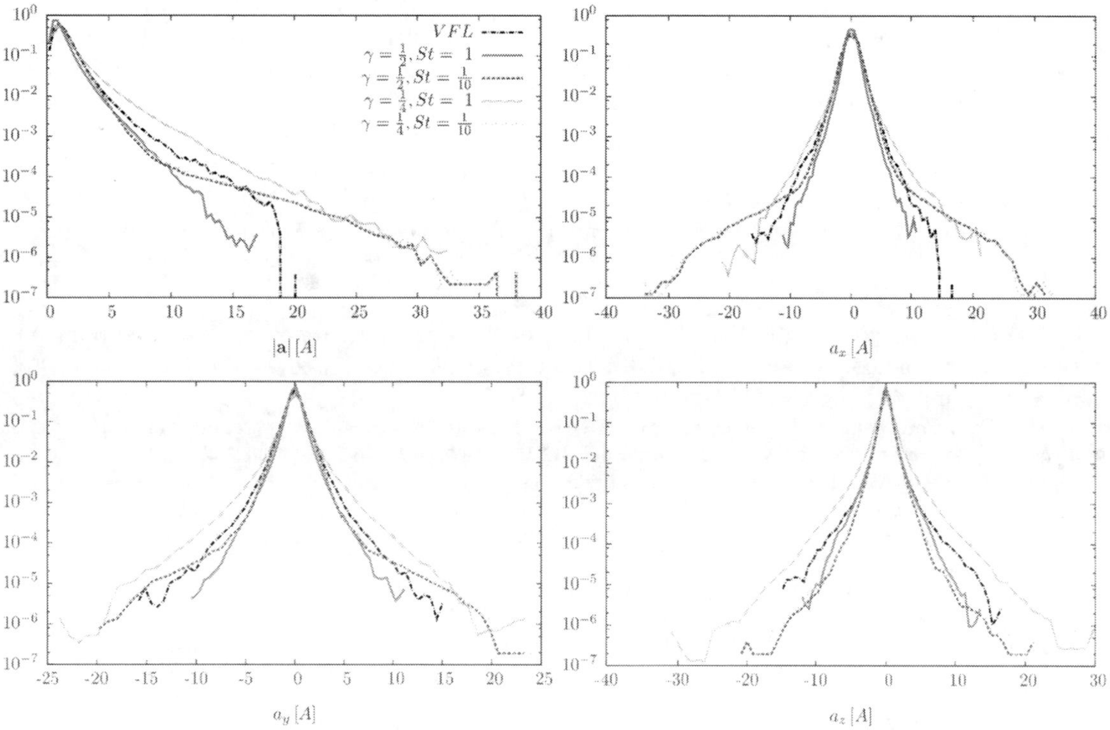

Figure 5. Acceleration PDF for VFL and CP

The probability distribution functions of the accelerations for VFL and CP are shown in figure 5. It is apparent that even though the drag term is present, there are important differences between charged particles and velocity field lines. The case $\gamma = 1/2$, $St = 1$ is the one that is closest to the behaviour of fluid particles. Note that the x and y components of the acceleration are sums of two different distributions. The long tailed distribution is generated in the initial relaxation phase (this is shown by results to be discussed elsewhere), whereas the central peak is generated in the quasistationary regime. It is however quite interesting that the second distribution with a longer tail is not observed for the z component of the acceleration.

3. Conclusions

The transport of charged test-particles in a fluid interacting with both the fluid velocity and the electromagnetic fields has been considered. The fluid flow is generated by the so-called Kolmogorov forcing which is known to induce persistent large-scale structures in the system. The fluid is assumed to be described by the incompressible MHD formalism.

Three types of trajectories have been considered. Test particles that follow the guiding center approximation and charged particles subject to a combination of the Stokes and the Lorentz forces have been compared to velocity field lines. Strong similarities between velocity field lines and charged particles can be observed. Also, the distributions of velocities for the guiding center trajectories are strongly affected by the constraint that the velocity is always directed along the local magnetic field with a constant amplitude. In all cases, the interaction of the particle trajectories with large-scale structures in the Kolmogorov flow generates anisotropic transport properties.

This preliminary study can be extended in various ways. First, more sophisticated guiding center approximations taken into account the presence of the electric field or the topology of the magnetic field could be investigated. The comparison with the charge particle trajectories would be a very valuable assessment of the guiding center approximations. Secondly, a more systematic parameter study could be initiated. It would definitively be interesting to evaluate the respective influence of the Stokes and Lorentz force on the diffusive properties of charged particles in the study of impurity transport in fusion plasmas. Finally, the influence of an imposed external magnetic field could also be investigated. Such a constant field would be an additional source of anisotropy that could compete with the large-scale structure influence.

3.1. Acknowledgements

D.C. is research director of the FRS-FNRS (Belgium). This work has been partly supported by the contract of association EURATOM-Belgian state. D.C. and C.C.L. would like to acknowledge support from COST Action MP0806.

[1] Lalescu C C, Carati D, Teaca B, Weyssow B and Grecu B D 2007 *Physics AUC* **17 (part II)** 1–10
[2] Priego M, Garcia O E, Naulin V and Rasmussen J J 2005 *Physics of Plasmas* **12** 062312
[3] Dubuit N, Garbet X, Parisot T, Guirlet R and Bourdelle C 2007 *Physics of Plasmas* **24** 042301
[4] Beyer P and Benkadda S 2001 *Chaos* **11** 774
[5] Sarris I E, Jeanmart H, Carati D and Winckelmans G 2007 *Physics of Fluids* **19** 095101
[6] Bec J, Biferale L, Cencini M, Lanotte A S and Toschi F 2010 *Journal of Fluid Mechanics* **646** 527–536
[7] Homann H, Bec J, Fichtner H and Grauer R 2009 *arXiv:0902.3212v1 [physics.flu-dyn]*
[8] Carati D *et al.* 2008 Solver for turbulent flows with periodic boundary conditions
[9] Lalescu C C, Teaca B and Carati D 2010 *Journal of Computational Physics* **229** 5862 – 5869
[10] Hairer E, Lubich C and Wanner G 2006 *Geometric Numerical Integration — Structure-Preserving Algorithms for Ordinary Differential Equations* 2nd ed (Springer)
[11] Homann H, Dreher J and Grauer R 2007 *Computer Physics Communications* **177** 560–565
[12] van Aartrijk M, Clercx H J H and Winters K B 2008 *Physics of Fluids* **20** 025104
[13] van Aatrijk M and Clercx H J H 2009 *Physics of Fluids* **21** 033304

Rayleigh-Bénard convection via Lattice Boltzmann method: code validation and grid resolution effects

V. Lavezzo[1], H.J.H. Clercx[1,3] and F. Toschi[1,2]

[1] Department of Applied Physics and J.M. Burgers Centre for Fluid Dynamics, Eindhoven University of Technology, Den Dolech, 2, 5612 AZ Eindhoven, The Netherlands
[2] Department of Mathematics and Computer Science and International Collaboration for Turbulence Research, Eindhoven University of Technology, Den Dolech, 2, 5612 AZ Eindhoven, The Netherlands
[3] Department of Applied Mathematics, University of Twente, Post Office Box 217, 7500 AE Enschede, The Netherlands

E-mail: v.lavezzo@tue.nl

Abstract. Thermal plumes, formed at the wall of turbulent natural convection cells, play an important role in the re-suspension and dispersion process of inertial particles. For this reason, a good resolution of the region close to the wall is necessary to correctly describe the plumes and, consequently, the particle dynamics. In this work, a Lattice Boltzmann Method (LBM) coupled with Lagrangian particle tracking is used to understand the effects of the filtering action exerted by the grid resolution on particle trajectories. A validation of the numerical method against the work of Kunnen (2009) and Schumacher (2009) is presented and, in this framework, mean and RMS statistics on fluid temperature are considered and analyzed in detail.

1. Introduction

The dispersion of inertial particles in turbulent convection has direct relevance for many industrial and environmental applications, where the fluid heat transfer can be modified by the presence and deposition of particles at the walls (e.g. nuclear power plants, petrochemical multiphase reactors, cooling systems for electronic devices, pollutant dispersion in the atmospheric boundary layer, aerosol deposition, etc.). Over the past years, turbulent convection has been the subject of extensive studies (see e.g. [1, 5, 7]), which attempted to determine the main flow features and the relation between the different parameters which characterize the fluid velocity and temperature fields, but only few of them focused on Lagrangian statistics. A Lagrangian tracking of tracer particles has been employed in [8] to study the properties of the mixing zone, the zone close to the wall dominated by rising and falling thermal plumes. The behaviour of heavy particles in stably stratified turbulence has been studied in [11] focusing on the relation between preferential concentration and flow stratification.

The focus of the present work is to evaluate the effects of the grid resolution on the flow characteristics and on particle behaviour in a turbulent Rayleigh-Bénard cell. To this aim a Lattice Boltzmann Method (LBM) coupled with a Lagrangian particle tracking has been chosen to solve for the fluid and particle velocity and temperature fields. Unlike the conventional numerical schemes based on discretization of macroscopic continuum equations, LBM is based on simplified kinetic models that incorporate the essential physics of mesoscopic scales, so that the macroscopic averaged properties obey the desired macroscopic equations (see e.g. the review articles [2, 4]). For what concerns single-phase turbulent flows, the LBM can be considered as an alternative numerical method with low algorithm complexity and good parallelization capabilities. However, if a local grid refinement is not considered, the LBM is restricted to a discretization onto a regular grid which might be not enough to capture correctly the physics in the close vicinity of the wall where the velocity and temperature boundary layers are formed. As in the present work the code is based on a uniform grid and, of course, we expect the LBM to perform as other discretization schemes, we need to understand which grid spacing is accurate enough for the purposes of this study. For this reason, the code has been validated against other two numerical works by [5] and [8] at different Rayleigh numbers. A good agreement between the LBM and the other methodologies has been found, confirming that the method provides an accurate description of the fluid momentum and temperature fields and, therefore, suitable for an analysis of particle resuspension dynamics in the domain. In Section 3, some preliminary results on tracer particles tracked in the same physical system described with three different domain discretizations, are also presented. Results showed that the grid resolution has a strong impact on particle behaviour exhibiting a lower dispersion if a coarser grid is used.

2. Methodology

The Lattice Boltzmann method is a well established tool described in detail in many previous publications (see e.g. [10] and [2]), so here only the main features characterizing this technique are reported. The simplest lattice Boltzmann equation, under the BGK approximation, can be written as follows:

$$f_i(x + \Delta t c_i, t + \Delta t) - f_i(x, t) = -\frac{\Delta t[f_i(x,t) - f_i^{eq}(x,t)]}{\tau} + F_i \Delta t \quad (1)$$

where $f_i(x,t) = f_i(x, v = c_i, t), i = 1, ..., n$ is the probability of finding a particle at lattice site x at time t, moving along the lattice direction defined by the discrete speed c_i during a time step Δt. The left-hand side of the equation represents the so-called streaming term, which corresponds to the evolution in time of the probability function f. The RHS describes the collision via a simple relaxation towards local equilibrium f_i^{eq} (a local Maxwellian distribution expanded to second order in the fluid speed) in a time lapse τ. This relaxation time fixes the fluid kinematic viscosity as $\nu = c_s^2(\tau - 1/2)$, where c_s is the speed of sound of the lattice fluid, ($c_s^2 = 1/3$ in the present work). Finally, F_i represents the effects of external forces. The set of discrete speeds must be chosen such that rotational symmetry is fulfilled to the required level of accuracy. The fluid density $\rho = \sum_i f_i$, and speed $u = \sum_i f_i c_i/\rho$ can be shown to evolve according to the (quasi-incompressible) Navier-Stokes equations of fluid dynamics. In this paper, we will refer to the nineteen speed D3Q19 model shown in Figure 1a. The temperature population g_i is evolved on the lattice according to the same formulation of the Boltzmann equation used for the fluid momentum. The temperature is modeled as a scalar $T = \sum_i g_i$, and it provides an external forcing via a buoyancy term, which is taken into account by introducing a suitable shift of the velocity and temperature fields entering the local equilibrium distributions.

The reference geometry, shown schematically in Figure 1b, is a Cartesian slab bounded by two horizontal walls having dimensions equal to the size of the computational grid employed for the calculations. Since in the present work a Lattice Boltzmann approach with a grid resolution equispaced in all directions has been chosen, it becomes useful to validate the code against other numerical works to check its capability in correctly capturing the flow field properties in the vicinity of the wall. The two works chosen as reference for this benchmark are: the finite-difference code by [5] and the pseudo-spectral code by [8]. The former has been selected for its relatively low Rayleigh number, the use of grid refinement close to the wall according to a log-normal distribution of the grid nodes and for the more realistic choice of no-slip boundary

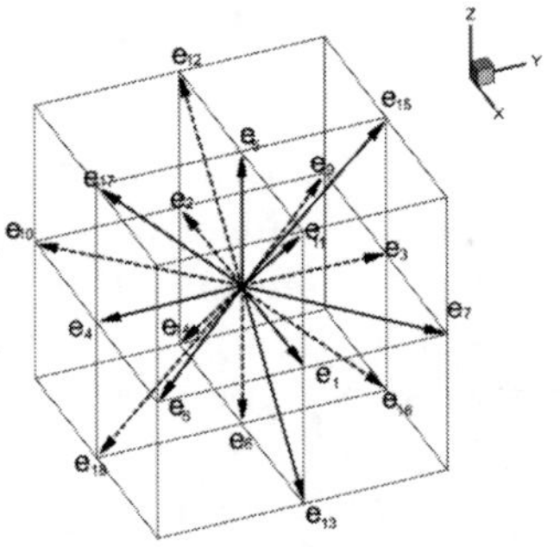

(a)

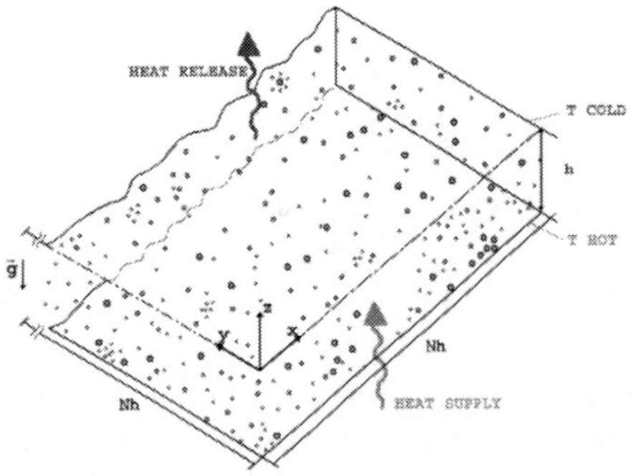

(b)

Figure 1. (a) Scheme of the D3Q19 model for the velocities in the LBM, (b) Sketch of the computational domain.

Table 1. The code has been validated against the numerical works by Kunnen et al. (2009) and Schumacher (2009) and for each test case two grid resolutions under the same physical parameters have been tested with the Lattice Boltzmann Method. The parameters used for the two benchmarks are here summarized.

	Grid	Grid type	b.c. vel/T	ΔT	Ra	Pr
Kunnen et al. (2009)	128x128x64	Refined	No-slip $/T_{fixed}$	1	$2.5 \cdot 10^6$	1
LBM	128x128x64 256x256x128	Regular	No-slip $/T_{fixed}$	1	$2.5 \cdot 10^6$	1
Schumacher (2009)	2048x2048x512	Regular	Stress-free $/T_{fixed}$	2π	$1.2 \cdot 10^8$	0.7
LBM	1024x1024x256 512x512x128	Regular	Stress-free $/T_{fixed}$	2π	$1.2 \cdot 10^8$	0.7

conditions. The latter because it employs a pseudo-spectral code, considered to be one of the most accurate Navier-Stokes solvers, on a regular grid at relatively high Rayleigh number using stress-free boundary conditions. To mimic the two numerical works as close as possible, the system properties, the grid resolution and the boundary conditions have been modified accordingly in the LBM. In Table 1 the parameters used in the validation tests are summarized. Periodic boundary conditions are imposed on both the velocity and the temperature field in the directions parallel to the walls. The no-slip and the stress-free conditions are implemented using a standard bounce back method which assure a second-order accuracy with the wall situated between two grid nodes. The temperature field is decomposed into a linear profile T_{lin} and fluctuations $T^{'}$ about the profile which are evolved in time. Since the total temperature is prescribed to be constant at the bottom ($Z = 0$) and top wall ($Z = 1$) the condition $T^{'} = 0$ follows there, so a mid-grid bounce back method is employed also for the temperature field.

The Rayleigh number is defined as $Ra = (\alpha g \Delta T H^3)/\nu\kappa$ where g is the gravitational acceleration, α the coefficient of thermal expansion, ΔT the temperature difference between the two walls, κ the thermal diffusivity and ν is the fluid kinematic viscosity. The Prandtl number is defined as the ratio of the kinematic viscosity to the thermal diffusivity of the fluid as $Pr = \nu/\kappa$. As by definition in LBM the grid resolution corresponds to the size of the computational domain, to maintain the same Ra and Pr number for different grid sizes i.e. for different values of H, the values of αg, ν and κ have been varied.

To understand the impact of the grid resolution on particle dynamics, a Lagrangian approach has been used to track a swarm of $1.6 \cdot 10^5$ tracer particles dispersed in the domain at $Ra = 5 \cdot 10^6$. No-slip boundary condition and constant temperature have been imposed on the velocity and temperature fields respectively. Three grid resolutions have been employed in this particular test case: $256 \times 256 \times 128$, $128 \times 128 \times 64$ and $64 \times 64 \times 32$. Particles are initially released at random positions on a plane placed at the same physical location from the bottom wall in all the grids considered. Specifically, the position H_0 of the plane in the lower resolution case is 0.5 lattice unit from the bottom wall (which sits 0.5 grid points below the first lattice node) and it is then multiplied by 2 and 4 to obtain the starting height in the other two grids. This particular configuration has been chosen to capture the effects of the plumes on particle resuspension at the early stages of the simulation, when hot plumes start to form at the bottom plate and particles are still non-homogeneously distributed in the domain. As particles have zero inertia, their equation of motion reduces to the simple kinematic relationship $d\mathbf{x}/dt = \mathbf{u}$, where $\mathbf{u}$ is the fluid velocity at the particle position and their temperature is equal to the fluid one calculated at the particle position. An explicit Adams-Bashforth scheme is used for the time integration of the equation of particle motion and a linear interpolation scheme is used to obtain both the fluid velocity and temperature at the point of the particle.

3. Results

3.1. Benchmark tests of the LBM code

In this section, results of the benchmark tests between the Lattice Boltzmann code and the works of [5] and [8] are presented.

In Figure 2a the mean temperature profile obtained with the Lattice Boltzmann method (symbols) is compared with the results obtained by [5] (black curve). All the curves are nicely overlapping on top of each other, thus suggesting that mean quantities are not really sensitive to both the numerical scheme and the employed grid resolution. It is possible to notice that the LBM, despite the regular grid, is capable to correctly capture the thermal boundary layer thickness δ_T in its geometrical extrapolation (see the inset of Figure 2a) if compared to the finite-difference code result. A value of about 0.045 in non-dimensional units is found. This value is consistent with the position of the peak in the temperature RMS profile visible in Figure 2b. However, some discrepancies in the RMS profile are still present in the bulk of the cell. This can be mainly due to two effects: (i) the different simulation time which is slightly longer in the finite difference case with respect to the present work, as it was checked by calculating the time-averaged statistics using different time windows and (ii) the different grid resolution used in the bulk region of the flow which is higher in the LBM with respect to the work of [5], thus providing a better description of this flow region. Since we are mostly interested in the dynamics of the boundary layer, this discrepancy can be considered negligible for the purposes of this study.

To quantify the total (conductive and convective) turbulent heat transport inside the domain, the volume and time averaged Nusselt number, normalized by the conductive heat flux obtained in absence of convection, has been calculated as:

$$Nu(z,t) = \frac{\langle u_z T \rangle_{A,t} - \kappa \partial_z \langle T \rangle_{A,t}}{\kappa \Delta T / H} \tag{2}$$

$$Nu = \frac{1}{H(t-t_0)} \int_0^H \int_{t_0}^t Nu(z,t)\,dz\,dt = 1 + \frac{H}{\kappa \Delta T} \langle u_z T \rangle_{V,t} \tag{3}$$

In [5] (see also [6]) a value of 11.08 for the total Nusselt number was found. In the present case a Nusselt number of 11.07 ± 1.03 and 11.16 ± 1.15 has been obtained for the low and high resolution

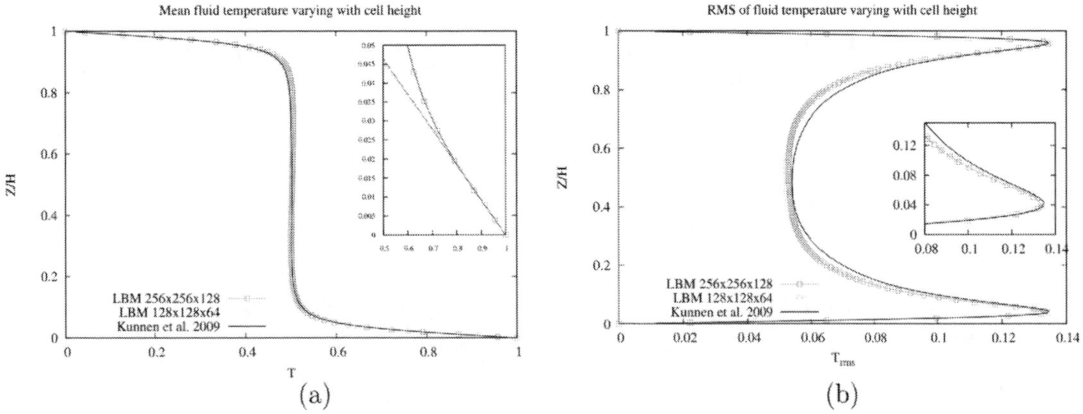

Figure 2. Benchmark between the LBM code using two grid resolutions (symbols) and the work by [5] (solid black line). (a) Mean temperature profile varying with the cell height with, in the inset, geometrical extrapolation of the thermal boundary layer thickness and (b) RMS of the temperature varying with the cell height with zoom in of the region where sharp temperature gradients are present.

case respectively. This result confirms the good agreement between the two codes and, as the introduced error between both approaches is around or less than 1%, the non-sensitivity of the results to the grid resolution. Using these values for the Nusselt number the boundary layer thickness can be calculated as:

$$\delta_T = \frac{H}{2Nu} \tag{4}$$

A value of 0.045 ± 0.005 has been found for both resolution, thus confirming the previous result obtained with the geometrical extrapolation.

In Figure 3a a comparison between the mean temperature profile varying with the cell height obtained with the LBM (symbols) and the pseudo-spectral code by [8] (black line) is reported. An overall good agreement among the curves is found. However, discrepancies are present in the close vicinity of the wall for the lower resolution case, as visible in the inset, where a small portion of the domain is considered for visualization. The black line indicates the value of the boundary layer thickness as in [8], whereas the colour dotted lines refer to the geometrical interpretation of δ_T obtained with the LB method. The lower grid resolution overestimates the value of δ_T and it is obviously not enough to correctly capture the dynamics in the boundary layer at this high Rayleigh number. In Figure 3b the temperature RMS varying with the cell height is presented. As expected, the lower grid resolution is overestimating the boundary layer thickness, but deviations are found in the bulk of the cell with both grid resolutions[1] of our LBM simulations. If we compare the value of the volume and time averaged Nusselt number a good agreement is found, also in this case. The obtained values of 57.0 ± 1.7 on the $1024 \times 1024 \times 256$ grid and of 55.3 ± 4.3 on the lower grid resolution compare well with the 56.4 ± 0.6 of [8]. The error in the Nusselt number, estimated by calculating the time averaged RMS of Nusselt number, is higher in the LBM case with respect to the pseudo-spectral code. This can be due to the different grid resolution employed in the two works which was limited in the LBM case by the computational resources available.

[1] As only a portion of the available computational resources was dedicated to the validation of the methodology, the low resolution case, being less expensive from a computational viewpoint, was run for a period longer than the higher resolution case. This allowed to obtaining slightly better results in the bulk of the cell with respect to the high resolution case.

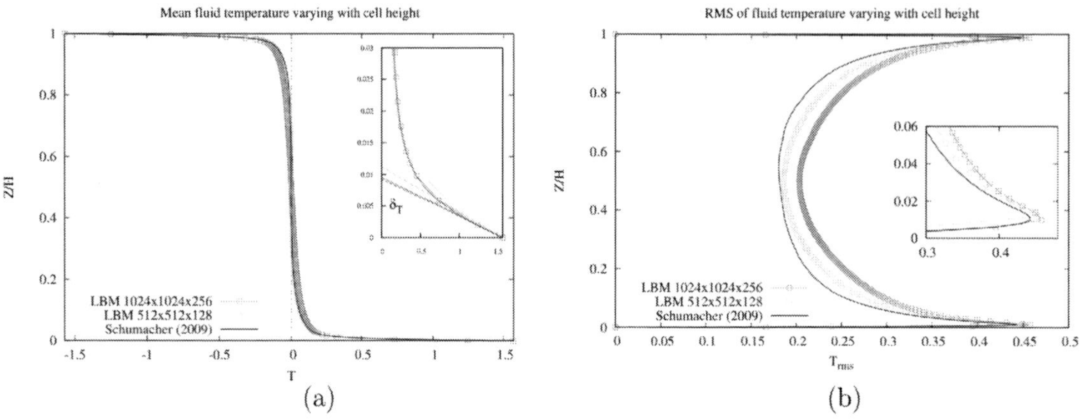

Figure 3. Benchmark between the LBM code and the work by [8]. (a) Mean temperature profile varying with the cell height with, in the inset, geometrical extrapolation of the thermal boundary layer thickness and (b) RMS of the temperature varying with the cell height with zoom in of the region of sharp temperature gradients.

3.2. Lagrangian insights

In this section some preliminary results on tracer particles dynamics are presented. The the reader should refer to Section 2 for details on particle tracking methodology and simulation parameters used in the following discussion.

The Lagrangian trajectories of tracer particles has put into evidence the direct impact that the computational grid has on the trajectory of the particles dispersed in the flow. In Figure 4a, the mean square displacement of particle position varying in time using three grid resolutions is shown. The mean square displacement has been calculated as the square of the difference between the particle initial position at time $t = 0$ and the particle position at time t as:

$$\langle \sigma(t)^2 \rangle = \frac{1}{N_p} \sum_{i=0}^{N_p} (\mathbf{x}_i(t) - \mathbf{x}_i(t = t_0))^2 \tag{5}$$

The statistics have been normalized in the vertical direction by the particle initial position H_0 and by the large circulation time scale on the horizontal axis. A coarser grid seems to under-predict single particle dispersion causing, with its implicit filtering action, particles to be confined in a smaller region of the domain. The reason for this can be found in the close relationship between the small-scale thermal plumes developing at the wall and particle dynamics. The sheetlike plumes close to the wall are responsible, in their morphological evolution, for the formation of mushroom-like plumes in a region confined to few δ_T [9, 7]. These three dimensional plumes are accountable, depending on their temperature, for the upward or downward particle transport. Changing the grid resolution modifies the statistical presence, the shape and the properties (velocity, temperature etc.) of sheetlike plumes and consequently affects the overall dispersion of particles in the cell. In Figure 4b an instantaneous snapshot of particles coloured with their temperature is shown. It is possible to notice that cold regions of fluid (green) are surrounded by warm (red) regions that correspond to the sheetlike structures. As we are focusing on the initial transient period when the convection starts to develop inside the domain, tracer particles are pushed at the edges of the sheetlike plumes, and bursts of particles are starting to rise due to the formation of mushroom-like type of structures, thus confirming the importance of these plumes in particle suspension.

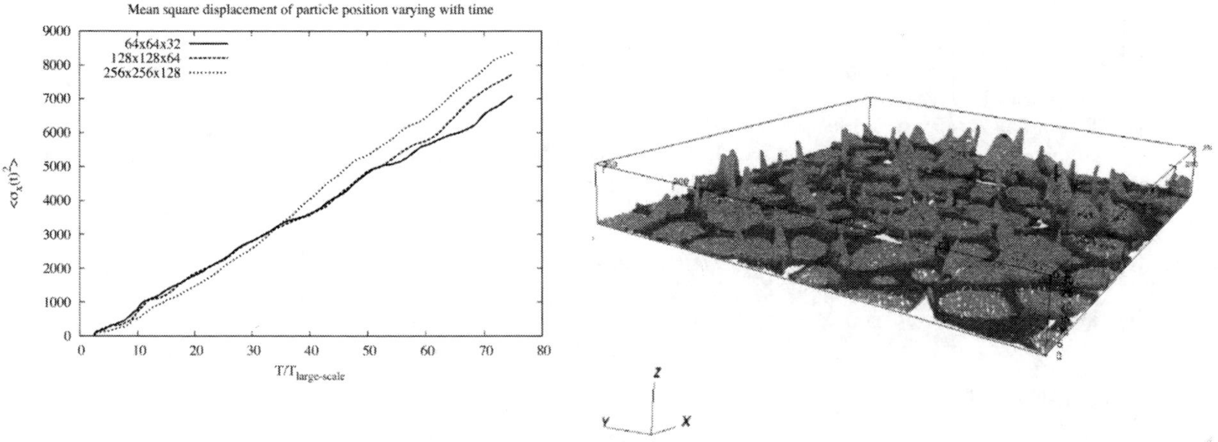

(a) (b)

Figure 4. (a) Mean square displacement of particle position varying in time obtained with three different grid resolutions ($256 \times 256 \times 128$, $128 \times 128 \times 64$ and $64 \times 64 \times 32$). (b) Snapshot of particles released in the flow obtained with the more refined grid.

4. Conclusions

A Lattice Boltzmann approach, coupled with Lagrangian particle tracking of passive tracers, is used in this work to understand the effects of the grid resolution on particle behaviour in a Rayleigh-Bénard cell. Results of a code validation against two other numerical works by [5] and [8] have been discussed. Good agreement in the mean temperature profile has been found in both cases. In the comparison with Kunnen and co-authors a good agreement, in the vicinity of the wall, is found also in RMS profiles for all grid resolutions considered, suggesting that at this low Rayleigh number the LBM results are independent from the grid resolution. Discrepancies are found in the bulk of the flow due to differences in both the simulation time lengths and the local grid resolutions. In the comparison with the work of [8] discrepancies are found in the temperature RMS profiles especially for the lower grid resolution which is not enough to correctly capture the thermal boundary layer thickness. Preliminary results on tracer particles suggest that particle motion is strongly influenced by the grid resolution as it impacts on the description of the small scales of the flow which are more effective in particle re-suspension and dispersion. Indeed, particles released in a flow obtained with the most refined grid are traveling longer distances with respect to those in the other two resolutions.

4.1. Acknowledgments

Authors wish to thank Dr. Rudie Kunnen (Eindhoven University of Technology, Eindhoven, The Netherlands) and Prof. Joerg Schumacher (Technical University of Ilmenau, Germany) for making their data accessible to us and for the useful discussions. This work is part of the research programme of the Foundation for Fundamental Research on Matter (FOM), which is part of the Netherlands Organisation for Scientific Research (NWO). The computational resources made available by SARA supercomputing center in Amsterdam under the DCCP project SH-176-10 and project SH-004-10 are gratefully acknowledged. The European COST Action MP0806 "Particles in Turbulence" is also acknowledged.

5. References

[1] Ahlers G, Grossman S and Lohse D 2009 *Rev. Mod. Phys.*, **81**, 503–537
[2] Benzi R, Succi S and Vergassola M 1992 *Phys. Rep.* **222**, 145–197
[3] Calzavarini E, Toschi F and Tripiccione R 2002 *Phys Rev E* , **66** 016304
[4] Chen S and Doolen G D 1998 *Annu. Rev. Fluid. Mech.* **30**, 329–364
[5] Kunnen R P J, Geurts B J and Clercx H J H, 2009 *Eur. J. of Mech. B/Fluids* **28**, 578–589
[6] Kunnen R P J, Clercx H J H and Geurts B J, 2006 *Phys. Rev. E* **74**, 056306
[7] Lohse D and Xia K 2010 *Annu. Rev. Fluid. Mech.*
[8] Schumacher J 2009 *Phys. Rev. E* **79**, 056301
[9] Shishkina O and Wagner C 2008 *J. Fluid Mech* **599**, 383–404
[10] Succi S 2001 *Oxford Science Publications, Oxford*
[11] van Aartrijk M, and Clercx H J H 2008 *Phys. Rev. Lett.* **100**, 254501 1-4

Nonlinear Algebraic Reynolds Stress Model for Two-Phase Turbulent Flows Laden with Small Heavy Particles in Circular Tube

R.V. Mukin, V.M. Alipchenkov, L.I. Zaichik, L.S. Mukina and V.F. Strizhov

Nuclear Safety Institute Russian Academy of Science, 115191, Bolshaya Tulskaya 52, Moscow, Russia

E-mail: mukin@ibrae.ac.ru

Abstract. The purpose of the study is to present an explicit algebraic Reynolds stress (nonlinear turbulent viscosity) model combined with modified $k-\varepsilon$ turbulence model taking into account particles effect on turbulence for calculating the main turbulent characteristics of two-phase flows. For calculating particles distribution in space we used diffusion-inertia model (DIM). The turbulence attenuating in the presence of particles is clearly observed, investigated and compared with the experimental data. The developed model adequately described turbulence anisotropy and the influence of particles inertia and concentration on the turbulence intensity.

1. Introduction

A great number of models intended for describing the turbulent characteristics of a fluid in the presence of particles can be found in literature. The most widespread model is the linear isotropic model of turbulent viscosity in combination with the equations for the kinetic turbulent energy and the rate of its dissipation. Thanks to its simplicity, this model is widely used in calculating two-phase flows of different types (see, for example, [1, 2, 3, 4, 5, 6, 7, 8]). However, in spite of its practical utility, the linear model has some well-known shortcomings. Thus, the model does not describe the turbulent stress anisotropy, as well as secondary flows in noncircular channels, and, moreover, leads to considerable errors in modelling strongly nonequilibrium flows with high velocity gradients due to sharp flow expansion or compression, strongly curved flows, etc. Rigorously speaking, the linear model of turbulent viscosity can be applicable only in calculating quasi-equilibrium flows characterized by an approximate balance between the generation and dissipation of the turbulence energy.

A more accurate and detailed modelling of the turbulent fluid characteristics can be based on the solution of the complete system of the Reynolds stress transport equation.This approach was realized in [9]; in this case, the empirical constants in the second moments equations are determined as a result of the calibration with the DNS data for a homogeneous shear flow. The transport equations for the Reynolds stress components of the continuous phase were also used in [9, 10, 11] in modelling axisymmetric jet and channel gas-disperse turbulent flows. However, modelling of complicated three-dimensional flows on the basis of the system of differential equations for all components of the turbulent stresses leads to a considerable increase in the computation time, as compared with the use of two-equation turbulence models of the $k - \varepsilon$ type. For this reason, for calculating single-phase flows the so-called nonlinear explicit algebraic Reynolds stress models (models of nonlinear turbulent viscosity) have received wide acceptance. These models, which possess almost the same accuracy as the differential models for the second moments of velocity fluctuations, make it possible to reduce considerably the computation time, increasing at the same time the numerical scheme stability.

In [12] turbulent flows and heat transfer in tubes were calculated using a very simple nonlinear model of turbulent viscosity with constant coefficients, without particles [13]. The original nonlinear model [14] is based on the theory of invariants and takes direct account of the particle effect. However, this model is not completely explicit and needs applying an iteration procedure to determine the turbulent energy production-to-dissipation ratio. As noted in [15], this iteration procedure considerably reduces the computational effectiveness of the algebraic model, since it can lead to nonuniqueness of solutions and, as a consequence, to the convergence to a non-physical solution. In [15, 16] a completely explicit, self-consistent algebraic Reynolds stress model for single-phase turbulence, which does not need an iteration procedure, was proposed. In spite of the fact that this model is rigorously valid only for two-dimensional homogeneous mean flows, this approach has gained a wide acceptance in modelling inhomogeneous and three-dimensional single-phase flows (see, for example, [17, 18]). In this study, a fully explicit, algebraic model for describing the Reynolds stresses of a fluid in the presence of small particles of the disperse phase is used.

The nature of the particles effect on the turbulent flow structure is not unique, small particles can attenuate fluid turbulence, while large particles can augment it, depending on their inertia and size. Thus, the presence of relatively small particles results in additional dissipation and an attenuation of turbulent fluctuations, owing to their decelerating (damping) action caused by their incomplete entrainment into the fluctuating fluid flow. With increase in the particle inertia the additional dissipation due to the fluctuating slip between the phases reduces and becomes unimportant for large particles. Apparently, it is the formation of an unsteady vortex structure (wake) due to flow separation behind a large particle in the flow that should be regarded as the main mechanism of the turbulence generation under the back action of the dispersed phase. Moreover, the turbulent characteristics of the continuous phase can considerably be influenced by the diffusive turbulent particle transport due to non-uniform dispersed phase distribution in space.

At the present time the vast amount of experimental data was gathered on the characteristics of particle-laden turbulent flows in channels, tubes, boundary layers etc. However there are always lots of factors in the experiment that make the flow more complicated, so that it's rather difficult to pick out the information about particular mechanisms of particles effect on turbulence. The impact of small heavy particles on the turbulent air flow in a vertical tube was studied in [19]. In that experiment the mean velocity profile was not distorted in the presence of particles, and no unsteady vortex structure (turbulent trace) due to flow separation behind the particle was formed. Thus the main way of the particles influence on turbulence is the damping effect related to incomplete entrainment of particles into the turbulent motion of the fluid, which causes additional dissipation and decrease in turbulent intensity. Taking into account the circumstances mentioned above, the experimental study performed in [19] is of a great value for validating theoretical models concerning the effect of small heavy particles on the turbulence intensity of the flow. In [20] fully explicit, algebraic model for describing the Reynolds stress of a flow in the presence of small particles of the dispersed phase is presented and validated for homogeneous shear flow. The main purpose of the present paper is to extend results of [20] to turbulent flow laden with small heavy particles in vertical tube and combine with DIM presented in our previous papers [21].

2. Mathematical formulation

2.1. Equations of the Turbulent Fluid

We consider the motion of a two-phase disperse medium consisting of an incompressible viscous fluid and small heavy particles. The volume concentration of the dispersed phase Φ is assumed to be sufficiently small so that inter-particle collisions can be neglected; however, the mass concentration $M \equiv \rho_p\Phi/\rho_f$ can be fairly large. The particle density ρ_p is assumed to be much greater than the fluid density ρ_f, while their diameter, d_p, is not greater than Kolmogorov length scale η. In this case, the equations for the continuous and disperse phases can be represented in the point-force approximation applied to the mass centres of individual particles. Moreover, the particle behaviour in a turbulent fluid and their effect on the flow is chiefly determined by the hydrodynamic drag, while the forces due to associated and displaced masses and the memory effect (Basset force) can be neglected. In this paper for taking into account particle effect on turbulence we used a single-point probability density function of particle velocity distribution [22, 23, 24]. Averaging Navier-Stokes equations for fluid flow with particles over the ensemble of random realizations of turbulent fields by representing the fluid velocity and the pressure in the form of the sums of mean and fluctuating components ($u = U + u'$ and $\mathrm{p} = \mathrm{P} + \mathrm{p}'$) leads to the Reynolds equation for the turbulent particle-laden fluid [22]:

$$\frac{\partial U_i}{\partial t} + U_j\frac{\partial U_i}{\partial x_j} = -\frac{1}{\rho_f}\frac{\partial \mathrm{P}}{\partial x_i} + \frac{\partial}{\partial x_j}\left(\nu_f\frac{\partial U_i}{\partial x_j} - \langle u_i'u_j'\rangle\right) + A_i, \tag{1}$$

$$U_{pi} = U_i + \langle u_i'\rangle_p,$$

where t is the time, x_i is the spatial coordinate, U_i is an average flow velocity, P is the pressure, ν_f is the kinematic viscosity coefficient, $\langle u_i'u_j'\rangle$ is the Reynolds stress tensor, τ_p is the the particle relaxation time, U_{pi} is the mean fluid velocity determined on a particle trajectory (the so-called fluid velocity seen by the particle), V_i is the mean velocity of the disperse phase.

Terms A_i and $\langle u_i'\rangle_p$ in (1) are describing the particle action on the the turbulent flow and the drift velocity between the fluid and the particles due to inhomogeneity of the disperse phase distribution [3]. Following [20] these terms written as follows:

$$A_i = \frac{\rho_p}{\rho_f\tau_p}\int\langle(v_i - u_i)p\rangle d\mathbf{v} = \frac{M}{\tau_p}\left(V_i - U_{pi}\right),$$

$$\langle u_i'\rangle_p = \frac{1}{\Phi}\int\langle u_i'p\rangle d\mathbf{v} = -\tau_p g_u\langle u_i'u_j'\rangle\frac{\partial \ln\Phi}{\partial x_j},$$

where p signifies the dynamic probability density of the particle velocity distribution in the phase space of the particle coordinates $\mathbf{x}$ and velocities $\mathbf{v}$.

In order to construct the fully explicit algebraic Reynolds stress model the following self-consistent method [15, 16] can be used, which applies the three tensor basis for Reynolds stress expression in the weak-equilibrium assumption. As a result of this approach being utilized to the Reynolds stress equation for turbulent gas laden by small heavy particles [25] we obtain:

$$\langle u_i'u_j'\rangle = \frac{2k}{3}\delta_{ij} - 2C_\mu^*\frac{k^2}{\varepsilon}\left\{S_{ij}^* - \frac{k}{\varepsilon}\left[B_1\left(S_{ik}^*S_{jk}^* - \frac{1}{3}S_{kn}^*S_{kn}^*\delta_{ij}\right) + B_2\left(S_{ik}^*W_{jk}^* + S_{jk}^*W_{ik}^*\right)\right]\right\}, \tag{2}$$

$$C_\mu^* = \frac{3\mathrm{A}_1\mathrm{A}_2}{3\mathrm{A}_1^2 - 2\mathrm{A}_3^2\overline{S}_{II}^* - 6\mathrm{A}_4^2\overline{W}_{II}^*}, \quad \mathrm{B}_1 = 2\mathrm{A}_3/\mathrm{A}_1, \quad \mathrm{B}_2 = \mathrm{A}_4/\mathrm{A}_1,$$

$$S_{ij}^* = \left(1 + Mf_{u1}\right)S_{ij}, \quad W_{ij}^* = \left(1 + Mf_{u1}\right)W_{ij},$$

$$S_{ij} = \frac{1}{2}\left(\frac{\partial U_i}{\partial x_j} + \frac{\partial U_j}{\partial x_i}\right), \quad W_{ij} = \frac{1}{2}\left(\frac{\partial U_i}{\partial x_j} - \frac{\partial U_j}{\partial x_i}\right)$$

$$\overline{S}^*_{II} = \overline{S}^*_{ij}\overline{S}^*_{ji}, \quad \overline{W}^*_{II} = \overline{W}^*_{ij}\overline{W}^*_{ji}, \quad \overline{S}^*_{ij} = \frac{k}{\varepsilon}S^*_{ij}, \quad \overline{W}^*_{ij} = \frac{k}{\varepsilon}W^*_{ij},$$

where S_{ij} and W_{ij} is a strain and rotation rate tensors for the fluid and S^*_{ij} and W^*_{ij} for the two-phase flow.

Together with linear by strain rate tensor S_{ij} term the Reynolds stress expression (2) includes also quadratic terms on the velocity deformations and rotation. Unlike the standard linear $k-\varepsilon$ model of turbulence, here C^*_μ is not a constant but depends on the strain and deformation tensor invariants for two-phase flow $\overline{S}^*_{II}$ and $\overline{W}^*_{II}$. For A_1 coefficient the following cubic equation is obtained:

$$\mathrm{A}_1^3 - \left(\mathrm{C}_1^0 - 2\right)\mathrm{A}_1^2 - \left\{\left[2\mathrm{A}_2\left(\mathrm{C}_1^1 + 2\right) + \frac{2\mathrm{A}_3^2}{3}\right]\bar{S}^*_{II} + 2\mathrm{A}_4^2\overline{W}^*_{II}\right\}\mathrm{A}_1 +$$
$$+ 2\left(\mathrm{C}_1^0 - 2\right)\left(\frac{\mathrm{A}_3^2\overline{S}^*_{II}}{3} + \mathrm{A}_4^2\overline{W}^*_{II}\right) = 0,$$

which has a solution:

$$\mathrm{A}_{1(1)} = \mathrm{M} + \mathrm{N} - \frac{\mathrm{a}}{3}, \mathrm{A}_{1(2,3)} = -\frac{\mathrm{M}+\mathrm{N}}{2} - \frac{\mathrm{a}}{3} \pm i\frac{\sqrt{3}(\mathrm{M}-\mathrm{N})}{2}, \tag{3}$$

where

$$\mathrm{M} = \sqrt[3]{-\frac{\mathrm{q}}{2} + \sqrt{\mathrm{Q}}}, \quad \mathrm{N} = \sqrt[3]{-\frac{\mathrm{q}}{2} - \sqrt{\mathrm{Q}}}, \quad \mathrm{Q} = \left(\frac{\mathrm{p}}{3}\right)^3 + \left(\frac{\mathrm{q}}{2}\right)^2, \quad \mathrm{p} = -\frac{\mathrm{a}^2}{3} + \mathrm{b},$$
$$\mathrm{a} = -\left(\mathrm{C}_1^0 - 2\right), \quad \mathrm{b} = -\left\{\left[2\mathrm{A}_2\left(\mathrm{C}_1^1 + 2\right) + \frac{2\mathrm{A}_3^2}{3}\right]\overline{S}^*_{II} + 2\mathrm{A}_4^2\overline{W}^*_{II}\right\},$$
$$\mathrm{c} = 2\left(\mathrm{C}_1^0 - 2\right)\left(\frac{\mathrm{A}_3^2\overline{S}^*_{II}}{3} + \mathrm{A}_4^2\overline{W}^*_{II}\right), \quad \mathrm{q} = 2\left(\frac{\mathrm{a}}{3}\right)^3 - \frac{\mathrm{ab}}{3} + \mathrm{c}.$$

If all the three roots (3) are real, then the largest root must be chosen [26]. Other coefficients in (2) equal:

$$\mathrm{A}_2 = \frac{4}{3} - C_2, \quad \mathrm{A}_3 = 2 - \mathrm{C}_3, \quad \mathrm{A}_4 = 2 - \mathrm{C}_4,$$

and according to [27] constants have the following values:

$$\mathrm{C}_1^0 = 3.4, \quad \mathrm{C}_1^1 = 1.8, \quad \mathrm{C}_2 = 0.36, \quad \mathrm{C}_3 = 1.25, \quad \mathrm{C}_4 = 0.4.$$

Compared to the original model [15, 16] the explicit algebraic Reynolds stress model (2) together with (3) contains the strain and deformation rate tensors and their invariants for two-phase flow instead of single-phase ones. The kinetic turbulent energy and its dissipation rate included in (2) can be calculated based on the $k-\varepsilon$ turbulence model for two-phase flows with small heavy particles [25]:

$$\begin{aligned}
(1 + Mf_{u1})\left(\frac{\partial k}{\partial t} + U_i\frac{\partial k}{\partial x_i}\right) &= \frac{\partial}{\partial x_i}\left\{\left[\nu + (1 + Mf_{u1})\frac{C_\mu k^2}{\sigma_k\varepsilon}\right]\frac{\partial k}{\partial x_i}\right\} - \\
&- (1 + Mf_{u1})\langle u'_i u'_j\rangle\frac{\partial U_i}{\partial x_j} - (\varepsilon + \varepsilon_p + \mathcal{G}_p) \\
(1 + Mf_{u1})\left(\frac{\partial \varepsilon}{\partial t} + U_i\frac{\partial \varepsilon}{\partial x_i}\right) &= \frac{\partial}{\partial x_i}\left\{\left[\nu + (1 + Mf_{u1})\frac{C_\mu k^2}{\sigma_\varepsilon\varepsilon}\right]\frac{\partial \varepsilon}{\partial x_i}\right\} - \\
&- \frac{\varepsilon}{k}\left[C_{\varepsilon1}(1 + Mf_{u1})\langle u'_i u'_j\rangle\frac{\partial U_i}{\partial x_j} + C_{\varepsilon2}(\varepsilon + \varepsilon_p + \mathcal{G}_p)\right]
\end{aligned} \tag{4}$$

where ε_p and $\mathcal{G}_p$ present relatively the additional dissipation in the presence of particles and the effect of particles nonuniform distribution in space. Equations (4) account for the particles contribution to the convection, diffusion, generation and dissipation of the main turbulent flow characteristics. The constants values in (4) are the same as for the single-phase flows in standard $k-\varepsilon$ turbulence model: $C_{mu} = 0.09$, $\sigma_k = 1.0$, $\sigma_\varepsilon = 1.3$, $C_{\varepsilon 1} = 1.44$, $C_{\varepsilon 2} = 1.92$. To set the boundary conditions when solving (4) we used the wall functions method [28], which is widespread for single-phase flow simulations at high Reynolds numbers.

2.2. Diffusion-inertia model

Transport and dispersion of the particles in two-phase turbulent flows a diffusion-inertia model (DIM) is used. This model was investigated in our previous papers [21, 29, 30, 31], combined with the simulation of the carrying turbulent flow within the framework of Reynolds-Averaged Navier-Stokes (RANS) equations. The model is based on the kinetic equation for the probability density function (PDF) of the particles velocity distribution [22, 23, 24], and is valid for two-phase flows with particles, which dynamic relaxation time does not exceed the Lagrangian integral timescale of the turbulence. The expression of the particle velocity in the form of the expansion in terms of carrying flow characteristics is a feature of this model and, thereby, causes a reduction of the problem of the particles transport and dispersion to the solution of a single equation for particle concentration:

$$\frac{\partial M}{\partial t} + \frac{\partial U_i M}{\partial x_i} + \frac{\partial}{\partial x_i}\left[\tau_p\left(F_i - \frac{DU_i}{Dt}\right)M\right] = \frac{\partial}{\partial x_i}\left[(D_B\delta_{ij} + D_{Tpij})\frac{\partial M}{\partial x_j}\right] + \\ + \frac{\partial}{\partial x_i}\left(M\frac{\partial q_u D_{Tpij}}{x_j}\right) \qquad (5)$$

$$D_{Tij} = \langle u_i' u_j' \rangle T_{Lp}, \quad q_u = \frac{\tau_p f_u}{T_{Lp}}, \quad \frac{D}{Dt} = \frac{\partial}{\partial t} + U_j\frac{\partial}{\partial x_j},$$

where F_i is the acceleration of a body force (for example, the gravity force) acting on a particle, D_B is the Brownian diffusion coefficient, D_{Tpij} is the particles turbulent diffusion tensor, q_u is the turbulent migration coefficient of particles, T_{Lp} is Lagrangian integral time of the liquid velocity fluctuations on the particle position.

The third term on the left-hand side of (5) describes the impact of gravity and other body forces, and the transport of by reason of the deviation of particle trajectories from the fluid streamlines. The first term on the right-hand side of (5) describes particles transport due to Brownian and turbulent diffusion and the second term quantifies turbulent migration (turbophoresis) due to the gradients of velocity fluctuations.

In DIM average velocity of particles can be expressed as an expansion in terms of the local characteristics of the turbulent fluid, with the particle response time as a small parameter:

$$V_i = U_i - (D_B\delta_{ij} + D_{Tpij})\frac{\partial \ln M}{\partial x_j} + \tau_p\left(F_i - \frac{DU_i}{Dt} - \frac{\partial q_u D_{Tpij}}{\partial x_j}\right). \qquad (6)$$

It should be noticed that using of anisotropic nonlinear Reynolds stress model compared with the standart isotropic model can improve the calculation results of particles transport and dispersion based on DIM in two points. Firstly, as a result of refined modelling of mean velocity field U_i, which determines the convective particles transport. And secondly because of the direct specifying of the Reynolds stress tensor components $\langle u_i' u_j' \rangle$, that is significant in modelling of particles transport due to turbulent migration (turbophoresis) and turbulent diffusion.

The eddy–particle interaction timescales are determined in an isotropic assumption $T_{Lp} = \left(T_{Lp}^{\ell} + 2T_{Lp}^{n}\right)/3$, where the superscripts ℓ and n stand for, respectively, the longitudinal (streamwise) and normal (spanwise) directions to the relative velocity vector $\mathbf{V_r} = \mathbf{V} - \mathbf{U_p}$. The integral timescales of particles interaction with turbulent vortices are determined from the model [32], that takes into account the particles inertia and different timescales of particle interaction with turbulent eddies in different directions - a phenomenon that arises due to "crossing-trajectories effect" [33],

$$T_{Lp}^{\ell} = \left\{ \frac{3\Im + m(2+3\gamma^2)}{3\Im\,(1+m\Im)^2}\left[1 - F(\mathrm{St}_E)\right] + \frac{F(\mathrm{St}_E)}{1+m\gamma} \right\} T_E,$$

$$T_{Lp}^{n} = \left\{ \frac{6\Im + m(4+3\gamma^2)}{6\Im\,(1+m\Im)^2}\left[1 - F(\mathrm{St}_E)\right] + \frac{2+m\gamma}{2\,(1+m\gamma)^2} F(\mathrm{St}_E) \right\} T_E,$$

$$\Im = \left(1+\gamma^2\right)^{1/2}, \quad \mathrm{St}_E = \frac{\tau_p}{T_E}, \quad \gamma = \frac{|\mathbf{V}_r|}{(2k/3)^{1/2}}, \quad T_E = \frac{3\,(1+m)^2}{3+2m} T_L, \quad m = \frac{T_E(2k/3)^{1/2}}{L},$$

$$F(\mathrm{St}_E) = \frac{\mathrm{St}_E}{1+\mathrm{St}_E} - \frac{0.9m\mathrm{St}_E^2}{(1+\mathrm{St}_E)^2\,(2+\mathrm{St}_E)},$$

where T_E is the Eulerian temporal integral scale of turbulence, L is the spatial integral scale of turbulence. The Lagrangian integral timescale is assumed to be equal to

$$T_L = \alpha \frac{k}{\varepsilon}, \qquad \alpha = \mathrm{const}$$

The quantities f_{u1} and g_u in equations (1), (2) and (4) are the coefficients of the particle entrainment into the turbulent fluid flow. Assuming these coefficients to be scalar are determined on the basis of the two-scale bi-exponential autocorrelation function and take the form [34]:

$$g_u = \frac{1}{\Omega} - f_u, \quad f_u = \frac{2\Omega + z^2}{2\Omega + 2\Omega^2 + z^2}, \quad f_{u1} = \frac{\left(2\Omega + z^2\right)^2 - 2\Omega^2 z^2}{\left(2\Omega + 2\Omega^2 + z^2\right)^2}, \quad \Omega = \frac{\tau_p}{T_{Lp}}, \quad z = \frac{\tau_T}{T_{Lp}},$$

where the Taylor differential timescale of fluid turbulence is determined as [21]:

$$\tau_T = \left(\frac{2\mathrm{Re}_\lambda \nu}{15^{1/2} a_0 \varepsilon}\right)^{1/2}, \quad a_0 = \frac{a_{01} + a_{0\infty}\mathrm{Re}_\lambda}{a_{02} + \mathrm{Re}_\lambda}, \quad \mathrm{Re}_\lambda = \left(\frac{20k^2}{3\varepsilon\nu}\right)^{1/2},$$

$$a_{01} = 11, \quad a_{02} = 205, \quad a_{0\infty} = 7.$$

The dynamic response time of a small particle is determined by the relation:

$$\tau_p = \tau_{p0}\left(1 + 0.15\mathrm{Re}_p^{0.687}\right)^{-1}, \quad \tau_{p0} = \frac{\rho_p d_p^2}{18\rho\nu},$$

where τ_{p0} is the particle response time, d_p is the particle diameter, and the particle Reynolds number Re_p is evaluated as

$$\mathrm{Re}_p = \frac{d_p\left[|\mathbf{V}_r|^2 + 2\,(1-2f_u)\,k\right]^{1/2}}{\nu}.$$

Average relative velocity between particles velocity and flow velocity seen by particles, according to (6) equals:

$$V_{ri} = V_i - U_{pi} = \tau_p\left(F_i - \frac{DU_i}{Dt}\right) - \frac{1}{M}\frac{\partial}{\partial x_j}\left[(D_B + g_u D_{Tpij})\,M\right].$$

3. Validation

In order to validate the presented nonlinear Reynolds stress model for two-phase turbulent flows the comparison was made with the experimental data published in [19]. In [19] the well-known method of laser anemometry was used to investigate the stationary upward fully developed flow in a vertical tube. Reynolds number based on the mean velocity and the channel diameter was Re = 25600. Particles were made of glass (SiO_2) with the diameter of 50 and 100 μm and density 2550 kg/m^3 as well as of alluminium oxide (Al_2O_3) with the diameter of 50 μm and density 3950 kg/m^3. Mass concentration of particles M varied from 0.12 to 0.39.

Fig. 1 shows the comparison of the calculated results for mean axial gas velocity U_x far away from the inlet section with the experimental data. These are the results in the universal coordinates: $u_+ = U_x/u_*$, $y_+ = yu_*/\nu$, where u_* is the friction velocity, y is the wall distance. Both the calculated and measured results reveal that the particles in use (of such mass concentration and inertia) don't affect the mean velocity distribution, which can be well described by the log-law [35]:

$$u_+ = \frac{\ln y_+}{\kappa} + \mathrm{B}, \quad \kappa = 0.41, \mathrm{B} = 5.2 \tag{7}$$

Fig.2 presents the distribution of the root mean square (rms) streamwise $\langle u_x'^2\rangle^{1/2}/U_{x0}$ and spanwise $\langle u_y'^2\rangle^{1/2}/U_{x0}$ components of velocity fluctuations, normalized by the mean velocity on the tube axis U_{x0}. The turbulence attenuating in the presence of particles is clearly observed. The attenuating effect increases as the mass concentration of particles in increased. Besides if the particle response time increases (within considerable range) the particle effect on turbulence is reduced due to decrease of additional dissipation ε_p. The comparison with the experimental data indicate the developed model to adequately describe turbulence anisotropy as well as effect of particle mass concentration and response time on the turbulence intensity. At the same time the velocity fluctuations predicted by the model are slightly lower than the experimental ones (particularly in the spanwise component of velocity pulsations and in the streamwise component near the wall). To improve the quantitative agreement with the experiment one should presumably use a low-Reynolds number model of turbulence instead of the high-Reynolds one to resolve the pulsations near the wall more accurate.

It should be mentioned that the particles concentration distribution in the cross-sections is considerably non-uniform, that is presented in Fig.3. However, authors in [19] observed uniform distributions of the particles and didn't observe particles concentration increasing near the wall of a tube presumably because measurement accuracy of particles concentration is decreasing near the wall. Presently, there is no doubt that the preferential concentration of particles laden in turbulent flow arises due to the so-called turbophoresis effect caused by the gradient of the turbulence intensity [36, 37]. But in the same time according to Fig.2 root mean square (rms) components of velocity fluctuations in the near-the-wall region do not depend on particle concentration. It can be explained by that the inertia of the particles (Stokes number St_L) strongly increases in the region near the wall, because the characteristic time T_L of carrying energy turbulent eddies becomes much less than that at the pipe axis. Therefore, the rate of particles involving in the fluctuation motion and the damping of the air turbulence intensity near the pipe wall are much smaller compared with the pipe axis region.

4. Conclusions

Nonlinear Reynolds stress turbulent model for two-phase flows laden with small heavy particles is presented. The model is based on the fully explicit algebraic Reynolds stress model, developed by Girimaji [15, 16] for single-phase turbulent flow. The presented model is valid for the particle-laden gas flows where particle relaxation time doesn't exceed the turbulent time macroscale. The model of nonlinear turbulent viscosity was implemented into a high-Reynolds number turbulence model for two-phase turbulent flow, including differential equations for kinetic turbulent energy k and turbulent energy dissipation rate ε taking into account particle influence on turbulence. The described nonlinear turbulent model was coupled together with the diffusion-inertia model for particles transport and dispersion in turbulent flow, developed by the authors earlier.

By means of comparison with experimental data in circular tube the presented model was shown to describe adequately the anisotropy of Reynolds stress tensor and influence of the particles concentration and inertia on turbulence intensity as well. The attenuation of turbulence increases as the particles concentration is increased. At the same time it is reduced as the particles response time τ_p is increased.

In the near wall region inertia of the particles (their Stokes number St_L) strongly increases, because the characteristic time $T_L = \alpha k/\varepsilon$ of carrying energy turbulent eddies becomes much less than that at the pipe axis. Therefore, the rate of particles involving in the fluctuation motion and the damping of the air turbulence intensity near the pipe wall are much smaller compared with the pipe axis region. It leads to the fact that the components of velocity fluctuations in the near wall region do not depend on particle concentration. In order to model velocity fluctuations in the near-the-wall region the low-Reynolds modification of the presented non linear ARSM model is to be suggested. However, proposed model can be applied to high-Re industrial problems on complex geometries

The work was done with the support of RFBR projects №09-08-00084 and №10-08-00210. It was started in a close collaboration and under the supervision of Professor Leonid Zaichik, and was finished in his light memory. This paper is dedicated to the memory of Professor Leonid Zaichik a great scientist, admirable man and a great human being. First author would also like to thank Dr. V.M. Alipchenkov for discussion on modelling issues and useful advices on the paper.

References

[1] Elghobashi S E and Abou-Arab T W 1983 *Phys. Fluids* **26** 931
[2] Rizk M A and Elghobashi S E 1989 *Int. J. Multiphase Flow* **15** 119
[3] Simonin O 1991 Second-moment prediction of dispersed-phase turbulence in particle-laden flows *Proc. 8th Symp. on Turbulent Shear Flows* vol 1 (Munich, Germany) pp 7–4–1
[4] Vinberg A A, Zaichik L I and Pershukov V A 1992 *Fluid Dynamics* **27** 353
[5] Squires K D and Eaton J K 1994 *Trans. ASME. J. Fluids Eng.* **116** 778
[6] Mando M, Lightstone M F, Rosendahl L, Yin C and Sorensen H 2009 *Int. J. Heat Fluid Flow* **30** 331
[7] Benavides A and van Wachem B 2009 *Int. J. Heat Fluid Flow* **30** 452
[8] Pakhomov M A and Terekhov V I 2009 *Fluid Dynamics* **44** 419
[9] Taulbee D B, Mashayek F and Barre C 1999 *Int. J. Heat Fluid Flow* **20** 368
[10] Zhou L X, Xu Y, Fan L S and Li Y 2001 *Powder Technol.* **116** 178
[11] Lain S and Sommerfeld M 2003 *Int. J. Heat Fluid Flow* **24** 616
[12] Boulet P and Moisette S 2002 *Int. J. Heat Mass Transfer* **45** 4201
[13] Speziale C G 1987 *J. Fluid Mech.* **178** 459
[14] Mashayek F and Taulbee D B 2002 *Numer. Heat Transfer. Pt. B* **41** 31
[15] Girimaji S S 1997 *Phys. Fluids* **9** 1067
[16] Girimaji S S 1996 *Theoret. Comput. Fluid Dynamics* **8** 387
[17] Wallin S and Johansson A V 2000 *J. Fluid Mech.* **403** 89
[18] Gatski T B and Rumsey C L 2002 Linear and nonlinear eddy viscosity models *Closure Strategies for Turbulent and Transitional Flows* (Cambridge: Cambridge Univ. Press)
[19] Varaksin A Y, Polezhaev Y V and Polyakov A F 2000 *Int. J. Heat Fluid Flow* **21** 562 – 567
[20] Alipchenkov V and Zaichik L 2010 *Fluid Dynamics* **45**(6) 909–923
[21] Zaichik L I, Drobyshevsky N I, Filippov A S, Mukin R V and Strizhov V F 2010 *Int. J. Heat Mass Transfer* **53** 154
[22] Zaichik L I 1997 *J. Appl. Math. Mech.* **61** 127
[23] Zaichik L 1999 *Phys. Fluids* **11** 1521
[24] Zaichik L I, Oesterle B and Alipchenkov V M 2004 *Phys. Fluids* **16** 1956
[25] Zaichik L, Alipchenkov V and Kozlov I 2010 *High Temp.* **48**(2) 205

[26] Jongen T and Gatski T B 1999 *J. Fluid Mech.* **399** 117
[27] Speziale C G, Sarkar S and Gatski T B 1991 *J. Fluid Mech.* **227** 245
[28] Launder B E and Spalding D B 1974 *Computer Methods in Applied Mechanics and Engineering* **3** 269–289
[29] Filippov A S, Alipchenkov V M, Drobyshevsky N I, Mukin R V, Strizhov V F and Zaichik L I 2010 *Journal of Engineering for Gas Turbines and Power* **132** 122901
[30] Drobyshevsky N I, Zaichik L I, Mukin R V, Strizhov V F and Filippov A S 2009 *Thermophysics and Aeromechanics* **16**(4) 521
[31] Drobyshevsky N I, Zaichik L I, Mukin R V and Filippov A S 2009 *Journal of Engineering Thermophysics* **18**(4) 271
[32] Zaichik L I and Oesterle B 2006 *Int. J. Multiphase Flow* **32** 838
[33] Csanady G T 1963 *J. Atmos. Sci.* **20** 201
[34] Zaichik L and Alipchenkov V 2007 *Statistical Models of Particle Motion within Turbulent Fluids* (Moscow: Fizmatlit)
[35] Pope S B 2000 *Turbulent Flows* (UK, Cambridge: Cambridge University Press)
[36] Reeks M W 1983 *Journal of Aerosol Science* **14** 729 – 739
[37] Eaton J K and Fessler J R 1994 *International Journal of Multiphase Flow* **20** 169 – 209

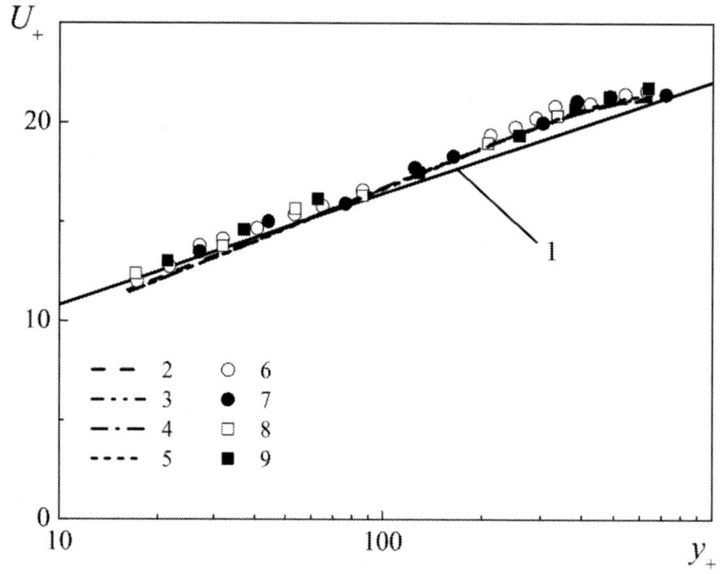

Figure 1. Mean velocity profiles: **1** – function (7); **2 - 5** – experiments by [19]; **6 - 9** – calculations; **2, 6** – $M = 0$; **3, 7** – Al_2O_3, $M_0 = 0.26$, $d_p = 50\ \mu m$; **4, 8** – SiO_2, $M_0 = 0.39$, $d_p = 50\ \mu m$; **5, 9** – SiO_2, $M_0 = 0.39$, $d_p = 100\ \mu m$.

Journal of Physics: Conference Series **333** (2011) 012012 doi:10.1088/1742-6596/333/1/012012

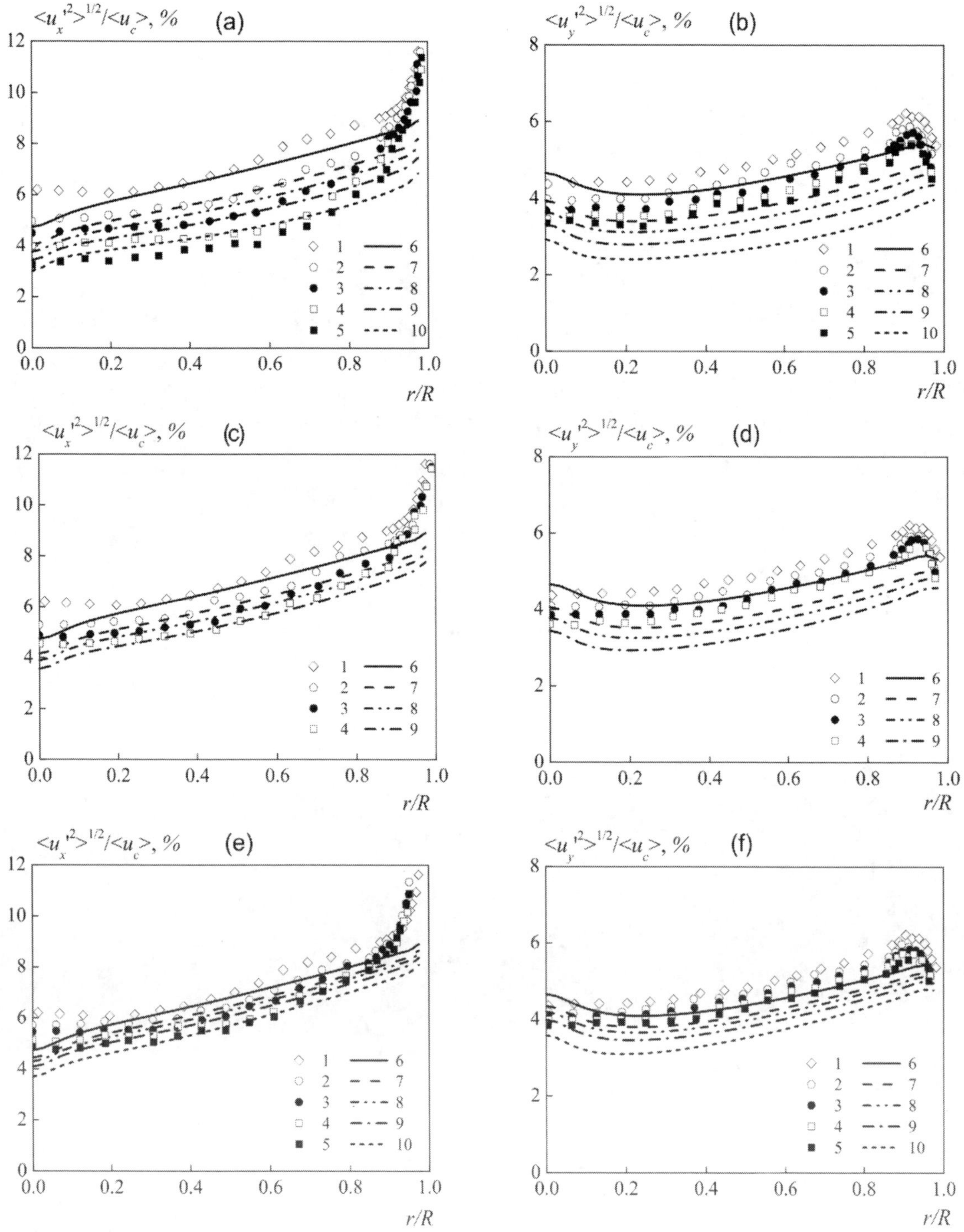

Figure 2. Streamwise rms velocity fluctuations – **a, c, e**; and spanwise rms velocity fluctuations – **b, d, f**; **a, b** – SiO_2, $d_p = 50$ μm, $\tau_p = 19.7$ ms; **c, d** – Al_2O_3, $d_p = 50$ μm, $\tau_p = 30.5$ ms; **e, f** – SiO_2, $d_p = 100$ μm, $\tau_p = 78.7$ ms; **1-5** – experiments by [19]; **6 - 10** – calculations; **1, 6** – $M_0 = 0$; **2, 7** – $M_0 = 0.12$; **3, 8** – $M_0 = 0.18$; **4, 9** – $M_0 = 0.26$; **5, 10** – $M_0 = 0.39$;

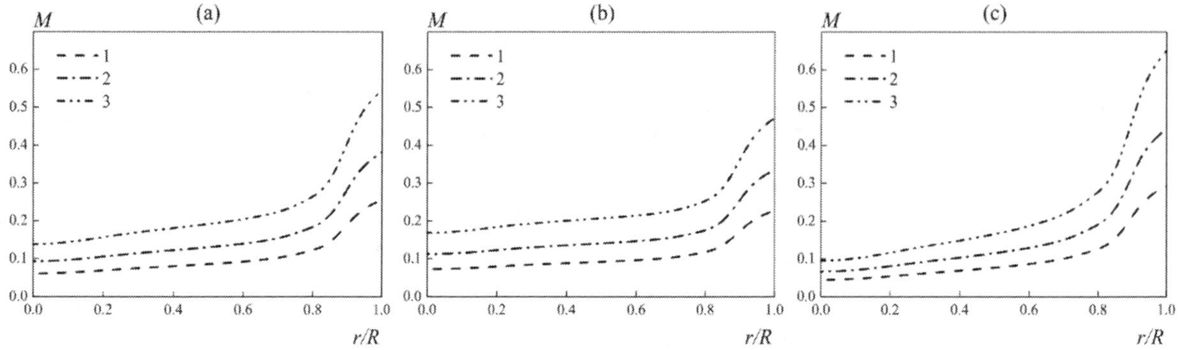

Figure 3. Mass concentration profiles: **a** – Al_2O_3, $d_p = 50$ μm, $\tau_p = 30.5$ ms; **b** – SiO_2, $d_p = 50$ μm, $\tau_p = 19.7$ ms; **c** – SiO_2, $d_p = 100$ μm, $\tau_p = 78.7$ ms; **1** – $M_0 = 0.12$; **2** – $M_0 = 0.18$; **3** – $M_0 = 0.26$;

Particles in Turbulence 2011 IOP Publishing
Journal of Physics: Conference Series **333** (2011) 012013 doi:10.1088/1742-6596/333/1/012013

Particle algorithms for population dynamics in flows

Prasad Perlekar[1], Roberto Benzi[2], Simone Pigolotti[3], and Federico Toschi[1]

[1]Department of Physics and Department of Mathematics and Computer Science, Eindhoven University of Technology, Eindhoven 5600MB, The Netherlands.
CNR-IAC, Via dei Taurini 19, 00185 Rome, Italy.
[2] Dipartimento di Fisica and INFN, Università "Tor Vergata", Via della Ricerca Scientifica 1, I-00133 Roma, Italy.
[3] Departament de Fisica i Enginyeria Nuclear, Universitat Politècnica de Catalunya Edif. GAIA, Rambla Sant Nebridi s/n, 08222 Terrassa, Barcelona, Spain.

E-mail: `f.toschi@tue.nl`

Abstract. We present and discuss particle based algorithms to numerically study the dynamics of population subjected to an advecting flow condition. We discuss few possible variants of the algorithms and compare them in a model compressible flow. A comparison against appropriate versions of the continuum stochastic Fisher equation (sFKPP) is also presented and discussed. The algorithms can be used to study populations genetics in fluid environments.

1. Introduction

Bacterial colonies growing on the surface of a Petri dish form intriguing patterns that can be mathematically modeled and studied in terms of a stepping stone model of population genetics [1]. Oceans cover a vast amount of the Earth's surface supporting an incredibly diversity of species, playing a key role since early stage of life on our planet. Many microorganisms must have found ways to survive in the flows generated in ocean. Therefore, though not much explored, it is important to understand the role of flows on models for the dynamic of populations. To understand this question two approaches can be taken: (a) Use continuum models and couple them with fluid flows [3, 4]. This is an approach used commonly in oceanography models to study plankton blooms [2]; (b) Use particle based models coupled to flows. The advantage of this approach is that it allows us to pose in a simpler way questions related to the populations to number fluctuations and the extinction of a mutant species, which in turn is central to the understanding of population genetics [1, 10]. A similar approach has been employed earlier to study the role of discrete effects in the propagation of a Fisher wave [5].

In this paper we discuss algorithms to study particle-based population dynamics in flows, following the above approach (b). An entity based algorithm with different variants is discussed in section 2. We then validate our algorithm for the Moran process in section 3, where we first present the results for the case of uniformly mixed populations with zero diffusivity and compare our results against theoretical predictions. Next we focus on the spatial diffusivity. The continuum limit of Moran model with spatial diffusivity is the stochastic Fisher-Kolmogorov-Petrovsky-Piscounov equation (sFKPP) [7, 8, 11]. We study the role of population fluctuations on the speed of a Fisher wave and compare our results against the asymptotic predictions of Ref. [11].

Finally, in section 4 we study the reaction $A \to A + A$, $A + A \to A$ in a flow, which also has the Fisher equation as a continuum limit, but with a different noise term [11, 15]. We show that in presence of a simple, deterministic, compressible flow the average population size (carrying capacity) decreases, in agreement with earlier investigations [4, 15].

2. Numerical algorithm

One of the example of population dynamics is the growth of bacteria on a Petri dish. The motion of the frontier of a bacterial colony can be understood as a solution of the stochastic Fisher equation. Another way to study frontier growth is by encoding the birth and death rules at an individual entity level. The advantage of this approach is that biologically relevant questions about the birth and extinction of species can be easily addressed. We now describe the two algorithms used by us to study population genetics with flows.

2.1. Algorithm 1

The numerical algorithm that is used to solve the population dynamics is similar to the algorithm used to study the front propagation of Fisher equation in Ref. [5]. We consider each individual in the population as a particle that is advected by the flow, diffused by the Brownian motion, and that can reproduce itself, die, or compete with other particles. Birth, death and competition can be modeled in terms of binary reactions. In order to efficiently implement these binary reactions (which necessarily happen when the two particles sufficiently close to each other) we introduce a spatial mesh. More precisely, we divide our one dimensional domain of size L into M subintervals of width $\delta = L/M$ and where δ is the interaction distance. As a typical starting condition one may consider particles uniformly distributed at positions $x \in L$. The population is evolved according to the following two main steps:

- *Advection and spatial diffusion:* the particles are advected and diffused according to $x_i(t + \Delta t) = x_i(t) + u\Delta t + \sqrt{2D\Delta t}\Gamma_i(t)$, $i = 1, 2, \dots n$. Here D is the spatial diffusion coefficient, u is the advecting velocity, Δt is the time step, n is the total number of particles, and $\Gamma(t)$ is a Gaussian white noise with zero mean and unit variance.
- *Population dynamics:* the particle population is assumed homogeneous inside the subdomain δ and interact with each other using a prescribed set of rules determining the population dynamics.

Given the framework that models advection, diffusion and locality of interaction, one must specify the proper reactions which encode for the biological processes in the populations. Here below we describe how to implement three of the building blocks at the base of the different reaction models used in this manuscript. Supposing that our ecosystem is composed by two distinct populations, A and B, we denote the number of particles within the interaction radius, δ, of type A as N_A and those of type B as N_B. The total number of particles in the interval δ is indicated with $N = N_A + N_B$ whereas the total number of particles in L is indicated with n.

(i) $A + B \xrightarrow{k} B + B$: In this type of reaction, a particle of one type, A, interacts with a particle of the other type, B, and at a rate k converts itself into B. To implement this rule we first count the number of B particles present in δ. If $kN_B\Delta t < r_t$ then the particle A converts itself into B otherwise nothing happens (r_t denotes a random number uniformly distributed in $[0, 1]$).

(ii) $A \xrightarrow{k} A + A$: In this reaction a particle of type A gives birth to another particle of the same type at a rate k. If $k\Delta t < r_t$ then the particle multiplies into two; otherwise nothing happens.

(iii) $A + A \xrightarrow{k} A$: In this reaction a particle of type A interacts with another particle of its own kind and one of the two dies with a rate k. Again we first count the number of A particles present in δ, if $k(N_A - 1)\Delta t < r_t$ then the particle annihilates; otherwise nothing happens.

2.2. Algorithm 2

This type of algorithm constitutes a variant over algorithm 1. Here the advection and diffusion parts are implemented in the same way. Two main differences are implemented in the population dynamics part of the algorithm:

Grid size At each timestep, the particle distribution is binned on a finer resolution. The bin size is chosen as δ/m, where δ is the interaction distance as before and m is an odd natural number (in the following we will choose $m = 11$). Binary reactions occur at a rate which depends on the total number of neighbors in the row of m cells centered on the bin where each particle is.

Rates The timestep Δt is taken small enough that no more than one reaction of each type can occur within each timestep. Then, a reproduction $A \to 2A$ occurs with a probability $kn(t)\Delta t$ where $n(t)$ is the total number of particles. A death by competition, $A + A \to A$ occurs with a rate $k\Delta t \sum_j N_j(t)$, where $N_j(t)$ is the total number of particles in the row centered in the bin where particle j is and the sum runs on all the particles present in the system. In the case of reproduction, the reproducing particle is chosen at random with uniform probability, while in the case of death by competition the choice is weighted with the number of neighbors, i.e. particle i has a probability to be chosen equal to $N_i/\sum_j N_j$.

The first modification allow to have competition in an area δ more precisely centered in the particle position. We found that the second modification slows down simulation considerably for our parameter values. Therefore, it should be implemented only in cases where the timescales of population dynamics are much longer than those of advection diffusion, so that one is not forced to choose a very small time step in order to avoid having more than one reaction per time step.

3. Validation

The validation is divided in two parts aimed at testing, respectively, the reaction algorithm alone and the diffusion-reaction implementation. We first focus on the reaction part and validate our population dynamics algorithm for the case of Moran process [6, 9] of population genetics. Here we compare our results with the theoretical prediction on the fixation probability from Kimura [9] (see Subsection 3.1).

In subsection 3.2 we add the effect of the spatial diffusion: with this addition the Moran process becomes equivalent to a stochastic Fisher equation [7, 8, 11]. We compare the results of our discrete simulations against the analytical predictions based on the stochastic Fisher equation, both in strong and weak noise limit [11, 14].

3.1. Moran model

We consider a homogeneous population of two species A and B with N_A individuals of type-A and N_B individuals of type B in a domain of size δ. When an individual of type A (B) interacts with an entity of the other type B (A) then B (A) converts into A (B) with a rate k_A (k_B). These reaction processes can be written in form the following form:

$$\begin{cases} A + B \xrightarrow{k_A} A + A \\ A + B \xrightarrow{k_B} B + B \end{cases}$$

During the above processes the total number of entities $N \equiv N_A + N_B$ remain conserved for all times. The continuum limit of the above process is the stochastic equation $dc =$

$\mu c(1-c)+\sqrt{\sigma^2 c(1-c)}dW$ (FKPP equation with noise) where $c = N_A/N$ is the concentration of the type-A species, $\mu \equiv (k_A - k_B)N/\delta$ and $\sigma^2 \equiv \frac{(k_A+k_B)}{\delta}$, dW is a Wiener process.

One of the central question in population genetics is the fixation probability of a mutation. For what concerns the Moran model this question can be formulated as: if B is the parent species and A is the mutant, what is the probability that the mutant converts all the parent species B into itself A (i.e. A gets fixated)? The two control parameters are the initial mutant population concentration $p = c_A(t=0) = N_A/N$ and the selective advantage μ. For convenience, in this section we set $\delta = 1$. For the case of the Moran process Kimura [9, 10] predicted that the fixation probability is:

$$P_{fix} = \frac{1-e^{-\alpha N p}}{1-e^{-\alpha N}} \tag{1}$$

where $\alpha \equiv \frac{k_A-k_B}{k_A+k_B}$ and p is the initial concentration of the mutant population. Note that for $k_A = k_B$ or $\mu = 0$ (case corresponding to no selective advantage), the fixation probability $P_{fix} = p$. The plot in Figure 1 shows the fixation probability for two different populations of size N with $p = 1/N$ and varying μ. The negative values of the selective advantage implies that the parent species has a selective advantage over the mutant. We find that, in agreement with the theoretical prediction, the fixation probability increases with the selective advantage of the mutant. For the special case of zero selective advantage ($k_A = k_B$) we find, in agreement with theory, $P_{fix} = p$. In our simulations the fixation probability, P_{fix} , is defined as one minus the fraction of realizations for which the mutant dies. Our measurements were obtained by averaging over an ensemble of 2000 realizations for each data point.

3.2. Stochastic Fisher equation

To study the effect of spatial diffusivity, Fisher, Kolmogorov *et al.* [7, 8] added a diffusion term to the Fisher equation:

$$dc = [\mu c(1-c) + D\nabla^2 c]dt + \sqrt{\sigma^2 c(1-c)}dW. \tag{2}$$

In absence of noise, for the case of a localized an initial condition, the Fisher equation exhibits a traveling wave solution with a front speed given by $v_F = 2\sqrt{D\mu}$ [7, 8]. More recently, it was shown that the speed of the Fisher wave is reduced in presence of noise. In the regime of weak noise [11, 12, 13, 14]:

$$v \sim \sqrt{D\mu}\left[2 - \frac{\pi^2}{(\log N)^2}\right] \tag{3}$$

where N is the total population within the interaction radius δ and $\sigma^2 \propto 1/N$. In the strong noise regime the Fisher speed is drastically reduced [11, 14]:

$$v \sim \frac{2D\mu}{\sigma^2}. \tag{4}$$

In order to validate our algorithm (1) we conducted a series of discrete particle simulations of the Moran process with spatial diffusion, with diffusivity D. The continuum limit is the sFKPP equation for different values of μ and σ^2. In Figure 2 we plot the normalized speed, v/v_F, of the Fisher wave versus the dimensionless noise strength $\sigma/(D\mu)^{1/4}$ as obtained from our discrete particle simulations. In the simulations the size of the domain was $L = 100$ with an interaction radius of $\delta = 1$. We observe, in agreement with [11], a cross-over in the Fisher wave speed around the value unity for the normalized noise strength. The asymptotic predictions for the weak and strong noise limits, see Equations (3) and (4), are also captured with our discrete particle simulations.

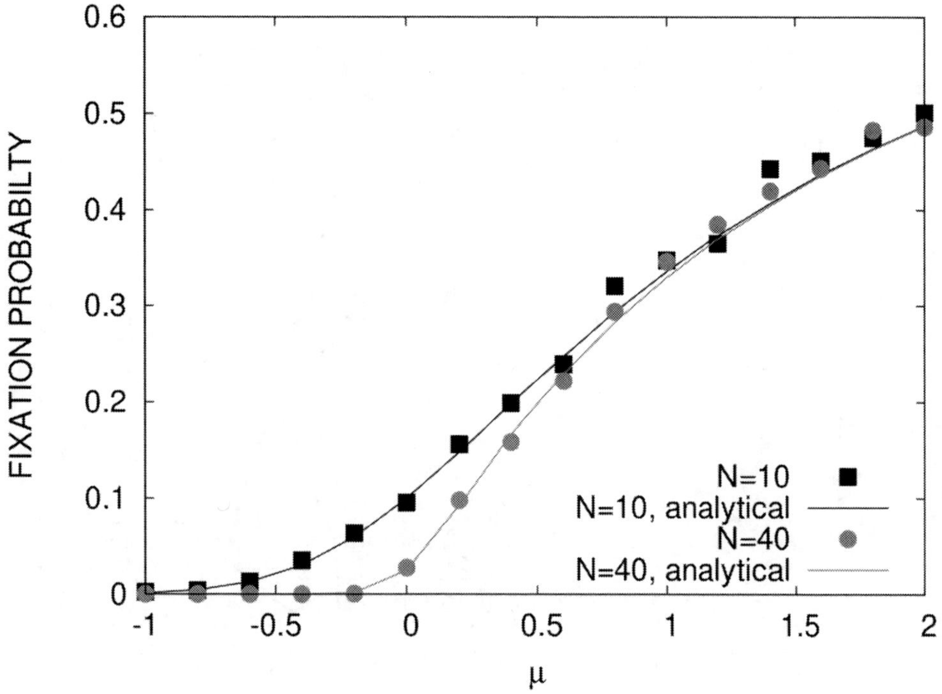

Figure 1. The behaviour of the fixation probability as a function of the selective advantage μ, for population sizes $N = 10$ and $N = 40$. A comparison with the analytical prediction is shown (see text for more details). We have chosen the initial allele frequency as $p = N_A/N = 1/N$ in all simulations.

4. Results on the carrying capacity in a model toy flow

A biologically relevant quantity is the carrying capacity (average population size in an ecosystem). In a recent paper [3] it was shown that a one-dimensional turbulent compressible velocity field leads to a reduction in the carrying capacity. This reduction of carrying capacity was later verified in a more realistic two-dimensional surface flow model of turbulence [4]. To investigate the behavior of our algorithms in presence of flow field we consider a very simple sinusoidal flow $u = U\sin(x)$. This flow is compressible and leads to a reduction in the carrying capacity on increasing the flow strength [15]. Below we consider a very simple model of population dynamics (the birth-coagulation process) whose mean-field limit also gives the Fisher equation.

$$\begin{cases} A + A \xrightarrow{\chi} A \\ A \xrightarrow{\gamma} A + A \end{cases}$$

where χ denotes the death (coagulation) rate and γ is the birth rate. The continuum limit of the birth-coagulation process gives:

$$\partial_t c + \partial_x(uc) = D\nabla^2 c + \mu c(1-c) + \sqrt{\sigma^2 c(1+c)}\Gamma(t), \tag{5}$$

where c is the local population concentration and $\Gamma(t)$ is a Gaussian white noise with zero mean and unit variance. We use the algorithm described in Section 2 to study the population dynamics

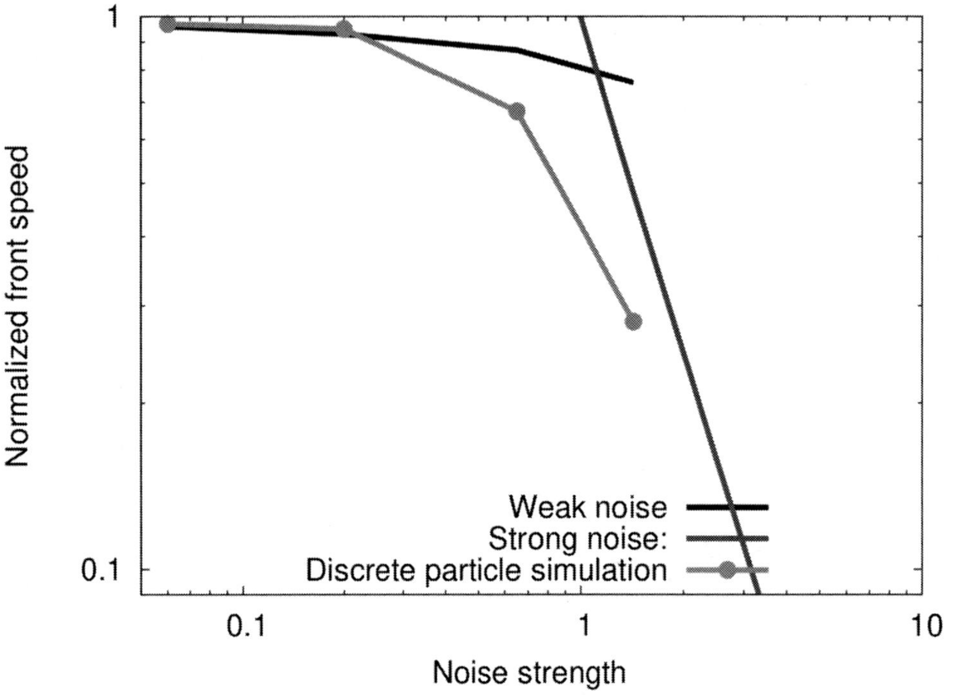

Figure 2. The normalized front speed of the Fisher wave (made dimensionless by v_F) versus the dimensionless noise strength. Data from our particles ssimulations are the red filled circles, while the asymptotic conjectures for the weak noise (blue line, $v \sim \sqrt{D\mu}\left[2 - \frac{\pi^2}{(\log N)^2}\right]$) and the strong noise (black line, $v \sim 2\frac{D\mu}{\sigma^2}$) limits are also shown. Each data point is obtained by doing an average over 50 independent noise realizations.

in this simple flow with the above population rules. An important point to notice is that this model, at variance with the Moran model, allows for fluctuations in the total population size. In absence of any flow the population reaches a steady state with a population size N_0, while the population size can changes drastically due to the presence of the sinusoidal flow field. We define the carrying capacity as:

$$Z_p(t) \equiv \frac{N_{av}}{N_0} \tag{6}$$

where N_{av} is the average population size in the presence of the flow. Figure 3 shows that the carrying capacity Z_p decreases strongly at increasing the forcing strength. This is consistent with earlier $1d$ and $2d$ simulations of population dynamics in turbulence flow field [3, 4] as well as with the stochastic simulations from [15].

5. Conclusion

We have introduced and discussed simple algorithms to study the dynamics of populations subject to an advecting flow. We benchmarked the numerical implementation amongst variants of the algorithm itself as well as against theoretical results. We show that the compressible flow reduces the total population size and thereby the carrying capacity. The proposed algorithms are a natural choice to investigate population genetics in flows. Future studies would look at the effect of population genetics in higher dimensional flows which are close to realistic settings

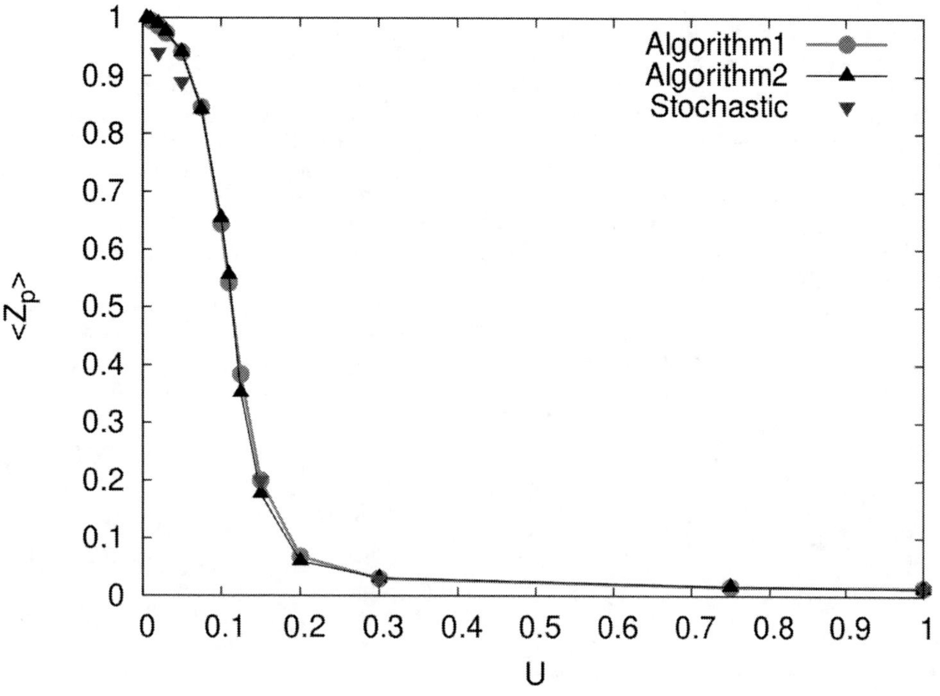

Figure 3. Carrying capacity versus the forcing strength of the sinusoidal flow field, U, for our one-dimensional compressible toy flow. We observe that the carrying capacity dramatically decrease with increasing forcing strength and saturates for $U \geq 0.3$. Algorithm 1 and algorithm 2 are the two algorithms described above, while stochastic refers to the results obtained from the numerical integration of Equation 5.

for e.g., surface flows. This would aid in our understanding of fixation of species in natural environments where fluid flow plays an important role.

6. Acknowledgement

We thank M. H. Jensen and D.R. Nelson for discussions. We acknowledge the COST Action MP0806 for support. FT and PP acknowledge the Kavli Institute of Theoretical Physics for hospitality. This research was supported in part by the National Science Foundation under Grant No. NSF PHY05-51164.

7. Reference

[1] K.S. Korolev, M. Avlund, O. Hallatschek, and D.R. Nelson, *Genetic demixing and evolution in linear stepping stone models*, Rev. Mod. Phys. **82**,1691 (2010).
[2] Mathias Sandulescu, Cristóbal López, Emilio Hernández-García, and Ulrike Feudel, *Biological activity in the wake of an island close to a coastal upwelling*, Ecological Complexity **5**, 228 (2008).
[3] R. Benzi and D.R. Nelson, *Fisher equation with turbulence in one dimension*, Physica D (Amsterdam) **238**, 2003 (2009).
[4] P. Perlekar, R. Benzi, D.R. Nelson, and F. Toschi, *Population dynamics at high Reynolds number*, Phys. Rev. Lett. **101**, 144501 (2010).
[5] S. Berti, C. Lopez, D. Vergni, and A. Vulpiani, *Discreteness effects in a reacting system of particles with finite interaction radius*, Phys. Rev. E **76**, 031139 (2007).
[6] P. Moran, *The Statistical Processes of Evolutionary Theory*. Clarendon Press (1962).

[7] R.A. Fisher, *The wave of advance of advantageous genes*, Ann. Eugenics **7**, 353 (1967).
[8] A. Kolmogorov, N. Petrovsky, N. Piscounov, *Etude de l'equation de la diffusion avec croissance de la quantite de la matiere et son application a un plobleme biologique* Moscow Univ. Math Bull. **1**, 1 (1937).
[9] M. Kimura, *On the probability of fixation of mutant genes in a population*, Genetics **47**, 713 (1962).
[10] O. Hallatschek and D.R. Nelson, *Population genetics and range expansion* **62**,42 (2009).
[11] C. R. Doering, C. Mueller, and P. Smereka, *Interacting particles, the stochastic Fisher-Kolmogorov-Petrovsky-Piscounov equation and duality* Physica A (Amsterdam) **325**, 243 (2003).
[12] E. Brunet, B. Derrida, *Shift in the velocity of a front due to a cutoff*, Phys. Rev. E **56** 2597 (1997).
[13] D.A. Kessler, Z. Ner, L.M. Sanderm *Front propagation: precursors, cutoffs, and structural stability*, Phys. Rev. E **58** 107 (1998).
[14] O. Hallatschek and K.S. Korolev, *Fisher Waves in the Strong Noise Limit*, Phys. Rev. Lett. **103**, 108103 (2009).
[15] S. Pigolotti, R. Benzi, M.H. Jensen, and D.R. Nelson, *Population genetics in compressible flows*, arXiv:1106.3506v1.

Effects of subfilter velocity modelling on dispersed phase in LES of heated channel flow

Jacek Pozorski, Maria Knorps, Mirosław Łuniewski

Institute of Fluid-Flow Machinery, Polish Academy of Sciences,
Fiszera 14, 80952 Gdańsk, Poland

E-mail: {jp, mknorps, mirmir}@imp.gda.pl

Abstract. A non-isothermal turbulent flow with the dispersed phase is modelled using the Large Eddy Simulation (LES) approach for fluid, one-way coupled with the equations of point-particle evolution. The channel is heated at both walls and isoflux boundary conditions are applied for fluid. Particle velocity and thermal statistics are computed. Of particular interest are the r.m.s. profiles and the probability density function of particle temperature in the near-wall region. We compare our findings with available reference data for particle-laden, heated channel flow. Moreover, an open issue in LES is the influence of non-resolved (residual) scales of fluid velocity and temperature fields on particles. In the present contribution, we apply a stochastic model for subfilter fluid velocity at the particle positions that aims at reconstructing the effects of the smallest scales of turbulence on particle dynamics. We analyse the impact of this model on particle thermal statistics.

1. Introduction

Channel flows are important benchmarks because of the modelling issues representative of inhomogeneous and wall-bounded turbulence. For model validation, comprehensive data sets are available from the Direct Numerical Simulations (DNS). Despite their simple geometry, turbulent channel flows exhibit complex physical features, including the near-wall vortical structures that affect the temperature field. The presence of particulate phase only increases this complexity [1]. From the practical point of view, exact predictions of particle dynamics in the near-wall region, their separation on the walls, and temperature distribution in non-isothermal cases are desirable.

We consider the issue of particle motion one-way coupled with the non-isothermal fluid flow that is determined from the Large Eddy Simulation (LES). The temperature is assumed to be a passive scalar, so it does not affect the fluid motion. We analyse here the effect of subfilter fluid velocity field on the dispersed phase, called the subfilter particle dispersion. To this aim, a stochastic model of the Langevin equation type is applied for the residual fluid velocity along particle trajectories. First, we report results from a regular LES (with no reconstruction of the residual velocity) of particle-laden channel flow. The LES corresponds to the computational conditions of the DNS reference data: for the flow case defined by the COST Action benchmark [2], for the heated flow [3], and for the particle thermal correlations [4]. Following our earlier work on particle velocity statistics [5], we analyse the impact of the subfilter particle dispersion model on particle temperature statistics. We discuss these issues and present results obtained for the particle-laden, turbulent channel flow with isoflux conditions at the walls.

2. Governing equations for fluid and dispersed phase

The fluid phase is governed by the filtered Navier-Stokes equations. In the momentum equation for the large-scale velocity field $\bar{\mathbf{U}}_f(\mathbf{x}, t)$, the resulting subgrid-scale (SGS) stress tensor is modelled through the subfilter viscosity ν_t, closed here with the dynamic (Germano-Lilly) approach [6]. The passive scalar equation for the filtered temperature $\bar{T}_f(\mathbf{x}, t)$ takes the form

$$\frac{\partial \bar{T}_f}{\partial t} + \bar{U}_{f,i}\frac{\partial \bar{T}_f}{\partial x_i} = \frac{\partial}{\partial x_i}\left[\left(\frac{\nu_f}{\mathrm{Pr}} + \frac{\nu_t}{\mathrm{Pr}_t}\right)\frac{\partial \bar{T}_f}{\partial x_i}\right] \tag{1}$$

and the SGS heat flux term is closed with the subfilter thermal diffusivity ν_t/Pr_t (cf. Table 1).

Particles in the flow are assumed to be smaller than the Kolmogorov length scale and are treated here as point-particles. The mass and volume load of the dispersed phase are taken much smaller than 1, therefore the one-way momentum and energy coupling with the fluid phase is sufficient. We note that some recent computational studies address the motion and heat transfer of finite-size particles with two-way coupling between the phases, cf. [7] and references therein.

Simplified equations of motion for particles are applied here. Only the drag force term is considered, which is acceptable for heavy particles, $\rho_p \gg \rho_f$. The collisions of particles with the walls are assumed elastic (the particle wall deposition case was studied in [5]). The equations of particle motion and temperature evolution are:

$$\frac{d\mathbf{x}_p}{dt} = \mathbf{U}_p\,, \qquad \frac{d\mathbf{U}_p}{dt} = f_D\frac{\mathbf{U}^* - \mathbf{U}_p}{\tau_p}\,, \qquad \frac{dT_p}{dt} = f_\theta\frac{T^* - T_p}{\tau_\theta}\,. \tag{2}$$

The fluid velocity $\mathbf{U}^*$ and temperature T^* along the particle trajectory are obtained from the second order interpolation of the LES field. The particle momentum relaxation time $\tau_p = \rho_p d_p^2/18\rho_f\nu_f$ and thermal relaxation time $\tau_\theta = \rho_p c_p d_p^2/12\lambda_f$ are assumed to be equal in the present study. The Stokes number (the particle inertia parameter) is conveniently defined in wall-bounded flows as $St = \tau_p^+$ (viscous scaling). The coefficients f_D and f_θ are well-known, semi-empirical correction factors for the drag force and heat transfer, respectively [8].

Lower-inertia particles are more responsive to small-scale eddy structures and, since the fluid velocity field is deprived of subfilter (higher) frequencies, an additional subgrid model for fluid velocity seen by particles is needed. Let us decompose it as $\mathbf{U}^* = \bar{\mathbf{U}}^* + \mathbf{u}^*$, where $\bar{\mathbf{U}}^* = \bar{\mathbf{U}}_f(\mathbf{x}_p, t)$ is the part resolved in LES and $\mathbf{u}^*$ is the SGS contribution. As proposed in [9], we use a stochastic Langevin equation, aiming to mimic the evolution of fluid velocity at particle's position. This is done by adding a random term, representing the effect of fluctuations in the SGS fluid velocity on the dispersed phase. The subfilter particle dispersion model states:

$$du_i^* = -\frac{u_i^*}{\tau_L^*}dt + \sqrt{\frac{2\sigma_{\mathrm{sg}}^2}{\tau_L^*}}dW_i, \tag{3}$$

where dW_i is an increment of the Wiener process. The subfilter velocity scale $\sigma_{\mathrm{sg}} = (\frac{2}{3}k_{\mathrm{sg}})^{1/2}$ is obtained from the estimated kinetic energy lost by filtration (e.g., the Yoshizawa assumption with the dynamic procedure [10]), and $\tau_L^* = C_{sg}\bar{\Delta}/\sigma_{\mathrm{sg}}$ is the subfilter time scale of turbulence, where $\bar{\Delta}$ is the filter width and C_{sg} is the model constant responsible for the time correlation of the subfilter velocity and, consequently, for the amount of kinetic energy added to particles [9]. A separate model for subfilter temperature at the particle position has not yet been included.

3. Numerical solution

A fully-developed turbulent channel flow is computed at the Reynolds number, based on the friction velocity, of $Re_\tau = 150$ (a benchmark test case of the COST Action LES-AID, [2]).

Table 1. Simulation parameters

Symbol	Value	Description
Re_τ	150	Reynolds number based on friction velocity
Pr	0.7	Prandtl number (air)
Pr_τ	0.98	turbulent Prandtl number
C_{sg}	0.01, 0.05	SGS Langevin model constant
St	1, 5, 25, 125	particle Stokes number τ_p^+

More numerical simulation parameters are presented in Table 1. The pressure-driven flow was assumed periodic in the streamwise and spanwise directions. The size of the flow domain in the streamwise (x), wall-normal (y) and spanwise (z) directions was $4\pi h \times 2h \times (4/3)\pi h$, discretised with $64 \times 84 \times 64$ FV meshes. For the fluid LES, we used a finite volume, academic solver of second-order accuracy (`FASTEST3D` code – courtesy of Prof. M. Schäfer, TU Darmstadt, Germany). The mesh size Δy^+ (in wall units) varied from 0.17 at the wall up to 10 at the centerline. In the periodic directions, the mesh was uniform with $\Delta x^+ = 29.5$ and $\Delta z^+ = 9.8$. At the walls, the isoflux b.c. were imposed. The present case complements therefore that of [8] where the heated-cooled channel with isothermal b.c. was studied.

Computations were performed with the time step $\Delta t^+ = 0.097$. The dispersed phase was added to the flow once the statistically-steady fluid thermal state was achieved ($t^+ \approx 60\,000$). The total simulation time for the particle-laden channel flow with $C_{sg} = 0.01$ was $t^+ \approx 34\,000$, and with $C_{sg} = 0.05$ it was about 10 000. The presence of the stochastic model for the particles affected the convergence time of the computations. The large particles ($St = 125$) achieved a stationary near-wall concentration faster with the SGS dispersion model (after $t^+ \approx 17\,000$). The small particles ($St = 1$) without the model were rapidly converged (after $t^+ \approx 1000$), while in the presence of the model the concentration slightly varied even at $t^+ \approx 30\,000$.

4. Results

First, the LES fluid solution of the channel flow case was compared to the DNS data of Kasagi et al. [11]. Figures 1 and 2 show the mean and r.m.s. profiles of fluid temperature, respectively. The mean temperature obtained in the LES does not differ from the DNS one. However, the r.m.s. fluctuations in LES, although smaller in the central part of the channel, are higher near the walls and the peak r.m.s. value is somewhat shifted closer to the wall. Such a behaviour may be due to the LES effect on small structures.

In order to estimate the impact of the Langevin equation-based model on the dispersed phase statistics, we compare the r.m.s. velocity fluctuations for of $St = 1$ particles (Fig. 3) to available DNS data [2]. Three cases of LES are shown: without the SGS particle dispersion model (denoted by 'LES'), with the SGS stochastic model and constant $C_{sg} = 0.01$, and with the model and constant $C_{sg} = 0.05$. Even without an additional model for particles, the LES gives good results for the streamwise velocity fluctuations, where the absolute error is small. For the wall-normal and spanwise directions (not shown, but qualitatively similar), the velocity fluctuations are underestimated as compared to the DNS. The presence of the model, with $C_{sg} = 0.01$, does not cover for energy of small scales, unresolved in LES. Another model constant tested was $C_{sg} = 0.05$ and the velocity fluctuation profiles are more consistent with DNS in the near-wall region, but according to [5] the higher the value of C_{sg}, the larger the overestimation of particle wall deposition velocity. Consequently, as the model is too dispersive in terms of preferential concentration patterns (high values of C_{sg}), despite the correct amount of energy added to the flow, the particle concentration is at odds with the physical situation.

The r.m.s. fluctuations of particle temperature for $St = 1$ and 25 are shown in Figs. 4 and 6.

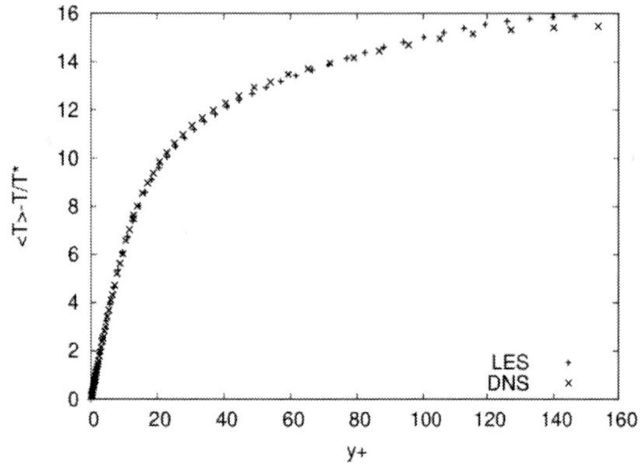

Figure 1. The mean fluid temperature: LES (+) and DNS (×) ref. data [11]

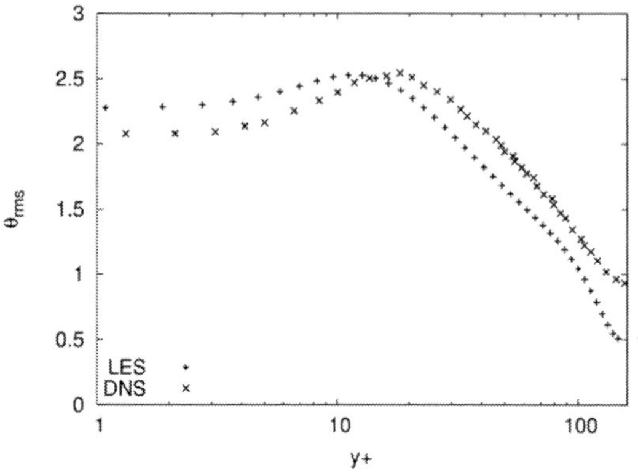

Figure 2. The r.m.s. fluid temperature: LES (+) and DNS (×) ref. data [11]

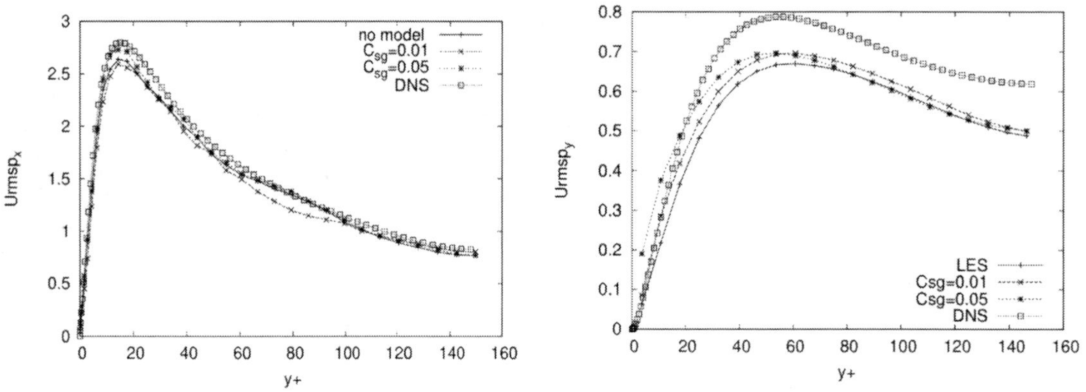

Figure 3. The r.m.s. of $St = 1$ particle velocity: a) streamwise, b) wall-normal. DNS (□), LES without model for particles (+), LES with SGS model with constant $C_{sg} = 0.01$ (×), LES with SGS model with constant $C_{sg} = 0.05$ (∗).

There is no significant difference between the temperature statistics of particles in LES with and without the model in the core of the flow, but near the walls ($y^+ < 10$) there is a change in the r.m.s. shape, depending on St. On the other hand, the effect of the subfilter particle dispersion model is not reflected in the probability density function (PDF) of particle temperature (Figs. 5 and 7); henceforth, we consider LES results without the SGS particle dispersion model.

The turbulent heat flux (Fig. 8) obtained from LES is in a qualitative agreement with the DNS data [4]. Also, a similar trend in both LES and DNS is observed, i.e., for increasing particle inertia the turbulent heat flux $\langle u\theta\rangle_p$ increases. The near-wall region is most affected by LES filtering and streamwise vortex rolls are not well resolved; therefore, the fluid velocity-temperature correlation coefficient $R_{u\theta} = \langle u\theta\rangle / u_{rms}\theta_{rms}$ is lower than in experiment (Fig. 9). For particles, the correlation increases with St, similarly as the turbulent heat flux.

The particle temperature PDF near the wall (Fig. 10) for $St = 1$ is in good agreement with the experimental data for fluid. Morover, a deviation from the normal distribution is observed. The "heavy tail" occurs in the region of lower temperatures, more pronounced, as the particle size increases. The effect may be due to the longer thermal relaxation time of heavy particles. They are more reluctant to transfer heat from the fluid in the near wall region, which results in

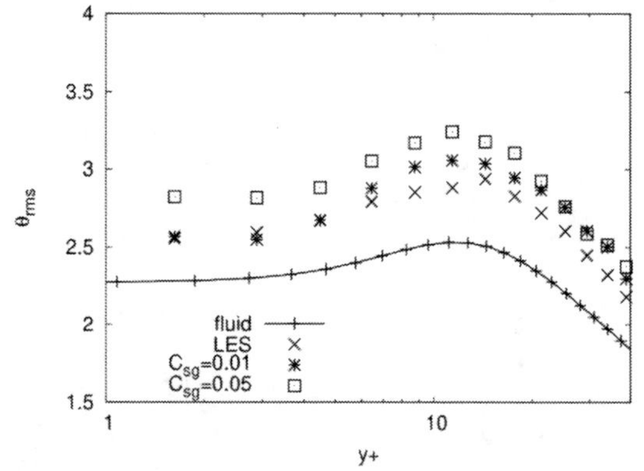

Figure 4. The r.m.s. particle temperature for St=1; DNS data [3]

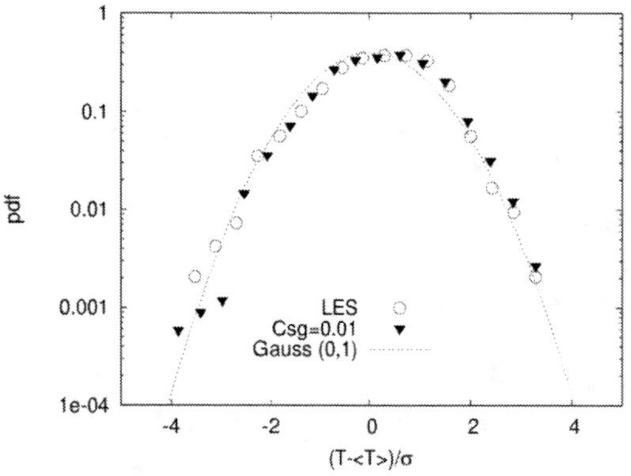

Figure 5. The PDF of particle temperature with and without the model at $y^+ = 3$; St=1

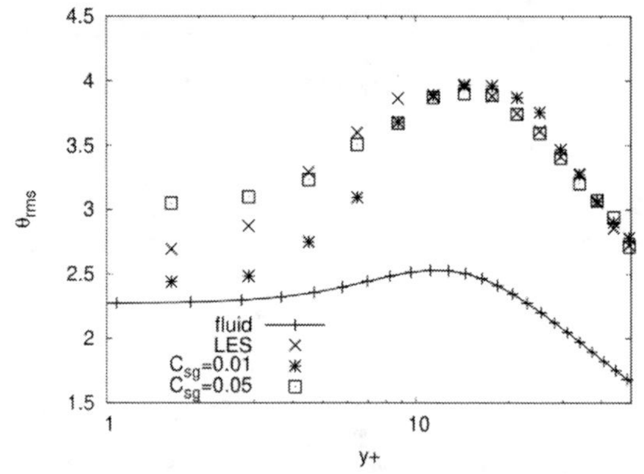

Figure 6. The r.m.s. particle temperature for $St = 25$; DNS data [3]

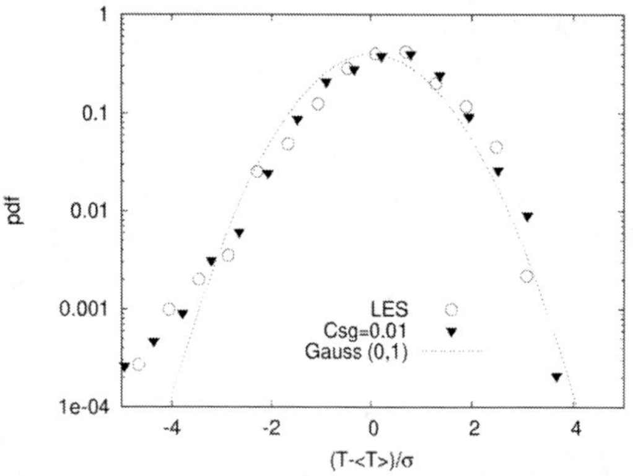

Figure 7. The PDF of particle temperature with and without the model at $y^+ = 3$; St=25

a greater number of cold particles. A similar situation is observed at the centre of the channel (Fig. 11), but now the "heavy tail" is on the hot particle side.

5. Concluding remarks

The LES simulation performed for turbulent heated channel flow with the dispersed phase is in a reasonable agreement with available DNS and experimental data. The results show some impact of the subfilter particle dispersion model. The presented model is of diffusion type, and although it recovers the kinetic energy of particles, it is known to alter the preferential concentration patterns. A solution to the problem may be in a structural model, enhancing near-wall structures of fluid velocity field. The r.m.s. particle temperature and the velocity-temperature correlation coefficients are found to be affected by the proposed model in the near-wall region. Moreover, the dispersed phase behaves similarily as in the DNS: with the increase of particle size comes a higher correlation of velocity and temperature. The PDF of particle temperature shows a non-gaussian behaviour, more pronounced for larger Stokes numbers.

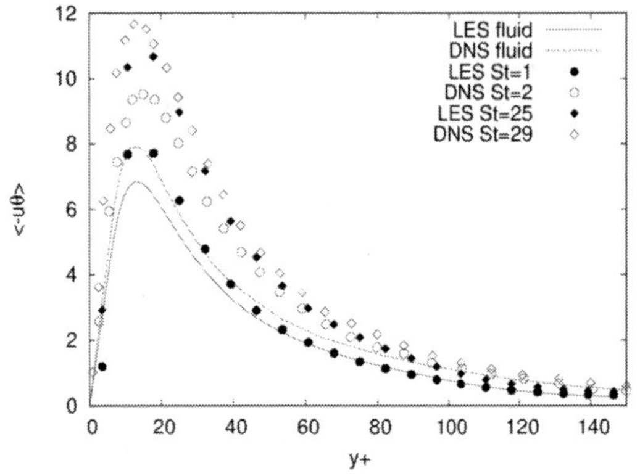

Figure 8. The turbulent heat flux; ref. DNS data [4]

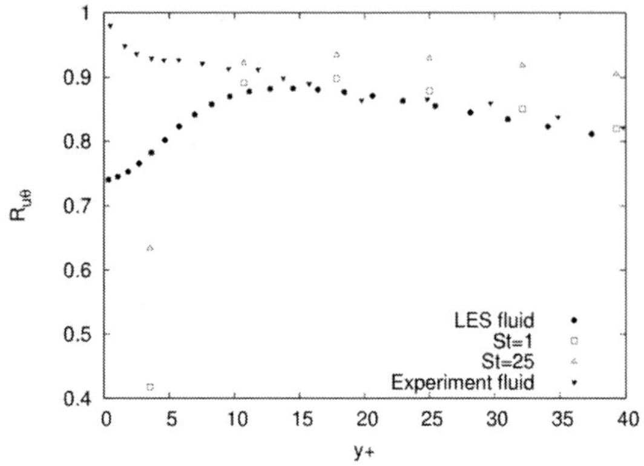

Figure 9. The correlation coefficient of temperature and streamwise velocity; experimental data [12]

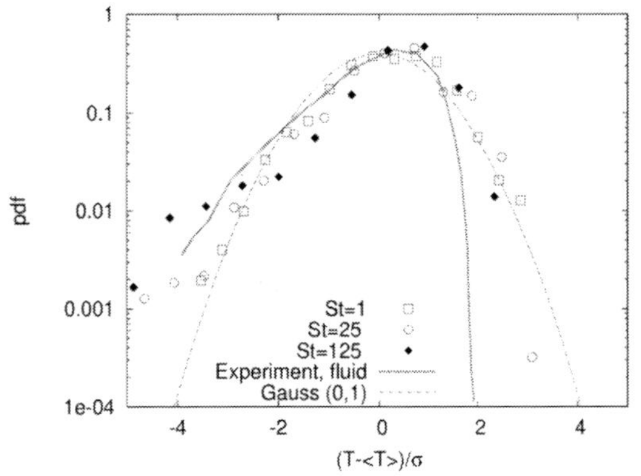

Figure 10. The particle temperature PDF at $y^+ = 3$; experimental data [12]

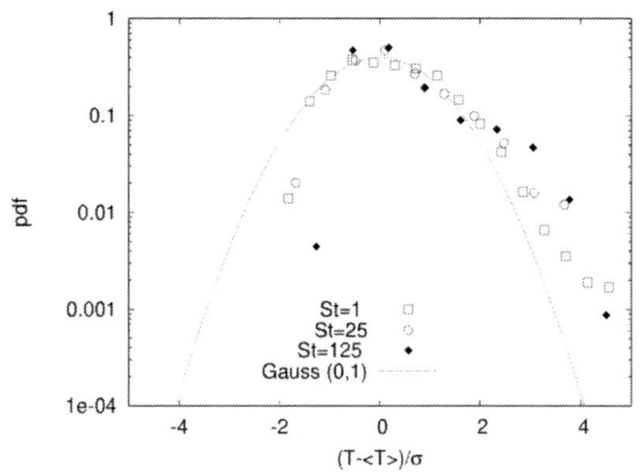

Figure 11. The particle temperature PDF at the centre of the channel

Acknowledgments

The work has been done in the framework of COST Action MP0806 "Particles in Turbulence". Calculations were carried out at the Academic Computer Center TASK in Gdańsk.

References

[1] Soldati A 2005 Appl. Math. & Mech. ZAMM **85** 683–699
[2] Marchioli C, Soldati A, Kuerten J, Arcen B, Tanière A, Goldensoph G, Squires K, Cargnelutti M and Portela L 2008 Int. J. Multiphase Flow **34** 879–893
[3] Kawamura H, Abe H and Matsuo Y 1999 Int. J. Heat Fluid Flow **20** 196–207
[4] Jaszczur M and Portela L 2008 Quality and Reliability of Large-Eddy Simulations ed Meyers J, Geurts B and Sagaut P (Springer) pp 343–354
[5] Pozorski J and Łuniewski M 2008 Quality and Reliability of Large-Eddy Simulations ed Meyers J, Geurts B and Sagaut P (Springer) pp 331–342
[6] Lilly D 1991 Phys. Fluids A **3** 1760
[7] Hagiwara Y 2011 Flow Turbulence Combust. **86** 343–367
[8] Pozorski J and Łuniewski M 2011 Quality and Reliability of Large-Eddy Simulations II ed Salvetti M, Geurts B, Meyers J and Sagaut P (Springer) pp 171–180
[9] Pozorski J and Apte S 2009 Int. J. Multiphase Flow **35** 118–128
[10] Yoshizawa Y 1986 Phys. Fluids **29** 2152–2164
[11] Kasagi N, Tomita Y and Kuroda A 1992 J. Heat Transfer **114** 598–606
[12] Antonia R, Krishnamoorthy V and Fulachier L 1988 Int. J. Heat Mass Transfer **31** 723–730

Statistical analysis of the influence of forces on particles in EM driven recirculated turbulent flows

M Ščepanskis[1], A Jakovičs[1] and E Baake[2]

[1] Laboratory for Mathematical Modelling of Environmental and Technological Processes, University of Latvia, 8 Zeļļu str., Rīga, LV-1002, Latvia

[2] Institute of Electrotechnology, Leibniz University of Hanover, 4 Wilhelm-Busch-str., Hannover, D-30167, Germany

E-mail: mihails.scepanskis@lu.lv

Abstract. The present paper contains an analysis of the statistical distribution of forces affecting non-conducting particles dispersed in an EM induced recirculated flow in induction furnaces. The simulation is conducted adopting the LES-based Euler-Lagrange approach in the limit of dilute conditions (one-way coupling). It is done by means of a development of *OpenFOAM* software code. The used Lagrange equation for particle tracking includes drag, EM, buoyancy, lift, acceleration and added mass forces. The relevant approximations for the forces are chosen on the basis of the statistical analysis of the non-dimensional parameters (particle Reynolds number, shear stress and acceleration parameter). The effect of force distribution on particle homogenization is described under different density ratios and particle sizes. The recommendations of the consistence of the Lagrange model for the simulation of the particle motion in the laboratory scale induction crucible furnace are given in conclusion.

1. Introduction

The present paper deals with the transportation of solid particles in a recirculated flow of metal melt in industrial electroheat equipment. Different types of induction furnaces are used, but all of them have the same physical principle of operation and consequently similar flow distribution in the induction furnaces, e.g. in a vertical cross section of an induction crucible furnace (ICF) [1] and in a cross section of a channel in a channel induction furnace (CIF) [2].

The sketch of an axis-symmetric induction system is shown on the figure 1. Alternating current in the inductor (j_{I} is the current density) creates a magnetic field $\boldsymbol{B}$ and induces an alternating current in the conductive liquid metal (j_{B} is the density of the induced current). The directions of j_{B} and j_{I} are opposite. The result of the interaction of j_{B} and $\boldsymbol{B}$ is the Lorentz force, which can be expressed as $f_{\mathrm{em}}(t)=[j_{\mathrm{B}}\times\boldsymbol{B}]$. This force drives the liquid away from the wall. So far as f_{Lorentz} has the maximum in the middle zone of the crucible (on the half-height of the inductor), the flow pattern consists of two eddies as it is shown on the figure 1 and in the case of cylindrical 3D geometry it means two toroidal vortices. Generally $f_{\mathrm{em}}(t)$ is harmonic, but as far as the frequency of j_{I} in the industrial electroheat equipment and density of metal are sufficiently high, the time for the metal to respond on the changes of the current is not enough. Therefore we can replace the harmonic force density with its average value (hereinafter we will type amplitude of the harmonic variables without any additional symbols):

$$f_{\mathrm{em}} = \frac{1}{2}\left[j_{\mathrm{B}} \times \boldsymbol{B}^{*} \right], \tag{1}$$

where $\boldsymbol{B}^{*}$ is complex conjugated magnetic field induction.

The power of ICFs and CIFs is sufficiently high enough to produce turbulent flow with high Reynolds numbers (e.g. for laboratory ICF it is about 10^5). The flow patterns and the turbulent properties of such systems were studied well experimentally and numerically during the last years (see e.g. [1,3-5]). As it is already discussed above the two toroidal vortices appear in the time averaged case, but in practice at each moment the turbulent flow consists of numerous eddies with different sizes. However, after the initial transition period the common structure of the flow becomes quasistationary (statistically stationary).

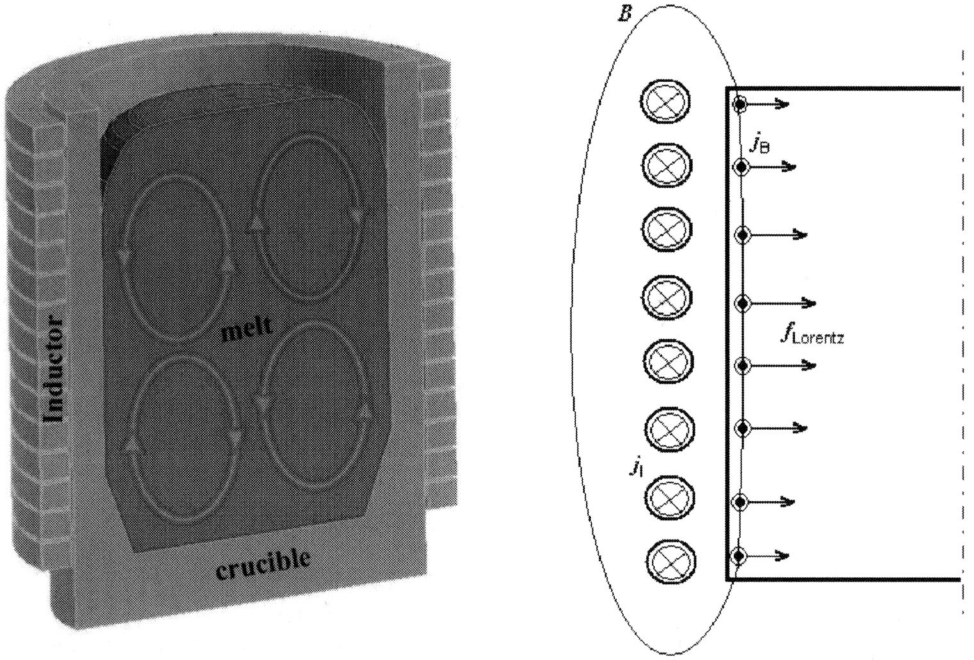

Figure 1. Sketch of axis-symmetric induction system on the example of induction crucible furnace.

The results produced by k-ε turbulence model, with a standard set of constants, show the highest values of turbulent kinetic energy in the eddy centers and the lowest between the eddies. Such distribution is characteristic for the k-ε model even in the case of 3D transient simulation [5]. Experimental results show that the maximum of the turbulent energy is between the vortices of the averaged flow and close to the wall of the crucible [1,5]. The average flow velocity in this area is close to zero, so turbulent pulsations of velocity are large there. It was shown [1,5] that the Large Eddy Simulation (LES) should be used for flow modeling in such equipment to achieve more realistic results. As far as the mass and heat transfer in the zone between the eddies depends on the resolution of the velocity pulsations, the LES has to be used for the objectives of this paper. The spectrum of the pulsations of the axial velocity between these eddies has a clear maximum, which corresponds to the oscillation with a period between 8 and 12 seconds depending on the inductor current [5] (these values are valid only for the particular installation that is simulated also in the present paper).

The flow inside induction furnaces is sufficiently complex due to the presence of the electromagnetic (EM) forces, which drive a turbulent flow with two main toroidal recirculated vortices. Therefore, the case described in this paper significantly differs from the well-studied turbulence in a pipe, a channel and other classical flows. It also differs from the turbulent flow in the gap between two counterrotating disks (the von Kármán flow) due to intensive axial pulsations between main eddies near the wall [1,5]. Therefore, the behaviour of the particles in such a kind of a flow is interesting from an engineering point of view (see paragraph 1.1) and also is still poor researched (see paragraph 1.2). Moreover, the EM field directly influences the non-conducting particles in the conducting liquid within the penetration depth and transport them to the wall, which is another specific aspect of considered system. The layer of significant EM field is sufficiently thin (about 20% of the radius of the crucible). Due to the non-slip boundary conditions the flow velocity is zero at the wall and increases dramatically in a radial direction until it achieves the maximal value. However, this maximum is in flows with high Reynolds numbers inside the layer of EM penetration. Obviously particles preferentially move in the streamline of maximal velocity. Therefore the major part of the particles frequently to the layer of significant EM field. The particle motion in turbulent flows without EM is well researched, but we can not separate the large interior zone without EM field from the thin layer near the wall and should consider the motion of the particles in the whole volume.

1.1. Motivation

EM heating and melting is one of the most effective methods for melting and processing of conducting materials. However there are some problems concerning the deposition of the impurities like dirty secondary metals as well as the erosion of the walls of the crucible. These processes can significantly reduce the efficiency of the equipment up to breakdown (e.g. channel clogging in CIF). This is one of the reasons for studying of the mass transfer processes in electroheat equipment.

In addition there is a problem of the homogenization of alloying particles, which are mixed in a steel melt to improve properties like strength, hardness and wear resistance. It is important to achieve homogeneous admixtures distribution to ensure a high quality of the alloy. Furthermore it is desirable to reduce the time of mixing to decrease the energy consumptions and prevent the melt from overheating. It is possible to investigate the optimal parameters (such as the size of alloying particles, the time of mixing and other parameters) for industrial processes on the basis of the numerical simulation.

The most impurities and alloying elements have higher melting temperatures than metal. Thereby the melt can contain admixtures as solid particles. However, metals often have very high melting temperatures and are not optical transparent. Therefore experimental investigations of particle motion and mass transfer processes are very difficult to achieve.

There is no experimental technique (known by authors) that gives a possibility to study the dynamics of the particle motion inside the melt. The only known method is developed by Taniguchi et al. [6,7], which gives the possibility to investigate experimentally the rate of the particle deposition in a turbulent flow of liquid metal under EM force. However, because the results are obtained by cutting solidified liquid, it is impossible to receive any information about the dynamics of the process inside the melt using such experimental technique. Moreover the presence of the solidification front has an influence on the particles during the solidification and it is not clear, if this effect is negligible.

Generally transparent liquids are poor electrical conductors, so thermal convection dominates in such systems. Therefore they can not be used to produce such flow patterns. On the other hand non-optical particle measurement techniques are not applicable due to the strong EM field around the equipment.

Therefore numerical simulations are an important way to study parameters of induction equipments and to investigate optimal parameters of admixture particles and thereby to improve energy efficiency of melting and homogenization processes. However, because of the absence of the experimental results, the only chance to verify the models is to compare the results with unpublished industrial reports, which contain some qualitative information.

1.2. Previous investigations

Particle behaviour has been widely investigated in turbulence without EM influence (see e.g. [8]). However, the experimental results received by Taniguchi et al. [6,7] show the significant influence of the EM field on the non-conductive particles in well-conductive liquid. In spite of probable inaccuracy of this experimental technique (a doubt about accuracy is already discussed above), the preferential concentration of the non-conductive particles near the wall at the height of the maximum of the EM force (1) corresponds to the industrial observations.

McKee et al. proposed the greatly simplified analytical model for the calculation of the non-conductive particle path in rotating magnetohydrodynamic flow within a long cylinder [9].

Kirpo et al. used Fluent software to investigate the behaviour of the particle cloud in the turbulent flows in induction furnaces [10]. This model was sufficiently rough: only drag, buoyancy and EM forces were taken into account for the calculations of the particle motion. The particles were initially distributed in a vertical plane, but in an industrial process alloying particles are injected on the top surface of the melt. However, general peculiarity of the particle behaviour was confirmed: the big non-conductive particles were settling down on the wall in the zone between the main vortices.

The model used by Kirpo was enchanced by Ščepanskis et al. by means of a development of *OpenFOAM* software code [11]: 1) the initial cloud of the particles was situated on the top surface of the melt, which corresponds to the industrial case of admixture injection, and 2) the Lagrange equation was supplemented with lift, acceleration and added mass forces. The calculations predicted the scheme of the particle motion during the industrial process of the admixture homogenization in the last paper. The general objective of the paper [11] was to investigate numerically process of the homogenization of non-conductive admixture particles in ICF, but not enough attention was paid to the analysis of the significance of all forces in the model. The importance of full set of the forces was proved as the separating trajectories of the particles with equal initial position but different set of forces. However the insignificant shift of the trajectory can lead to the significant changes in turbulent flows. Therefore evidently the statistical analysis of the forces and the non-dimensional parameters, which is done in the present paper, is important to substantiate the model for the particle motion for this industrially essential application.

It was also stated in [11] that the homogenization takes place between the zones of the main vortices and has an oscillating nature. The rate of the homogenization depends on the size and relative density of the particles. It was also observed that the processes, which occur in the middle zone near the wall, define the characteristic type of the particle motion. The balance between EM, drag and buoyancy forces defines the rate of the homogenization. The last paper contains only approximate theoretical analysis of the force influence on the process of homogenization, so it became clear that additional simulations should be carried out to investigate the contribution of different forces to the homogenization process. The present paper contains such results of numerical simulation.

2. Parameters of laboratory scale equipment and the mesh

A cylindrical ICF is a typical object for inductive melting processes. The sketch of an ICF is shown on the figure 1 and its parameters are shown on the table 1.

Table 1. Parameters of the laboratory scale ICF.

Parameter	Value
inductor frequency	365 Hz
inductor effective current	2000 A
melt and inductor height	570 mm
crucible radius	158 mm
gap between crucible and inductor	38.5 mm
number of inductor turns	12

As far as the flow consists of two main toroidal eddies, the size of the main eddy is approximately equal to the radius of the crucible. The calculations show that we have to use at least 5 cells per eddy to resolve these vortices. To simulate directly the vortices of one order of magnitude smaller we should refine the cells until its linear size meets a size of 3 mm (2% of the size of the main eddy). The size of the elements decreases near the wall (the width of the smallest element is 0.1 mm). The characteristic velocity of the flow in the main eddies is about 0.15 m/s [5], therefore the approximate time of the circulation is about 3.3 s. The LES calculation was carried out using a time step of 5 ms, which is 0.15% of the typical circulation time.

3. Mathematical model

The present model includes the simulation of the flow and the particle motion. The turbulent flow is calculated using the LES method with the isotropic Smagorinsky subgrid viscosity model:

$$\nu_{\mathrm{SGS}} = C_{\mathrm{k}} \cdot \sqrt{k} \cdot \delta, \qquad k = 2\frac{C_{\mathrm{k}}}{C_{\mathrm{e}}} \cdot \delta^2 \cdot \left|D_{ij}\right|^2, \qquad \delta = \sqrt[3]{V_{\mathrm{c}}},$$

where D_{ij} is the deformation tensor, V_{c} is the volume of the computational cell, $C_{\mathrm{k}} = 0.07$ and $C_{\mathrm{e}} = 1.05$. No additional boundary treatment is applied. The slip boundary condition (BC) on the top surface and the non-slip BC on the walls are applied.

The flow is driven by EM force and thermal buoyancy force in the Boussinesq approximation. The thermal BC are convective on the top surface and adiabatic on the walls of the crucible.

At each time step the results of the flow simulation gives the field of flow velocities, which is used for further calculation of the particle trajectories. The Lagrange equation describes the motion of the inertial particles. In the LES framework we know only the filtered velocity and assume the isotropy of the subgrid part. A statistical approach to the analysis of the force distribution will be discussed below. In spite of the calculation of the individual trajectories of particles, the analysis will be done for the cloud, so the unfiltered isotropic part of velocity should not significantly influence the statistical results. As far as the homogenization criterion is also based on the statistics, the subgrid part should not affect the results of the homogenization process too.

3.1. Lagrange equation

Let us first consider a movable spherical particle in a still liquid. If we solve the stationary Navier-Stokes equation, we will obtain the force that acts on the spherical particle (the Stokes force) as follows:

$$\frac{1}{m}\boldsymbol{F} = \frac{18\nu}{d^2}\frac{\rho_{\mathrm{f}}}{\rho_{\mathrm{p}}} \cdot \left(\boldsymbol{u}_{\mathrm{f}} - \boldsymbol{u}_{\mathrm{p}}\right), \tag{2}$$

where m is the mass of the particle, ν is the kinematic viscosity of the liquid, d is a diameter of the particle, ρ_{f} and ρ_{p} are liquid and particle densities respectively, $\boldsymbol{u}_{\mathrm{f}}$ and $\boldsymbol{u}_{\mathrm{p}}$ are fluid and particle velocities respectively. If we solve the non-stationary Navier-Stokes equation, then the force (2) will be generalized by adding the non-stationary terms [12,13]:

$$\frac{1}{m}\boldsymbol{F} = \frac{1}{2}\frac{\rho_{\mathrm{f}}}{\rho_{\mathrm{p}}}\frac{\mathrm{d}\left(\boldsymbol{u}_{\mathrm{f}} - \boldsymbol{u}_{\mathrm{p}}\right)}{\mathrm{d}t} + \frac{\mathrm{d}\boldsymbol{u}_{\mathrm{f}}}{\mathrm{d}t} + \frac{18\nu}{d^2}\frac{\rho_{\mathrm{f}}}{\rho_{\mathrm{p}}} \cdot \left(\boldsymbol{u}_{\mathrm{f}} - \boldsymbol{u}_{\mathrm{p}}\right) + \frac{9}{d} \cdot \left(\frac{\nu}{\pi}\right)^{1/2} \frac{1}{\rho_{\mathrm{p}}} \cdot \int_0^t \frac{\mathrm{d}\left(\boldsymbol{u}_{\mathrm{f}} - \boldsymbol{u}_{\mathrm{p}}\right)}{\mathrm{d}\tau}\frac{\mathrm{d}\tau}{\sqrt{t-\tau}}. \tag{3}$$

We can compare this result to the result obtained for a potential liquid and generalize it by replacing the simple derivation $\mathrm{d}\boldsymbol{u}_{\mathrm{f}}/\mathrm{d}t$ with the material derivation $\mathrm{D}\boldsymbol{u}_{\mathrm{f}}/\mathrm{D}t$ [13,14].

The contribution to the force made by the summand, that contains the integral, is not so significant because of the oscillating flow acceleration [15]. In addition it is complex for numerical modeling and will be neglected.

It is necessary to note that the flow in our case is not the Stokes flow. Therefore the drag coefficient C_{D} [16] should correct the third summand in (3) and the first summand should be multiplied by the acceleration coefficient C_{A} [17]:

$$C_{\mathrm{D}} = \frac{18\nu}{d^2}\frac{\rho_{\mathrm{f}}}{\rho_{\mathrm{p}}}\left(1+0.15\cdot \mathrm{Re}_{\mathrm{p}}^{0.687}\right), \qquad C_{\mathrm{A}} = 2.1 - \frac{0.132}{0.12+Ac^2}, \tag{4}$$

where $\mathrm{Re}_{\mathrm{p}} = dU/\nu$ is the particle Reynolds number, $Ac = U^2 d^{-1}(\mathrm{d}U/\mathrm{d}t)^{-1}$ is acceleration parameter, $U = u_{\mathrm{f}} - u_{\mathrm{p}}$ is relative particle velocity. It is clear from (4) that if $\mathrm{Re}_{\mathrm{p}} > 1$, than C_{D} should be expressed in Schiller-Naumann form.

As it was mentioned above some additional forces should be taken into account in the Lagrangian equation.

Lift force appears due to the flow circulation around the particle [18]:

$$\frac{1}{m}\boldsymbol{F}_{\mathrm{lift}} = \frac{\rho_f}{\rho_p}\cdot C_{\mathrm{L}}\cdot[\boldsymbol{U}\times[\nabla\times\boldsymbol{U}]] \equiv \frac{\rho_f}{\rho_p}\cdot C_{\mathrm{L}}\cdot\boldsymbol{\xi}, \tag{5}$$

where C_{L} is lift coefficient that depends on Re_{p} and non-dimensional shear stress $Sr = |\partial \boldsymbol{u}_{\mathrm{f}}/\partial \boldsymbol{s}|\cdot d/U$, where $\boldsymbol{s}$ is a unit vector perpendicular to $\boldsymbol{U}$. Legendre & Magnaudet [19] generalized the analysis performed by Saffman [20] and McLaughlin [21] for the flows with the low particle Reynolds number, and C_{L} can be approximated as follows (the Saffman approximation) [18,19,21]:

$$C_{\mathrm{L}}^{\mathrm{low\,Re}}(\mathrm{Re}_{\mathrm{p}}, Sr) = \frac{6}{\pi^2}\frac{1}{\sqrt{\mathrm{Re}_{\mathrm{p}}\cdot Sr}}J'(\varepsilon), \qquad J'(\varepsilon) = \frac{J(\infty)}{\left(1+0.2\cdot\varepsilon^{-2}\right)^{3/2}}, \qquad \varepsilon = \left(\frac{Sr}{\mathrm{Re}_{\mathrm{p}}}\right)^{1/2},$$

where $J(\infty) = 2.255$. This approximation is also known as the Saffman force. Legendre & Magnaudet propose following approximation for the high particle Reynolds numbers [18]:

$$C_{\mathrm{L}}^{\mathrm{high\,Re}}(\mathrm{Re}_{\mathrm{p}}) = \frac{1}{2}\frac{1+16\cdot\mathrm{Re}_{\mathrm{p}}^{-1}}{1+29\cdot\mathrm{Re}_{\mathrm{p}}^{-1}}.$$

For the particle Reynolds number between the approximation (McLaughlin-Legendre-Magnaudet) is as follows [18]:

$$C_{\mathrm{L}}(\mathrm{Re}_{\mathrm{p}}, Sr) = \left(\left(C_{\mathrm{L}}^{\mathrm{low\,Re}}(\mathrm{Re}_{\mathrm{p}}, Sr)\right)^2 + \left(C_{\mathrm{L}}^{\mathrm{high\,Re}}(\mathrm{Re}_{\mathrm{p}})\right)^2\right)^{1/2}. \tag{6}$$

As far as C_{L} depends on two parameters (Re_{p} and Sr), it is useful to plot down these dependences (figure 2). The range of Re_{p} and Sr for the particular case will be obtained below and the suitable approximation will be chosen.

The densities ρ_{p} and ρ_{f} are different, so we have to take into account the buoyancy force. As it was mentioned above we should also include the EM force in the equation. A great transition resistance appears on the surface between the particles and the metal, hence, we can consider force acting on the non-conducting particles, which can be expressed as follows [22]:

$$\frac{1}{m}\boldsymbol{F}_{\mathrm{em}} = -\frac{3}{4}\frac{1}{\rho_{\mathrm{p}}}\boldsymbol{f}_{\mathrm{em}}.$$

Thus taking into account all above-mentioned forces, the following Lagrange equation for the non-conductive spherical particle motion is used:

$$\underbrace{\left(1+\frac{C_{\mathrm{A}}}{2}\frac{\rho_{\mathrm{f}}}{\rho_{\mathrm{p}}}\right)\cdot\frac{\mathrm{d}\boldsymbol{u}_{\mathrm{p}}}{\mathrm{d}t}}_{\mathrm{d}\boldsymbol{u}_{\mathrm{p}}/\mathrm{d}t+\text{added mass force}} = \underbrace{C_{\mathrm{D}}\cdot\boldsymbol{U}}_{\text{drag force}} + \underbrace{\left(1-\frac{\rho_{\mathrm{f}}}{\rho_{\mathrm{p}}}\right)\cdot\boldsymbol{g}}_{\text{buoyancy force}} \underbrace{-\frac{3}{4}\frac{1}{\rho_{\mathrm{p}}}\boldsymbol{f}_{\mathrm{em}}}_{\text{EM force}} + \underbrace{\frac{\rho_{\mathrm{f}}}{\rho_{\mathrm{p}}}C_{\mathrm{L}}\boldsymbol{\xi}}_{\text{lift force}} + \underbrace{\left(1+\frac{C_{\mathrm{A}}}{2}\frac{\rho_{\mathrm{f}}}{\rho_{\mathrm{p}}}\right)\cdot\frac{\mathrm{D}\boldsymbol{u}_{\mathrm{f}}}{\mathrm{D}t}}_{\text{acceleration+added mass}},$$

where vector $\boldsymbol{\xi}$ is defined in (5).

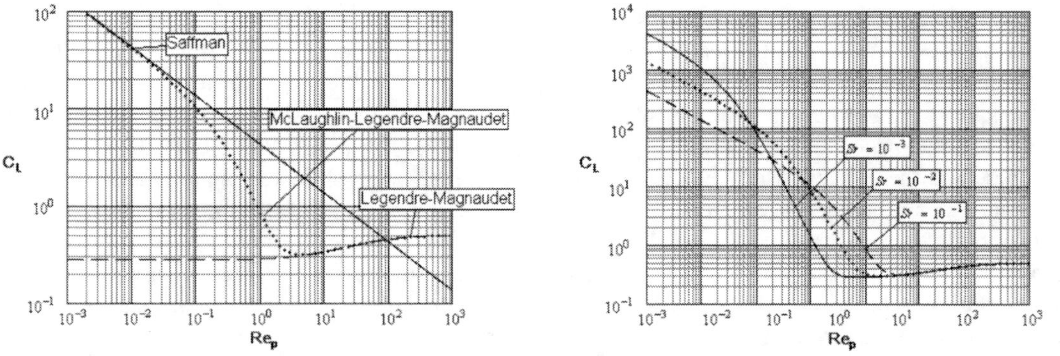

(a) (b)

Figure 2. (a) Different approximations for the lift coefficient C_L (here $Sr = 0.1$); (b) McLaughlin-Legendre-Magnaudet approximation for lift coefficient C_L for different Sr.

We assume that the particle-particle interaction is negligible and the particles do not affect the pattern and the velocities of the flow. This assumption is possible because the volume of the particles do not exceed 1% of the liquid volume.

The particle tracking library in *OpenFOAM* software is supplemented with EM, lift, acceleration and added mass forces. Thereby the Lagrange equation becomes non-linear. To ensure the convergence the hydrodynamic time step was split into various Lagrange time steps (LTS). The non-linear factors in the Lagrange equation are approximated with the values that correspond to the previous LTS (Pikar's method). The implicit scheme is used to solve the equation.

3.2. Particle collision with the wall

The standard *OpenFOAM* algorithm simulates a particle collision with the wall as follows:

$$u_p^{n\,'} = \varepsilon \cdot u_p^n,$$

$$u_p^{t\,'} = (1-\mu)\cdot u_p^t,$$

where u_p^n and u_p^t are normal and tangential components of the particle velocity before collision; $u_p^{n\,'}$ and $u_p^{t\,'}$ are normal and tangential components of the particle velocity after collision; The coefficients of restitution and friction have been set to $\varepsilon = 0.8$ and $\mu = 0.2$ in the present simulation.

However the *OpenFOAM* algorithm does not take into account the collision of a particle with the wall in a situation when the size of the particle is larger than the size of a mesh element. If the distance between the particle center and the wall is smaller than the radius, than the present model moves the particle to the distance of the radius from the wall at the end of the each LTS.

4. Statistical analysis of the forces

Particles with a diameter less than 300 μm are well admixed in the melt of the laboratory scale ICF (the parameters of the furnace are shown on the table 1), but larger particles conglomerate at the wall in the middle zone of the crucible due to relative increase of EM force per particle mass unit. It should be underlined that this limit in a particle size and following results are relevant only for ICF of the mentioned laboratory scale, while the obtained dependences should be universal. Mentioned effect can be prevented by increasing the flow velocity (see theoretical investigation in [11]). It is also clear that the size of the main eddies influences the balance of the forces in the middle zone. If the linear velocity of the circular motion is invariable, than the greater is the area of the vortex and the lower is the affect of centrifuging the light particles to the center of the eddy [8]. Therefore the EM force, which moves the particles to the wall, will be more competitive with respect to the centrifuging effect for the large vortices. The ratio of the liquid and particle densities $S = \rho_f / \rho_p$ affects the particle motion likewise through the EM/centrifuging balance. The dependence of the flow velocity on inductor current and frequency is also known [1]. Apparently there are a lot of dimensional parameters, an attempt to create a composite non-dimensional parameter was not successful. The EM parameters and geometry of the equipment are fixed in the present paper, subsequently the size of the main eddies and the characteristic velocity are also defined; the purpose of the paper is to investigate the dependence on the particle size and their density of the distribution of the forces in ICF of such laboratory scale.

The aim of the following paragraphs is to determine the distribution of the forces that has influence on the particles with different sizes and densities in whole volume. We inject 75 particles in a horizontal plane near the top surface of the crucible. During the period of 10 s the initial cloud of the particles is distributed more or less homogeneous inside the zones of the main vortices (see figure 1), but the number of particles is different between the zones of the upper and the lower main eddies [11]. So we skip this transitional period and start to collect data after that. The values of non-dimensional parameters and the magnitude of the forces are saved for each particle every 50 ms for a period of 20 s, which corresponds approximately to 6 periods of a particle circulation in the main vortex. Therefore the statistical database consists of 75 x 20 / 0.05 = 30000 information units. Thereby we assume that we collect the data from the whole volume and for all states of quasistationary turbulence.

First the statistical analysis of the non-dimensional parameters ($\mathrm{Re_p}$, Sr and Ac) will be done to choose the approximation of the forces. Then the distribution of the force densities will be evaluated.

The particles with the liquid-to-particle density ratio $S = \rho_f / \rho_p$ equal to 1 and 1.5 and the diameter d from 10 μm to 200 μm (that is from 6 ppm to 0.1% of the radius of the main vortex) are analyzed.

4.1. Non-dimensional parameters and approximation of the forces

The acceleration parameter $Ac = U^2 d^{-1} (\mathrm{d}U/\mathrm{d}t)^{-1}$ for all examined cases is between 10^2 and 10^5. Thereby the acceleration coefficient (4) can be simply approximated as $C_A = 2.1$.

The half width of the calculated statistical distribution of the particle Reynolds number $\mathrm{Re_p} = dU/\nu$ is shown on the figure 3. We can see, that the distribution of $\mathrm{Re_p}$ is more or less invariable for the particles with a density equal to the liquid density (figure 3b), except particles with

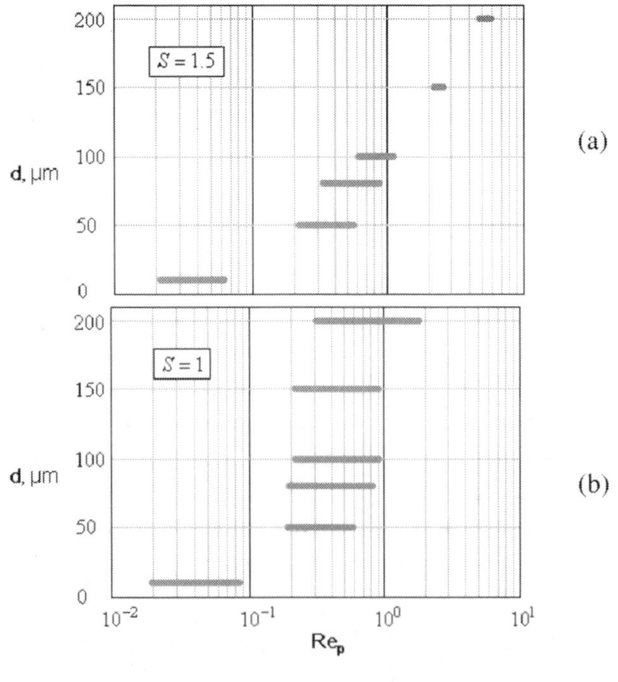

Figure 3. The half width of the calculated statistical distribution of $\mathrm{Re_p}$. (a) $S = \rho_f / \rho_p = 1.5$; (b) $S = 1$.

small diameter - 10 μm. For the particles with $S = 1.5$ $\mathrm{Re_p}$ increases with the particle increase in the size. This increase is connected with the significant role of buoyancy force that produces an axial velocity and will be explained below. For $S = 1.5$ and $d > 100\,\mu\mathrm{m}$ $\mathrm{Re_p} > 1$, therefore the Schiller-Naumann approximation (4) for drag coefficient should be used, in other cases the Stokes form for C_D is applicable.

The calculations show that the non-dimensional shear stress Sr varies between 10^{-3} and 10^{-1}. Taking also into account the figure 3 and the behaviour of the lift coefficient (figure 2), we can conclude that C_L should be defined within the McLaughlin-Legendre-Magnaudet approximation (6).

4.2. Forces

Apparently EM force is localized near the wall due to the limited penetration depth of the EM field. Therefore there is no reason to investigate this force statistically.

The half width of the statistical distribution of drag, lift, acceleration and added mass force densities for the particles with the diameters from 10 μm to 200 μm and $S = 1;1.5$ is shown on the figure 4. The mentioned densities approximately correspond e.g. to the densities of steel – niobium and steel – grey tin alloys respectively.

We can notice that added mass and acceleration force densities are relatively independent of the particle size in the both cases ($S = 1$ and $S = 1.5$). On the other hand, the lift force density decreases with the increase of the particle size. The behaviour of drag force density is different for various S. It decreases with the decrease of the particle diameter like the lift force density for the case $S = 1$. But in the case $S = 1.5$ drag force density decreases only until it achieves the value of buoyancy force density, and then it still remains distributed approximately around the buoyancy force point. The buoyancy leads to the axial motion and moves the particles to the surface. In the most cases the drag force opposes the buoyancy and mixes the particle deeper in the melt (e.g. in the zone of the upper eddy near the wall). As far as the drag coefficient C_D decreases with an increase of the particle size (4), buoyancy force and axial motion become relatively stronger for the larger particles. So the particles axial velocity increases and consequently the difference between the flow and particle velocities $\boldsymbol{U} = \boldsymbol{u}_\mathrm{f} - \boldsymbol{u}_\mathrm{p}$ also increases. Thereby we can observe such a coupling of the drag and buoyancy force for the big light particles.

Figure 4 shows that for the particles with $d \geq 80\,\mu\mathrm{m}$ and $S = 1$ we should take into account all forces, but for $d < 80\,\mu\mathrm{m}$ only drag and lift forces. For the particles with $S = 1.5$ and $d \leq 100\,\mu\mathrm{m}$ drag, lift and buoyancy forces should be taken into account, but for the particles with $d \geq 150\,\mu\mathrm{m}$ only drag and buoyancy forces dominate. So we find for the separating size: $d = 80\,\mu\mathrm{m}$ for $S = 1$ and $d = 100\,\mu\mathrm{m}$ for $S = 1.5$. These values separate the sizes, when different sets of forces are applicable. Apparently the separating size for $S > 1$ rises with an increase of S.

Generally the significance of acceleration and added mass forces decreases with the increase of S. The lift force also becomes unimportant for big particles. These regularities should be true also for such systems with the different parameters.

5. Modelling of particle homogenization

After the statistical estimation of the force distribution in the flow of the laboratory scale ICF is done, it is interesting to analyze their influence on the macroscopic particle transportation process such as the process of the alloying particle homogenization in ICF. Generally the homogenization process is already described in [11], but the aim of this paragraph is to study the influence of the different forces on the particle distribution during the homogenization process. Apparently the theoretical analysis of the particle behaviour in the middle zone of the crucible [11] is not enough to understand the role of the forces during this process. Thereby the simulations with different sets of forces were carried out and the results are discussed below.

Like it was done in [11], the homogenization process in the present paper is analyzed by plotting the values of $\Delta N/N$, where ΔN is the difference between the number of the particles in the zones of the upper and the lower eddies (see figure 5) and N is the number of the particles in the both zones. The magnitude of the average axial velocity is low in the middle zone of the crucible near the wall, but the turbulent pulsations have a maximum between the eddies [1]. Thus the intensive turbulent

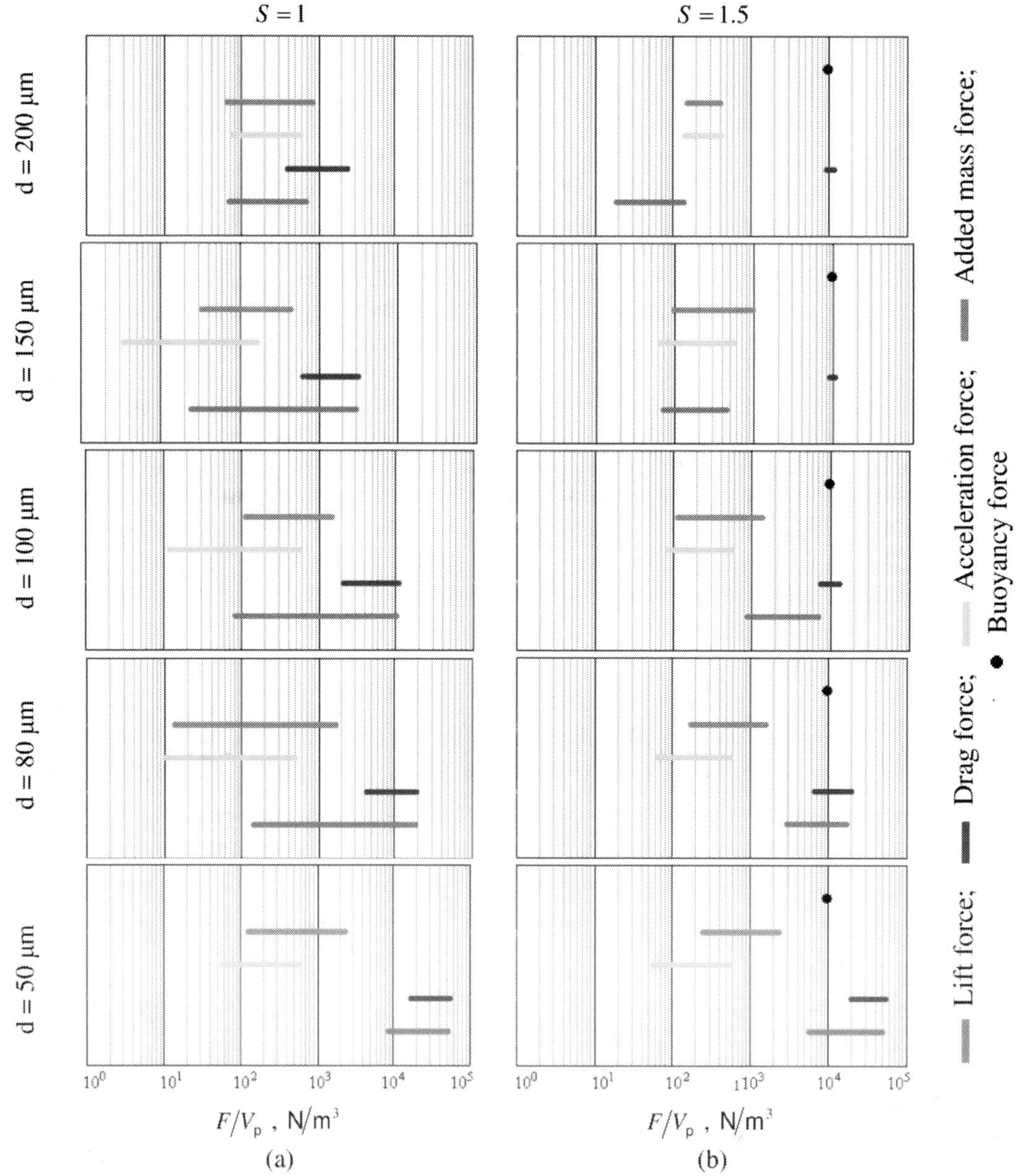

Figure 4. The half width of the calculated statistical distribution of the drag, lift, acceleration and added mass force densities F/V_p, where V_p is particle volume, for the particles with (a) $S = \rho_f/\rho_p = 1$, (b) $S = 1.5$.

pulsations in this area improve the particle exchange and the decrease of ΔN. The results of the simulation show that the radial and the angular distributions of the particles inside the crucible rapidly homogenize (except the thin layer near the wall). Therefore $\Delta N/N$ parameter was chosen for the analysis of the homogenization process.

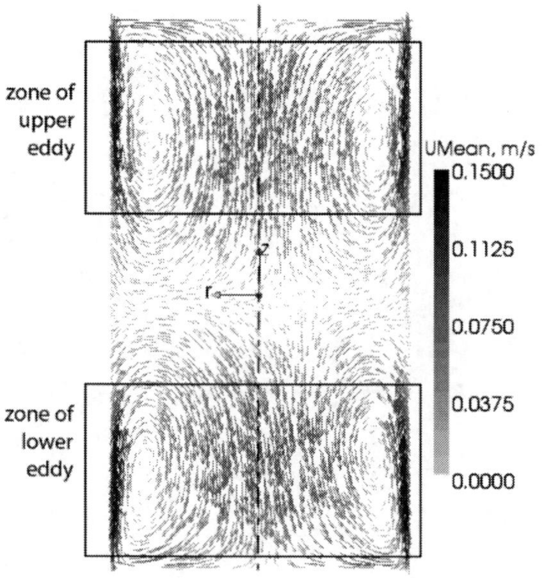

Figure 5. Average flow velocity in the central vertical plane in the ICF. The frames show the zones of the upper and the lower eddies. [11]

It has been already discussed [11] that in spite of the quasistationary flow pattern, the behaviour of the homogenization factor $\Delta N/N$ in time is not quasistationary, but it monotonously decreases. The exchange of the particles between the eddies takes place due to the axial flow pulsations. So each particle in the zone of the upper eddy can reach the lower eddy in only one time per period of the main circle motion. Consequently the process of homogenization is connected with the period of the particle circulation which is approximately 3.3 s (see paragraph 2). The figure 6 illustrates the process of the homogenization during the period of 50 s. It means that the period of the monotonous homogenization on the figure 6 is about 40 s (first 10 s can be counted out as the transitional period), that is about 12 periods of the main circular motion. $\Delta N/N$ tends asymptotically to a constant value which is zero for the case $S=1$ and to some positive value if $S>1$ [11]. $\Delta N/N$ reaches the asymptotic value in approximately 55 s if $S=1$ and in 75 s if $S=1.5$ [11]. However, the figure 6 shows that the absence of some forces in the model leads to a shift of the curve of $\Delta N/N$ and consequently to the misrepresentation of the time for achieving the asymptotic value.

We can see that for the particles with $d=50\mu\text{m}$ the difference between the curves that corresponds to the diverse sets of the forces is not so significant. On the contrary the curves for $d=100\mu\text{m}$ show a significant divergence. The largest divergence is between the curve that takes into account all forces and the curve without acceleration and added mass forces. This difference increases with an increase of the particle size and leads to a time delay in the process of the homogenization. At the initial moment, when the unmovable particle cloud should be accelerated, acceleration and added mass forces play a great role. Therefore the process of the homogenization is slower without acceleration and added mass forces. The absence of lift force also changes the curve but in the contrast with acceleration forces, it speeds up the homogenization. The divergence between the curve without lift force and the curve with all forces decreases with an increase of S.

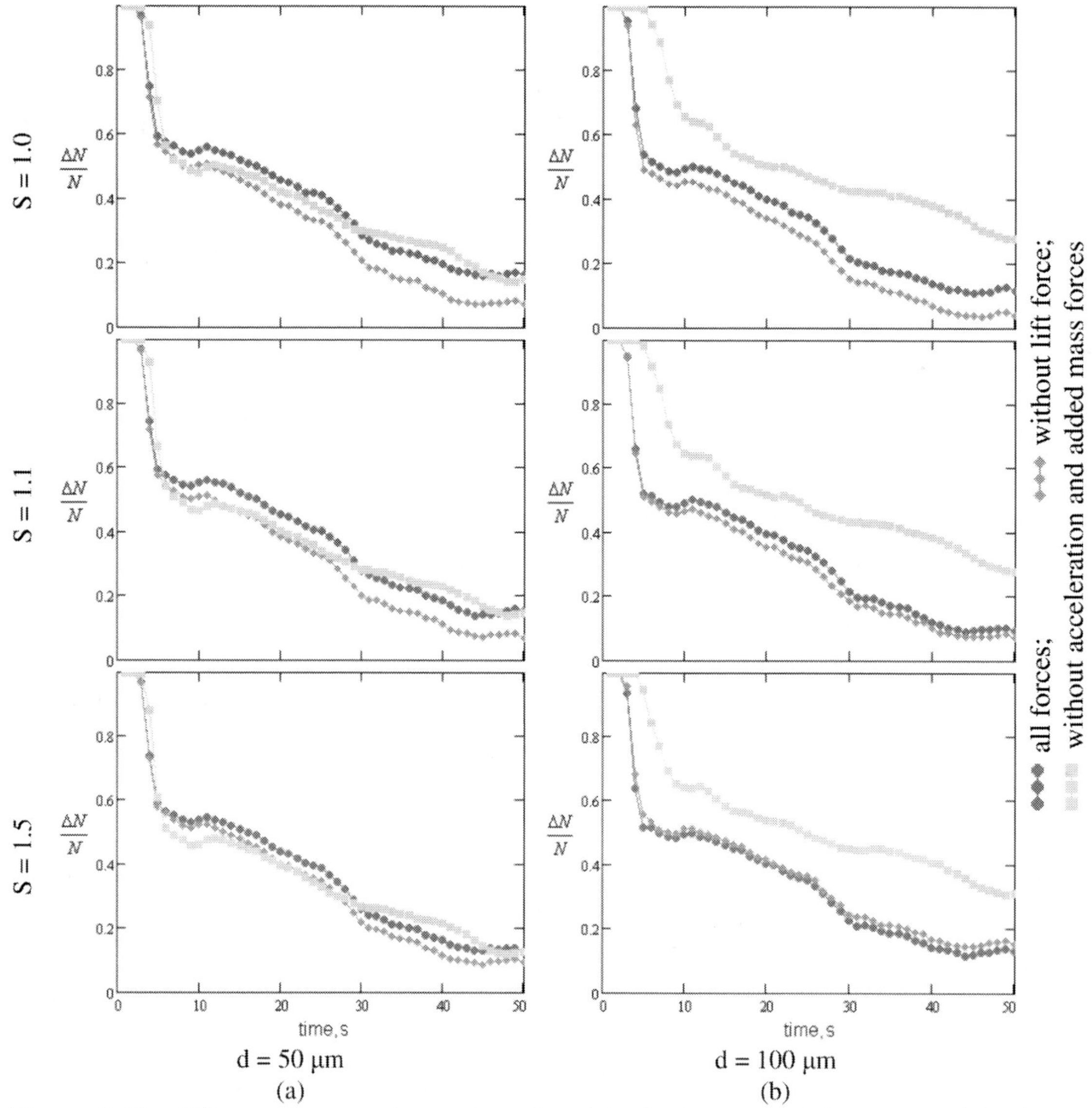

Figure 6. The difference between the number of particles in the zones of the upper and the lower eddies normalized with the respect to the number of the particles in the both zones $\Delta N/N$. (a) $d = 50\,\mu\text{m}$, (b) $d = 100\,\mu\text{m}$

So the simulation of the homogenization process shows that the absence of the forces in the Lagrange model also influences the process of the redistribution of the particles in the volume of the recirculated flow. And the separating values of the particle size for the significance of the forces are approximately equal to the separating values discovered in the paragraph 4.2.

6. Conclusions

The analysis of the significance of the different forces in the Lagrange equation for particle motion and the homogenization process is done in this paper on the basis of the statistical research. The present work proves the significance of drag, lift, EM, acceleration and added mass forces and forms the Lagrange equation that should be used for modelling of the particle transfer in the flow inside an ICF. However in the particular cases of a laboratory scale ICF we can use the short list of the forces:

$S = \rho_f / \rho_p = 1$, $d < 80\,\mu m$ - drag, lift and EM forces;

$S = 1.5$, $d \leq 100\,\mu m$ - drag, lift, buoyancy and EM forces;

$S = 1.5$, $d > 100\,\mu m$ - drag, buoyancy and EM forces.

Generally the significance of acceleration and added mass forces decreases with an increase of S. Lift force also becomes unimportant for large particles. The Schiller-Naumann approximation for the drag coefficient, the McLaughlin-Legendre-Magnaudet approximation for lift force and the extreme case of the Odar-Hamilton approximation for the acceleration coefficient should be chosen in the Lagrange equation.

The presence of these forces is also important for the simulation of the homogenization process. The absence of acceleration and added mass forces increases significantly the homogenization time for particles with $d > 50\,\mu m$. The presence of lift force becomes important when $S \approx 1$.

In spite of the particular relevance of the recommendations above (valid only for a laboratory scale ICF), the general dependences should be similar for different parameters of ICFs. Moreover the authors hope that the present recommendations for the Lagrange model will be useful in the engineering studies.

7. Acknowledgement

This work has been partly supported by the European Social Fund within the project «Support for Doctoral Studies at University of Latvia».

References

[1] Kirpo M, Jakovičs A, Baake E and Nacke B 2007 Analysis of experimental and simulation data for the liquid metal flow in a cylindrical vessel *Magnetohydrodynamics* **43** 161-72

[2] Baake E, Jakovičs A, Pavlovs S and Kirpo M 2010 Long-term computations of turbulent flow and temperature field in the induction channel furnace with various channel design *Magnetohydrodynamics* **46** 461-74

[3] El-Kaddah N, Szekeley J, Taberlet E and Fautrelle Y 1986 Turbulent recirculating flow in induction furnaces: a comparison of measurements with predictions over a range of operating conditions *Metallurgical Transactions B* **17B** 687-93

[4] Bojarevics A, Bojarevics V, Gelfgat Yu M and Pericleous K 1999 Liquid metal turbulent flow dynamics in a cylindrical container with free surface: experiment and numerical analysis *Magnetohydrodynamics* **35** 205-22

[5] Umbrashko A, Baake E, Nacke B and Jakovics A 2006 Modeling of the turbulent flow in induction furnaces *Metallurgical and Materials Transactions B* **37B** 831-8

[6] Taniguchi S and Brimacombe J K 1994 Application of pinch force to the separation of inclusion particles from liquid steel *ISIJ International* **34** 722-31

[7] Takahashi K and Taniguchi S 2003 Electromagnetic separation of non-metallic inclusion from liquid metal by imposition of high frequency magnetic field *ISIJ International* **43** 820-7

[8] Toschi F and Bodenschatz E 2009 Lagrangian properties of particles in turbulence *Annu. Rev. Fluid Mech.* **41** 375-404

[9] McKee S, Watson R, Cuminato J R and Moor P 1999 Particle-tracking within-turbulent cylindrical electromagnetically-driven flow *Int. J. Numer. Meth. Fluids* **29** 59-74

[10] Kirpo M, Jakovičs A, Baake E and Nacke B 2009 LES study of particle transport in turbulent recirculated liquid metal flows *Magnetohydrodynamics* **45** 439-50

[11] Ščepanskis M, Jakovičs A and Nacke B 2010 Homogenization of non-conductive particles in EM induced metal flow in a cylindrical vessel *Magnetohydrodynamics* **46** 413-23

[12] Basset A B 1888 *A treatise on hydrodynamics with numerous examples* vol 2 (Cambridge: Deighton, Bell and Co)

[13] Brennen C E 2005 *Fundamentals of multiphase flow* (New York: Cambridge University Press)

[14] Crowe C T, Sommerfeld M and Tsuji Y 1998 *Multiphase flows with droplets and particles*

(Boca Raton, Florida: CRC Press)
[15] Kirpo M, Jakovičs A, Baake E and Nacke B 2006 Modeling velocity pulsations in a turbulent recirculated melt flow *Magnetohydrodynamics* **42** 207-18
[16] Schiller L and Naumann Z 1933 Uber die grundlegenden Berechnungen bei der Schwerkrafttaufbereitung (in German) *Ver. Deut. Ing.* **77** 318-20
[17] Odar F and Hamilton W S 1964 Forces on a sphere accelerating in a viscous fluid *J. Fluid Mech.* **18** 302-14
[18] Legendre D and Magnaudet J 1998 The lift force on a spherical bubble in a viscous linear shear flow *J. Fluid Mech.* **368** 81-126
[19] Legendre D and Magnaudet J 1997 A note on the lift force on a bubble or a drop in a low-Reynolds-number shear flow *Phys. Fluids* **9** 3572-4
[20] Saffman P G 1965 The lift on a small sphere in a slow shear flow *J. Fluid Mech.* **22** 385-400
[21] McLaughlin J B 1991 Inertial migration of small sphere in linear shear flows *J. Fluid Mech.* **224** 261-74
[22] Leenov D and Kolin A 1954 Theory of electromagnetophoresis I Magnetohydrodynamic forces experienced by spherical and symmetrically oriented cylindrical particles *J. of Chem. Phys.* **22** 683-89

Search for possible relationship between volcanic ash particles and thunderstorm lightning activity

A Várai[1], M Vincze[1], J Lichtenberger[2], and I M Jánosi[1]

[1] Department of Physics of Complex Systems;
[2] Space Research Group, Eötvös Loránd University, Pázmány Péter sétány 1/A, H-1117 Budapest, Hungary

E-mail: janosi@lecso.elte.hu

Abstract. Explosive volcanic eruptions that eject columns of ash from the crater often generate lightning discharges strong enough to be remotely located by very low frequency radio waves. A fraction of volcanic ash particles can stay and disperse long enough to have an effect on weather phenomena days later such as thunderstorms and lightnings. In this work we report on lightning activity analysis over Europe following two recent series of volcanic eruptions in order to identify possible correlations between ash release and subsequent thunderstorm flash frequency. Our attempts gave negative results which can be related to the fact that we have limited information on local atmospheric variables of high enough resolution, however lightning frequency is apparently determined by very local circumstances.

1. Introduction

The atmosphere below 50 km is conducting because of the presence of ions created primarily by cosmic rays and background radioactivity. The contribution of free electrons in this region can be neglected due to their short lifetimes [1, 2]. Significant charge separation in the lower atmosphere appears to require strong updrafts which carry water droplets upward causing at the cloud tops the accumulating of increasing positive charge. The exact mechanism by which charge separation happens is still a subject of research [3, 4, 5, 6]. Similarly, the mechanism by which the lightning discharge begins is not well known [8, 9]. Measured electric fields in thunderclouds are typically not large enough to directly initiate a discharge [1, 2].

The explanation of lightning activities is further complicated by the fact that very similar meteorological situations, such as tropical cyclones of the same strengths and trajectories, can yield to extremely different flash rates [10, 11]. Nevertheless the increasing efficiency of remote detection on global scales provides a key to a better understanding of many aspects of related phenomena [12, 13, 14, 15].

One of the existing global collaborations is the World Wide Lightning Location Network (WWLLN, see http://wwlln.net) operating over 50 sensors at universities and research institutes. The physical basis of remote observation is the strong electromagnetic emission over a wide spectrum coupled to a lightning discharge. Radio waves in the Very Low Frequency (VLF) range (3–30 kHz) are damped very slowly in the atmosphere, therefore a network of synchronized detectors can locate remote lightning strikes when the vertical electric field exceeds the background noise [16].

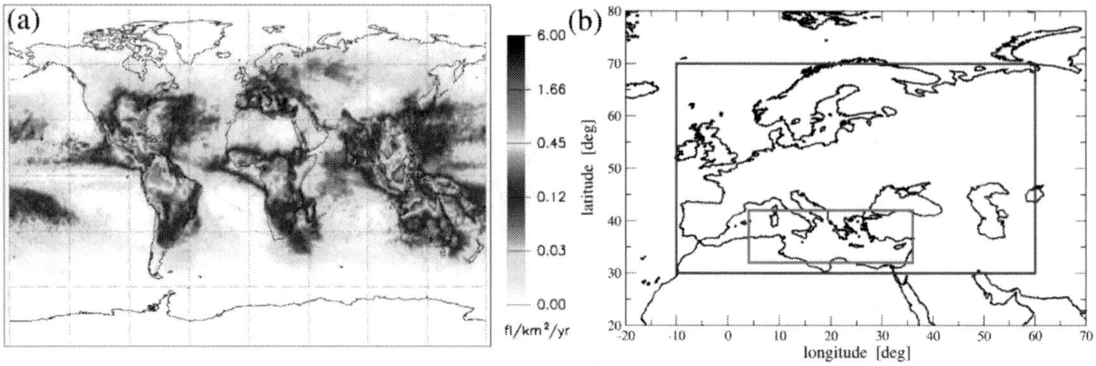

Figure 1. (a) Global mean lightning activity determined from WWLLN records, in the units of flash-number/km^2/year. (b) Geographic location of the test areas: Mediterranean Sea (MS, red rectangle, 32-42°N, 4-36°E), and the continental Europe (EU, blue rectangle, 30-70°N, 10°W-60°E).

It is well known that volcanic eruptions that eject columns of ash above a critical height (approximately 7 km) from the crater often generate strong lightning activity [1, 2, 17]. This observation motivated to initiate the WWLLN Ash Cloud Monitor experiment (http://wwlln.net/volcanoMonitor.html), where the lightning activity is evaluated in every minute over more than 1500 active volcanos. The primary goal of this project is to monitor volcanic activities and issue early alerts (e.g., about ash clouds endangering aviation) when it is necessary.

Besides triggering lightning discharges in the volcanic ash clouds, small aerosol particles staying longer in the troposphere can also change lightning activities over distant areas, too. In the present work we attempt to find evidences supporting this hypothesis. Two recent eruptions are analyzed over Europe. The main finding in both cases is that an apparent increase in the integrated lightning frequency over a given area cannot be explained directly by the possible presence of ash clouds, other factors appear to have stronger effects resulting in marked spatial and temporal fluctuations.

2. Data and methods

Global WWLLN lightning data are evaluated in the period of 28/10/2004 – 30/06/2010. The annual mean flash rate has a very uneven geographic distribution illustrated clearly in Fig. 1a. Approximately 70 % of lightnings is located in the tropics where the majority of thunderstorms occur. In spite of the common perception, only 20-25% is the so called cloud-to-ground flash, the rest is either cloud-to-cloud or intra-cloud discharge.

In order to evaluate possible effects of volcanic ash on lightning frequency, we concentrated on a limited area over the continental Europe. The reason is clearly seen in Fig. 1a: Europe has a moderate exposure compared to the tropical regions, nevertheless the strongest lightning activity is around the Mediterranean Sea (MS) [18]. Therefore a statistical comparison is performed over two rectangles: one covers the whole geographic area of Europe, a smaller embedded region is located to the Mediterranean, see Fig. 1b.

The time series shown in Fig. 2 exhibit a marked seasonal behavior, as expected at mid-latitude climate. The most active thunderstorm period is the middle of Summer (June-July) over the continent, and a characteristic delay is apparent over the MS region (October-November). Orange ellipses in Fig. 2 indicate unusually strong lightning activity over the continent during

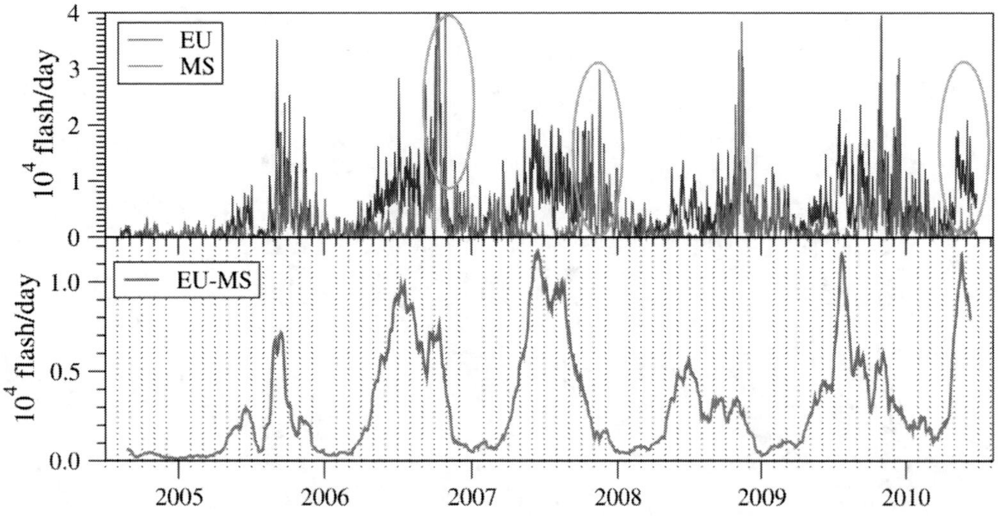

Figure 2. Top: time series of the lightning activity detected by the WWLLN network over the two rectangles denoted in Fig. 1b, see the legends. Bottom: the difference between daily flash numbers of the EU and MS data smoothed by a 31-day running mean procedure. Orange ellipses indicate the periods of closer analysis. (Vertical dotted lines denote the end of months.)

Autumn months in 2006 and 2007, and a very sharp increase in Spring of 2010. Here we pose the question whether this marked increase could be related to volcanic eruptions occurring shortly before these periods, or it is only an illusory temporal correlation.

3. Case study I: Mt. Etna, 2006-2007

One of the biggest producers of volcanic plumes in the world is Mount Etna in Sicily, Italy [19, 20]. In the year of 2006, a longer eruption lasted from 31 August to 15 December, and the ash spread from the source area to a distance of several kilometers away from the volcano, representing a serious danger for aviation. Andronico *et al.* published a very detailed analysis on ash plume dispersal by using ground based and satellite observations [19]. They introduced five plume categories and evaluated their duration, eruption typology, tephra fallout, and observable extent. Apart from class 1 (degassing plumes), all the others are characterized by light to continuous ash fallout and a limited spatial dispersion, the largest detected distances were only around 250 kilometers at moderate (3-4 km) heights.

An essential point is that the volcanic activity during the eruptive period (31/09 - 15/12) was more or less continuous with the emission of relatively weak and localized plumes [19]. However, the lightning activity exhibits very large fluctuations in the same period. As an example, we show the flash locations (Fig. 3) recorded by the WWLLN on four consecutive days, where a very large positive fluctuation occurred on 30 October 2006. According to Table 1 in Ref. [19], the whole month of October was rather quiet, only a few and very small (3-40 km long) plumes were recorded, which hardly explains the large temporal fluctuations in lightning activity.

Etna's activity was limited to the volcano's summit area in 2007-2008 [21]. While three of the four summit craters were quiescent during the entire period, the mild passive degassing activity of the Southeast crater has been repeatedly interrupted by a sequence of explosive–effusive eruptive episodes. Two remarkable events are listed in Ref. [21] in the Autumn period of 2007: both episodes (4-5 September, and 23-24 November 2007) started with mild discontinuous ash emissions (rich in lithic fragments), which after a few days (hours) were followed by mild

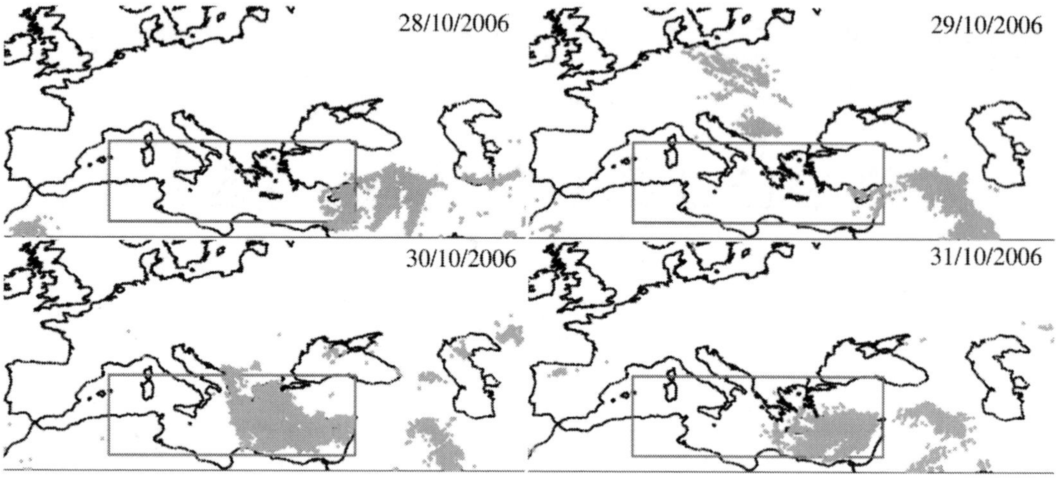

Figure 3. Lightning activity on four consecutive days over the rectangle EU (see Fig. 1), 28-31 October, 2006. Note the large isolated peak at the same time in Fig. 2, top curve. The total number of recorded events on 30 October was 66354 flashes.

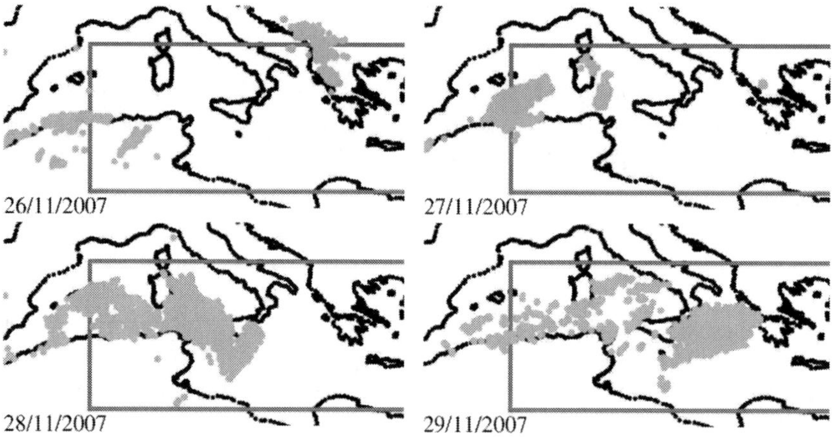

Figure 4. Lightning activity on four consecutive days around the MS region (see Fig. 1), 26-29 November, 2007. Note the large isolated peak in Fig. 2 top curve on the same days. The total number of recorded events on 29 November was 16755 flashes.

Strombolian explosions.

It is tempting to suggest that the localized and intense lightning activity shown in Fig. 4 has a causal link with the volcanic eruption on 23-24 November. Unfortunately, the figure sequence clearly shows that the thunderstorm zone on 26-29 November had a definite tendency to move from West to East, and it is difficult to imagine that the volcano ash cloud could do the opposite a few days earlier and contribute to triggering lightning discharges.

4. Case study II: Eyjafjallajökull 2010

On 26 February 2010, unusually strong seismic activity and rapid expansion of the Earth's crust were registered by the Meteorological Institute of Iceland [22, 23]. This gave strong indication that magma was pouring from underneath the crust into the magma chamber of

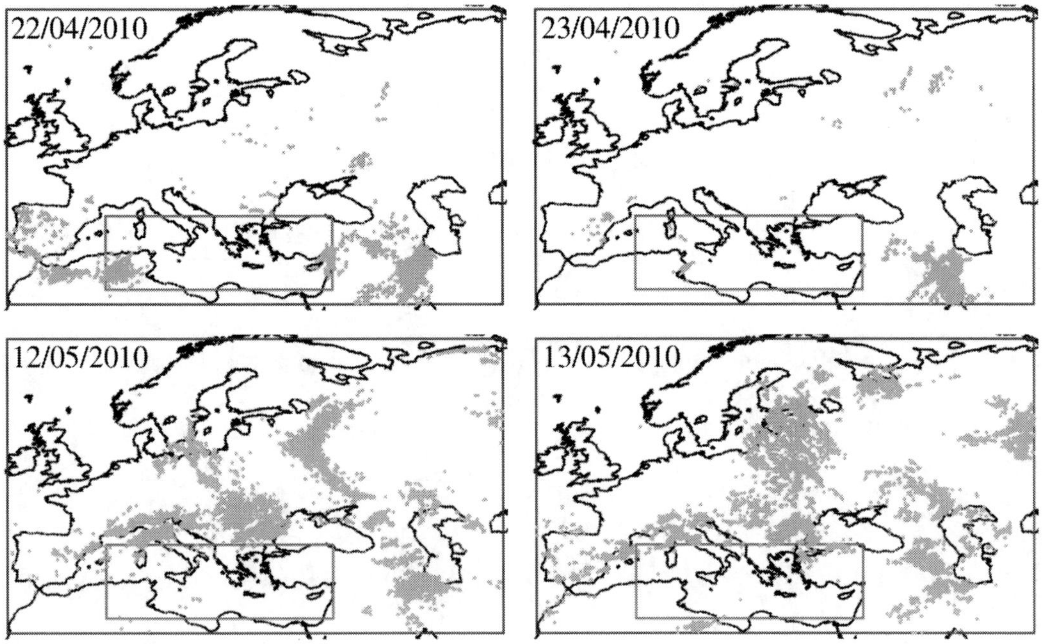

Figure 5. Lightning activity on four days over Europe (see Fig. 1), April and May 2010, when airport closures are reported.

the Eyjafjallajökull volcano. On 14 April 2010, a strong explosive eruption series occurred in the central crater causing meltwater floods and requiring over 800 people to be evacuated. The eruptions threw volcanic ash several kilometers up in the atmosphere which led to air travel disruption in northwest Europe for days in April and in May 2010.

Recent detailed numerical analysis [24] showed that a rapid spread of volcanic ash is possible, with all countries in Europe facing the possibility of an airborne ash concentration exceeding security limits within 24 hours of an eruption. Probabilities of significant concentrations of ash are highest to the east of Iceland, with probabilities exceeding 20 % in most countries north of 50°N. Deposition probabilities were highest at Scottish and Scandinavian airports. Ash concentrations usually remain higher for longer periods during Summer when the mean wind speeds are lower.

Haszpra and Tél report numerical studies on volcanic ash dispersal by using reanalysis wind-field data for the critical time period [25]. In spite of the fact that the overall lightning activity exhibits a quick increase after the main eruption on 14-15 April 2010 (see Fig. 2), the maps determined for the same days as shown in Fig. 2 in [25] exhibit very weak spatial correlations with Fig. 5 here.

5. Discussion

We are fully aware of the fact that the presence of volcanic ash or other aerosols at elevated concentrations would mean only a single factor contributing to trigger lightnings. Thunderstorms and other forms of intense atmospheric convection depend on so many local factors that a detection of direct connection is expected to be extremely difficult. As a last attempt to explain flash number variability, we show the result of correlation analysis between the lightning activity in the Mediterranean region (MS in Fig. 1b) and a daily index of North Atlantic Oscillation.

The North Atlantic oscillation (NAO) is the fluctuations in the difference of atmospheric

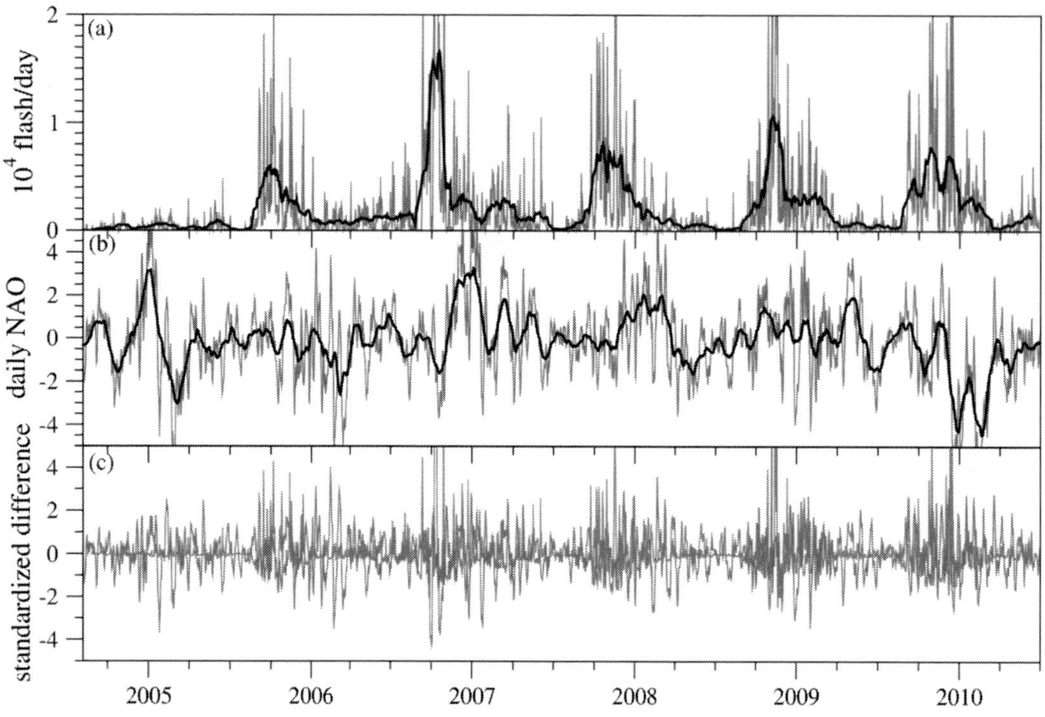

Figure 6. (a) WWLLN lightning counts over the Mediterranean Sea (red), and 31-day running mean (black). (b) Daily NAO index (green) fro the same period as in (a), and 31-day running mean (black). (c) Lightning anomaly (red) and NAO anomaly (green) computed as the difference between the daily value and the 31-day running mean.

pressure at sea level between the Icelandic low and the Azores high. A large positive difference between the pressure at the stations in Reykjavik and Lisbon implies increased westerlies, cool summers and mild and wet winters in Central Europe. In the opposite case (negative NAO phase), westerlies are suppressed, these areas suffer cold winters and storm tracks tend to move toward the Mediterranean Sea. The motivation for checking possible correlations is given by two recent papers where the authors claim to find an increased lightning frequency at low NAO index values [26, 27].

The NAO index is originally represented by monthly or seasonal mean patterns (and also used in this representation in [26, 27]). However, it is increasingly becoming evident that its dynamics is associated with shorter (intraseasonal) time scales. NAO pressure differences at daily resolution are available from 1948 (http://www.lasg.ac.cn/staff/ljp/) [28], therefore a direct comparison with WWLLN statistics is not difficult.

A clear technical problem at the proper comparison is that the lightning activity over the MS area has a marked seasonality (see Fig. 2), while the NAO signals lacks this feature. The standard way is to compute anomalies based on climatological mean values, however the limited length of WWLLN records and the changing detection efficiency are sources of large inaccuracies. (Changing detection efficiency is a consequence of the fact that the WWLLN project is in a developing phase and different stations operate at different periods.) Therefore we made an attempt to define daily anomalies by determining differences between the original signals and their 31-day running means. The result is shown in Fig. 6.

Figure 6c shows no sign of any significant correlation between the NAO signal and lightning rate over the MS region. It is not presented here, but we tested the correlations between WWLLN time series over Europe and NAO with negative result as well, also other related quantities such es running standard deviation or "volatilities" determined by different methods and different window lengths.

The negative results here do not mean that there is no relationship between aerosol concentrations, cloud formation and lightning discharges [29, 30, 31]. A proper understanding probably requires to collect more detailed local information about the atmospheric variables, such as local pressure, wind speed, vertical wind component, humidity, temperature, various gradients, etc.

Acknowledgments

The authors wish to thank the World Wide Lightning Location Network (http://wwlln.net), a collaboration among over 50 universities and institutions, for providing the lightning location data used in this paper. The authors thank Péter Kiss for developing codes to evaluate WWLLN data. This work was supported by the Hungarian Science Foundation (OTKA) under Grant No. NK72037 and NK100296, by the European Commission's COST Action MP0806, and by the European Union and the European Social Fund under the grant agreement no. TÁMOP 4.2.1./B-09/KMR-2010-0003.

References

[1] Rakov V A and Uman M A 2003 *Lightning: physics and effects* (New York: Cambridge University Press) p 679
[2] Betz H D, Schumann U and Laroche P (eds) 2009 *Lightning: principles, instruments and applications* (New York: Springer Science) p 641
[3] Berdeklis P and List R 2001 *J. Atmos. Sci.* **58** 2751
[4] Lang T J and Rutledge S A 2002 *Mon. Wea. Rev.* **130** 2492
[5] Yair Y 2008 *Planet. Atmos. Electricity* **30** 119
[6] McNutt S R and Williams E R 2010 *Bull. Volcanology* **72** 1153
[7] Bakhtiari M 2009 *J. Phys. D: Appl. Phys.* **42** 015209
[8] Colman J J, Roussel-Dupré R A and Triplett L 2010 *J. Geophys. Res.* **115** A00E16
[9] Silfvast W 2010 *Opt. Photon. News* **21** 32
[10] Price C, Asfur M and Yair Y 2009 *Nature Geosci.* **2** 329
[11] Abarca S F and Corbosiero K L 2011 *Mon. Wea. Rev.* **139** 175
[12] Orville R E and Huffines G R 2001 *Mon. Wea. Rev.* **129** 1179
[13] Lagouvardos K, Kotroni V, Betz H D and Schmidt K 2009 *Nat. Hazards Earth Syst. Sci.* **9** 1713
[14] Abreu D, Chandan D, Holzworth R H, and Strong K 2010 *Atmos. Meas. Tech.* **3**, 1143
[15] Golkowski M, Kubicki M, Cohen M, Kulak A and Inan U S 2011 *Acta Geophys.* **59** 183
[16] Collier A B, Lichtenberger J, Clilverd M A, Rodger C J and Steinbach P 2011 *J. Geophys. Res.* **116** A03219
[17] Lane S J, James M R and Gilbert J S 2011 *J. Phys.: Conf. Ser.* **301** 012004
[18] Altaratz O, Levin Z, Yair Y and Ziv B 2003 *Mon. Wea. Rev.* **131** 2060
[19] Andronico D, Spinetti C, Crsitaldi A and Buongiorno M F 2009 *J. Volcanol. Geotherm. Res.* **180** 135
[20] Scollo S, Prestifilippo M, Spata G, D'Agostino M and Coltelli M 2009 *Nat. Hazards Earth Syst. Sci.* **9** 1573
[21] Aiuppa A *et al.* 2010 *Geochem. Geophys. Geosyst.* **11** Q09008
[22] Sigmundsson F *et al.* 2010 *Nature* **468** 426
[23] Matoza R S *et al.* 2011 *Geophys. Res. Lett.* **38** L06308
[24] Leadbetter S J and Hort M C 2011 *J. Volcanol. Geotherm. Res.* **199** 2030
[25] Haszpra T and Tél T 2011 *J. Phys.: Conf. Ser.*, this issue.
[26] Soriano L R, De Pablo F and Tomas C 2004 *Int. J. Climatol.* **24** 1241
[27] De Pablo F and Soriano L R 2007 *Mon. Wea. Rev.* **135** 2810
[28] Li J and Wang J 2003 *Adv. Atmos. Sci.* **20** 661
[29] E. Williams E and Stanfill S 2002 *C. R. Physique* **3** 1277
[30] Kirkby J 2007 *Surv. Geophys.* **28** 333
[31] Kristjánsson J E, Stjern C W, Stordal F, Fjaeraa A M, Myhre G and Jónasson K 2008 *Atmos. Chem. Phys.* **8** 7373

Particles in Turbulence 2011 IOP Publishing
Journal of Physics: Conference Series **333** (2011) 012017 doi:10.1088/1742-6596/333/1/012017

On the optimum fields and bounds for heat and mass transport in two turbulent flows

Nikolay Vitanov[1,2]

[1]Institute of Mechanics, Bulgarian Academy of Sciences, Akad. G. Bonchev Str., Bl. 4, 1113 Sofia, Bulgaria,

[2] Max-Planck Institute for the Physics of Complex Systems, Nöthnitzerstr. 38, 1187 Dresden, Germany

E-mail: vitanov@imbm.bas.bg; vitanov@pks.mpg.de

Abstract. The optimum theory of turbulence is one of the few tools for obtaining analytical results for transport of heat, mass or momentum by turbulent flows. This is achieved by asymptotic theory which is valid for large values of the characteristic numbers of the investigated fluid system. For small and intermediate values of the Reynolds, Rayleigh or Taylor numbers we have to solve numerically the Euler-Lagrange equations of the corresponding variational problems. Below we discuss numerical results from the application of the Howard-Busse method of the optimum theory of turbulence to two problems: convective heat transport in non-rotating and rotating fluid layer and mass transport in pipe flow. We obtain profiles of the optimum fields and discuss the evolution of the thickness of the boundary layers as well as present our first results about the lower bound on the mass transport in a pipe flow.

1. Introduction

Turbulent flows are important for fluid motions of different scales ranging from micrometer droplet sizes in turbulent emulsification up to the large-scale motions in the planetary atmospheres and Earth's oceans. There are few theoretical tools for obtaining rigorous analytical results for the turbulent flows. One of these tools is the optimum theory of turbulence [1, 2, 3, 4, 5]. This theory leads to estimates for the bounds on turbulent quantities by means of a variational problem obtained directly from the governing Navier-Stokes equations, Boussinesq equations, etc. [6, 7, 8]. There exist two methods of the optimum theory of turbulence: the method of Howard-Busse [9, 10, 11, 12] and the method of Doering-Constantin [13, 14, 15, 16]. Below we apply the Howard-Busse method in order to study the evolution of the thickness of the boundary layers for case of thermal convection as well as to obtain numerical lower bounds on the mass transport in a pipe flow. This research continues our previous research on the optimum theory of turbulence [17, 18, 19, 20, 21].

The remaining of the paper is organized as follows. In Sect. 2 we formulate 3 variational problems from the optimum theory of turbulence. The problem for the upper bounds on the convective heat transport in a horizontal non-rotating fluid layer is formulated in details. In order to save space the other two problems (for the upper bounds on heat transport in a rotating fluid layer and for the lower bounds on the mass transport in a pipe flow) are formulated as briefly as possible. Sect. 3 presents results from the numerical investigation of the above problems. Several concluding remarks are summarized in Sect. 4.

2. Mathematical formulation of the problems

2.1. Upper bounds on the convective heat transport for the horizontal non-rotating fluid layer

We consider a horizontally infinite layer of fluid heated from below and we use the equations of the Boussinesq approximation [22] to model it. The thickness of the layer d is unit for length and the time scale is d^2/κ where κ is the thermal diffusivity of the fluid. The fixed temperatures T_1 and T_2 at the upper and lower boundaries of the fluid layer and the temperature scale is $(T_2 - T_1)/Ra$. $Ra = (\gamma(T_2 - T_1)gd^3)/(\kappa\nu)$ is the Rayleigh number. g is the acceleration of gravity, ν is the kinematic viscosity, and γ is the coefficient of thermal expansion. Denoting the vertical unit vector as $\mathbf{k}$, Prandtl number as $Pr = \nu/\kappa$, and the deviation from the static temperature distribution in dimensionless form as Θ we can write the Boussinesq equations as follows

$$\frac{1}{Pr}\left(\frac{\partial \mathbf{v}}{\partial t} + \mathbf{v}\cdot\nabla\mathbf{v}\right) = -\nabla p + \mathbf{k}\Theta + \nabla^2\mathbf{v}; \quad \nabla\cdot\mathbf{v} = 0; \quad \frac{\partial\Theta}{\partial t} + \mathbf{v}\cdot\nabla\Theta = Ra\mathbf{k}\cdot\mathbf{v} + \nabla^2\Theta \quad (1)$$

We consider rigid fluid boundaries: $v_z = \Theta = \frac{\partial v_z}{\partial z} = 0$ and denote the averages over the planes and over the fluid layer as $\overline{q}$ and $\langle q\rangle$. Θ is divided into averaged and fluctuation part $\Theta = \overline{\Theta} + T$. The variational problem is obtained on the basis of two relationships for the fluctuation quantities. We assume that the fluid layer is observed long after the last changes of the averaged characteristic quantities in the time. Thus the temporal derivatives of these quantities are assumed to be 0. In addition we consider all horizontally averaged quantities to be time independent and think in terms of a set of many identical experiments which allows us to replace the averages over a single layer by ensemble averages over an infinite set of layers. Then we can obtain the following two relationships

$$Ra\langle v_z T\rangle = \langle|\nabla T|^2\rangle + \langle(\overline{v_z T} - \langle v_z T\rangle)^2\rangle; \quad \langle v_z T\rangle = \langle|\nabla\mathbf{v}|^2\rangle \quad (2)$$

The convective part of the heat transport is described by the term $\langle v_z T\rangle$. The goal of the optimum theory of turbulence in this case is to obtain upper bound μ on this transport. μ is a monotonous function of R and the problem for obtaining maximum of μ is equivalent to the following variational problem:
Find the minimum $R(\mu, w, \theta)$ of the functional

$$\mathcal{R}(\mu, w, \theta) = \frac{\langle|\nabla\theta|^2\rangle\langle|\nabla\mathbf{u}|^2\rangle}{\langle w\theta\rangle^2} + \mu\frac{\langle(\overline{w\theta} - \langle w\theta\rangle)^2\rangle}{\langle w\theta\rangle^2} \quad (3)$$

among all vector fields $\mathbf{u} = (u, v, w)$ and scalar fields θ which satisfy the continuity equation and the rigid boundary conditions $w = \theta = \frac{\partial w}{\partial z} = 0$ at $z = 0, 1$.

The continuity equation is taken into account by means of the general representation of a solenoidal vector field $\mathbf{u} = \nabla\times(\nabla\phi\times\mathbf{k}) + \nabla\psi\times\mathbf{k}$. ϕ and ψ are determined by $\mathbf{u}$ and we can impose the condition $\overline{\phi} = \overline{\psi} = 0$ without loss of generality. ψ enters the variational functional only through positive terms in the numerator and w is determined solely by ϕ: $w = -\nabla_1^2\phi$ where ∇_1^2 is the horizontal Laplacian $\nabla_1^2 = \nabla^2 - \partial^2/\partial z^2$. Thus the functional reaches its minimum when $\psi = 0$ and the Euler-Lagrange equations of the variational problem have the form

$$\langle|\nabla_1\nabla^2\phi|^2\rangle\nabla^2\theta + w[Ra\langle w\theta\rangle - \mu(\overline{w\theta} - \langle w\theta\rangle)]; \quad \langle|\nabla\theta|^2\rangle\nabla_1^2\nabla^4\phi - \nabla_1^2\theta[R\langle w\theta\rangle - \mu(\overline{w\theta} - \langle w\theta\rangle)] = 0 \quad (4)$$

The above functional is homogeneous in both $\mathbf{u}$ and θ and we can impose the two normalization conditions: $\langle w\theta\rangle = \langle|\nabla\times\mathbf{k}\nabla^2\phi|^2\rangle = \mu$ and $\langle|\nabla\theta|^2\rangle = \gamma\mu$. The special form of the nonlinearity of the Euler-Lagrange equations allows us to seek solutions of following form

$$\phi = \sum_{n=1}^{N}\phi_n(x, y)\alpha_n^{-2}w_n(z); \quad \theta = \sum_{n=1}^{N}\phi_n(x, y)\theta_n(z) \quad (5)$$

where $\nabla_1^2 \phi_n = -\alpha_n^2 \phi_n$ and $\overline{\phi_n \phi_m} = \delta_{nm}$ and δ_{nm} is the Kronecker delta symbol. $w_n(z)$ and $\theta_n(z)$ $(n = 1, 2, \dots, N)$ are chosen as

$$w_n(z) = \sum_{m=1}^{M} a_{nm}[\cos(2(m-1)\pi z) - \cos(2m\pi z)]; \quad \theta_n(z) = \sum_{m=1}^{M} b_{nm} \sin[(2m-1)\pi z] \tag{6}$$

which satisfy the boundary conditions. The value of the parameter M must be sufficiently large such that the integral properties of the solution become independent on M. We choose it in such a way that the value of μ changes by less than 0.1% when M is replaced by $M - 5$. This ensures that the numerical solutions of the Euler-Lagrange equations do not depend in any significant way on this parameter.

When we introduce the relationships (5) and (6) in the Euler-Lagrange equations we obtain equations that have several parameters: the Rayleigh number Ra and the wave numbers α_n. We fix the Rayleigh number and calculate the Nusselt number as function of the wave number. In order to obtain the value of the upper bound we have to find the optimum value of the wave numbers ,i.e., the values for which the Nusselt number has a maximum. Our numerical experience shows that this maximum is a single maximum.

2.2. Upper bounds on the convective heat transport for a rotating horizontal fluid layer of infinite Prandtl number

In this case the heated from below horizontal layer of fluid rotates about the vertical axis with a constant angular velocity Ω. The layer sizes are infinite in horizontal directions. The non-dimensional Boussinesq equations are

$$\frac{1}{Pr}\left(\frac{\partial \mathbf{u}}{\partial t} + \mathbf{u} \cdot \nabla \mathbf{u}\right) = -\frac{\sqrt{Ta}}{2}\nabla p + \nabla^2 \mathbf{u} + RaT\mathbf{k} + \sqrt{Ta}\mathbf{u} \times \mathbf{k} \tag{7}$$

$$\frac{\partial \Theta}{\partial t} + \mathbf{u} \cdot \nabla \Theta = \nabla^2 \Theta; \quad \nabla \cdot \mathbf{u} = 0 \tag{8}$$

The boundary conditions are stress-free at $\pm z = 1/2$: $u_3 = \partial^2 u_3/\partial z^2 = T = 0$. $Pr = \nu/\kappa$ is the Prandtl number, $Ta = (2\Omega d^2/\nu)^2$ is the Taylor number, $Ra = (\gamma g \Delta T d^3)/(\kappa \nu)$ is the Rayleigh number, γ is the coefficient of thermal expansion, p is the pressure, and $\mathbf{k}$ is the unit vector in the direction opposite to the gravity.

The variational problem is formulated again by two moment equations obtained on the basis of the Boussinesq equations. When Prandtl number is infinite the Navier-Stokes equation becomes linear and we can include it as a constraint in the variational problem. The equation of continuity is taken into account by the general representation of a solenoidal field $\mathbf{u}$ in terms of a poloidal and a toroidal component as in Sect 2.1. Further we perform the rescaling $\mathbf{u} = \mu^{1/2}\langle w\theta \rangle^{-1/2}\mathbf{v}$; $T = \mu^{1/2}\langle w\theta\rangle^{-1/2}Ra^{-1}\theta$, where the $z-$component of the rescaled velocity field $\mathbf{v}$ is denoted as w. and introduce the toroidal - poloidal decomposition into the Navier-Stokes equation $(P = \infty)$. Taking the $z-$component of the horizontal curl and $z-$component of the double curl of the result we obtain the relationships

$$\nabla^2 f + \sqrt{Ta}\frac{\partial w}{\partial z} = 0; \quad \nabla^4 w + \nabla_1^2 \theta - \sqrt{Ta}\frac{\partial f}{\partial z} = 0 \tag{9}$$

where $f = -\nabla_1 \psi$ is the vertical component of the vorticity. By means of Lagrange multipliers p^* and q^* we include (9) in the variational problem which can be formulated as follows
Find the minimum $R(\mu)$ of the variational functional

$$\mathcal{F} = \frac{\langle |\nabla\theta|^2\rangle}{\langle w\theta\rangle} + \mu\frac{\langle(\overline{w\theta} - \langle w\theta\rangle)^2\rangle}{\langle w\theta\rangle^2} - \left\langle p^*\left(\nabla^2 f + \sqrt{Ta}\frac{\partial w}{\partial z}\right)\right\rangle - \\ - \left\langle q^*\left(\nabla^4 w + \nabla_1^2\theta - \sqrt{Ta}\frac{\partial f}{\partial z}\right)\right\rangle = 0 \tag{10}$$

among all fields w, θ, f that satisfy the boundary conditions $w = \theta = \frac{\partial^2 w}{\partial z^2} = \frac{\partial f}{\partial z} = 0$ at $z = \pm 1/2$.

The Euler-Lagrange equations for the above functional are

$$\langle | \nabla\theta |^2 \rangle \theta + 2\theta[\mu(\overline{w\theta} - \langle w\theta \rangle) - Ra\langle w\theta \rangle] - \langle w\theta \rangle^2 \left(\nabla^4 q^* - \sqrt{Ta}\frac{\partial p^*}{\partial z} \right) = 0 \tag{11}$$

$$\langle | \theta |^2 \rangle w - 2\langle w\theta \rangle \nabla^2 \theta + 2w[\mu(\overline{w\theta} - \langle w\theta \rangle) - Ra\langle w\theta \rangle] - \langle w\theta \rangle^2 \nabla_1^2 q^* = 0 \tag{12}$$

$$\nabla^2 p^* + \sqrt{Ta}\frac{\partial q^*}{\partial z} = 0; \quad \nabla^4 w + \nabla_1^2 \theta = \sqrt{Ta}\frac{\partial f}{\partial z}; \quad \nabla^2 f + \sqrt{Ta}\frac{\partial w}{\partial z} = 0 \tag{13}$$

2.3. Lower bounds on the mass transport in a pipe flow

Busse [23, 24, 25] derived numerous analytical upper bounds on mass and momentum transport in shear flows. Below we shall discuss the lower bounds on the mass transport in a uniform pressure gradient driven flow of incompressible fluid in an infinitely long pipe of radius r_0. The model equations are the Navier-Stokes equations (in a cylindrical coordinate system (r, φ, x)), non-dimensionalized by using r_0 and r_0^2/ν as length and time scales respectively. The boundary condition on the velocity is $\mathbf{u} = 0$ at $r = 1$. The velocity is divided in two components: $\mathbf{u} = \overline{\mathbf{u}} + \mathbf{v}^*$ where $\overline{\mathbf{u}}$ is the average of the velocity field over the surfaces $r =$ const and fluctuation field $\mathbf{v}^*$. For statistically stationary turbulent flow (for which the average of any quantity over the surfaces $r =$ const is time independent) the dimensionless mass transport through the pipe is given by the relationship

$$M = \frac{G}{8} - \frac{\langle r\sqrt{2} v_r^* v_x^* \rangle}{\sqrt{8}} \tag{14}$$

where G is the uniform pressure gradient along the pipe and $\langle \ldots \rangle$ denotes averaging over the total volume of the pipe. The last term in Eq. (14) is positive and thus in order to obtain a lower bound on the mass transport we search for the upper bound on the quantity $\langle r\sqrt{2} v_r v_x \rangle$ among all fields $\mathbf{v}$ that satisfy the continuity equation $\nabla \cdot \mathbf{v} = 0$ and the boundary condition $\mathbf{v} = 0$ at $r = 1$. This problem is equivalent to the problem for finding the minimum $P(\mu)$ of the functional

$$P^*(\mu, \mathbf{v}) = \frac{\langle | \nabla \times \mathbf{v} |^2 \rangle}{\langle r\sqrt{2} v_r v_x \rangle} + \mu \frac{\langle \overline{v_r v_x}^2 + \overline{v_r v_\varphi}^2 \rangle - \langle r\sqrt{2} v_r v_x \rangle^2}{\langle r\sqrt{2} v_r v_x \rangle^2} \tag{15}$$

for $\mu > 0$ for all bounded fields $\mathbf{v}$ that satisfy the continuity equation, the boundary condition and the requirement $\langle r v_r v_x \rangle > 0$.

Below we shall discuss solution of the Euler-Lagrange equations for the variational problem (15) for the case of single wavenumber optimum field. In the general case of $N-$wavenumber solution of the Euler-Lagrange equations the solution is searched in the form (below $v_r = w$ and $v_x = \theta$)

$$w = \sum_{n=1}^{N} F_n(\varphi)\frac{w_n(r)}{\alpha_n^2}; \quad \theta = \sum_{n=1}^{N} F_n(\varphi)\theta_n(r); \quad \langle F_n^2 \rangle = 1; \quad \frac{d^2 F_n}{d\varphi^2} = -\alpha_n^2 F_n \tag{16}$$

3. Results from the numerical investigation

3.1. Effectivity of the Galerkin method

In absence of rotation (Sect 2.1) the increasing Rayleigh number leads to increasing convective heat transport and to thinning of the boundary layers of the optimum fields w and θ. This leads to increasing number of modes M. M shows a tendency to follow approximately the power law $M = 0.128 Ra^{0.415}$. For the system of equations correspondent to the solved in Sect.

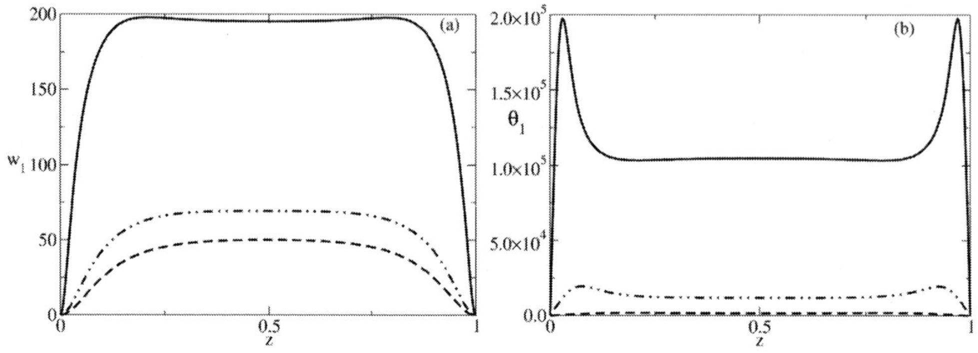

Figure 1. Behavior of the optimum fields for the case of thermal convection in non-rotating fluid layer. Left-hand side figure: the optimum field $w(z)$. Right-hand side figure: the optimum field $\theta(z)$. The values of the Rayleigh numbers are as follows. ——: $R = 10^6$; — · · —: $R = 10^5$; - - - -: $R = 10^4$.

2.1. variational problem we can proceed with the Galerkin method up to $M = 200$ which approximately corresponds to Rayleigh number $Ra = 5 \cdot 10^7$. The situation is different in presence of rotation (Sect. 2.2). The strong rotation leads to decrease of M and this decrease follows approximately a power law too. For $Ra = 10^8$ and for Taylor numbers in the interval between $Ta = 5 \cdot 10^8$ and $Ta = 10^{10}$ the power law is $M = 1.2 \cdot 10^6 Ta^{-0.474}$.

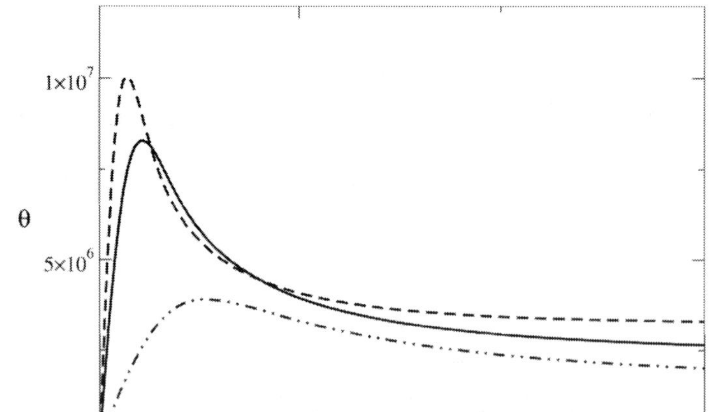

Figure 2. Erosion of the boundary layers of the optimum field $\theta(z)$ by rotation. The value of the Rayleigh number is $R = 10^8$. The values of the Taylor number are as follows. - - - -: $Ta = 10^9$;——: $Ta = 2.5 \times 10^9$; — · · —: $Ta = 10^{10}$;

3.2. Changes in the optimum fields

3.2.1. Formation of boundary layers with increasing Rayleigh number. Fig. 1. shows development of the optimum fields $w_1(z)$ and $\theta_1(z)$ of the single-wavenumber solution of the

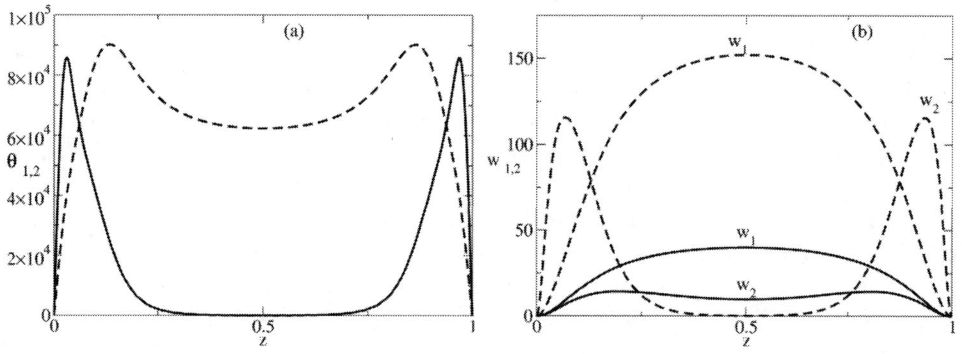

Figure 3. Figure (a): optimum fields $\theta_{1,2}$ connected to the two-wavenumber solution of the variational problem for the case without rotation. ——: θ_1; - - - -: θ_2. The value of the Rayleigh number is : $Ra = 5 \cdot 10^5$. Figure (b): Separation of the components connected to different wave numbers. The optimum fields are for the two - wavenumber solution for the case of thermal convection without rotation. ——: the components $w_{1,2}(z)$ for $Ra = 3.5 \cdot 10^4$; - - - -the components $w_{1,2}(z)$ and $\theta_1(z)$ of the two - wavenumber solution for $Ra = 5 \cdot 10^5$.

variational problem for the case of convection in a non-rotating fluid layer. From top to the bottom of the Figs. 1(a) and 1(b) the Rayleigh number increases and the profiles of the optimum fields change. What we observe is a slow formation of boundary layers of the field w_1 whereas the formation of the boundary layers of the field θ_1 is faster. At $Ra = 10^6$ the boundary layers of θ_1 are well formed whereas the formation of boundary layers of w_1 just begins. This happens by increasing value of w_1 close to the boundaries of the fluid layer and arising of a slight depression of w_1 in the middle of the layer of fluid (see Fig. 1(a)). This result of the numerical investigation reflects the fact that the thermal boundary layers for the discussed case of convection are formed much faster and are thinner than the boundary layer connected to the velocity field.

The presence of rotation changes the picture of forming of boundary layers which become thinner with increasing Rayleigh number. The nature of the changes depends on the strength of the rotation (the value of the Taylor number Ta). When the Rayleigh number is fixed we can distinguish three regions with respect to influence of rotation:

(1.) weak rotation: the rotation does not influence significantly the development of the boundary layers

(2.) intermediate rotation: the rotation slows the formation of boundary layers

(3.) strong rotation: the rotation stops the development of the boundary layers. If the rotation is strong enough it begins to erode the boundary layers and the very strong rotation inhibits and even stops the convection

Fig. 2 illustrates case **(3.)**. The increase of Taylor number from $Ta = 10^9$ to $Ta = 10^{10}$ leads to increasing thickness and decreasing peak height of the boundary layer for the optimum field θ. If the Taylor number increase further one observes vanishing of the boundary layer and convective motion stops.

3.2.2. Space separation of the components of the optimum fields connected to different wave numbers. When the number of the wave numbers connected to the solution of the Euler - Lagrange equations of the variational problem is larger than 1 then the phenomenon of space separation of components of the optimum fields must arise. The existence of the space separation is a basic assumption of the theory of Busse [2] of the multi-wavenumbers solutions of the variational problems of the optimum theory of turbulence. The numerical investigation confirms the assumption of Busse. One example for optimum fields connected to two-wavenumbers solution of the variational problem for the case of convection without rotation is shown in Fig. 3 (a). The separation phenomenon is shown in Fig. 3(b) for the components w_1 and w_2 of the two - wavenumber solution of the same problem. We observe that with increasing Rayleigh number the fields w_2 concentrates its non-zero values in the area close to the boundaries of the layer. In the same area with increasing Rayleigh number the values of w_1 become smaller and smaller in comparison to the values of w_2. The opposite concentration is observed at the central part of the fluid layer. There w_1 increases strongly with increasing Rayleigh number and the values of w_2 tend to be close to 0.

The presence of rotatition influences the process of separation too. Again three cases are observed which are identical to the three cases of influence of rotation on the boundary-layers formation as discused above. The intermediate rotation can slow the process of spatial separation of the components of the optimum field. The strong rotation can even reverse the process of separation.

3.3. Behavior of the thickness of boundary layers

A basic result from the analytical asymptotic solution of the Euler-Lagrange equations connected to the upper bounds on convective heat transport in a fluid layer is that the thickness of the optimum field $\overline{w\theta}$ is closely connected to the upper bound on the Nusselt number and follows a power law for large values of the Rayleigh and Taylor numbers. Numerical investigation of the Euler-Lagrange equations confirms this result and in addition exhibit another interesting feature: the thicknesses of the boundary layers of the fields w and θ can follow power law too when the values of the Rayleigh and Taylor numbers are large enough.

Let us denote the thickness of the boundary layers of the optimum fields as δ. An important property of the Galerkin method is that we can calculate very accurately the profiles of the optimum fields. Because of this we can determine precisely the thicknesses of the layers of the optimum fields. The boundary layer of an optimum field is in the region from the boundary of the layer (where the correspondent fields have value 0 in order to satisfy the boundary conditions) to the first maximum of the field. For the case of thermal convection in a non-rotating layer δ decreases with increasing Rayleigh number which leads to fast increase of the value of the truncation parameter M. The thinning of the boundary layers is much slower for the case of the field w which corresponds to the velocity field in the Boussinesq equation. The size δ of the boundary layer follows approximately a power law. For an example for the case of single-wavenumber solution of the Euler-Lagrange equations of the variational problem $\delta \approx 15 \cdot Ra^{-0.3}$. When the upper bound of the convective heat transport is given by the two-wavenumbers solution of the Euler-Lagrange equations then the field w_1 does not develop a boundary layers at the expense of the field w_2 which develops such layers and their thickness follows approximately the power law $\delta \approx 5.8 \cdot Ra^{-0.33}$. As we see the thickness of this boundary layer decreases faster than the thickness of the boundary layer connected to the single - wavenumber solution. Similar is the picture for the case of the field θ which corresponds to the temperature field in the Boussinesq equations. For this field the thickness of the boundary layers decreases faster that for the field w. For an example for the single - wavenumber solution of the variational problem and for large enough Rayleigh numbers the thickness of the boundary layer of the field θ_1 follows approximately the power law $\delta \approx 7.9 \cdot Ra^{-0.4}$.

The presence of rotation can reverse the process of the thinning of the boundary layers of the optimum fields. When we fix the Rayleigh number and increase the Taylor number the thickness of the boundary layers of the components of the optimum fields begins to increase and this increase is correlated with the inhibition of the convection by the rotation. The rotation destroys relatively fast the boundary layers of the field w. The boundary layers of the field

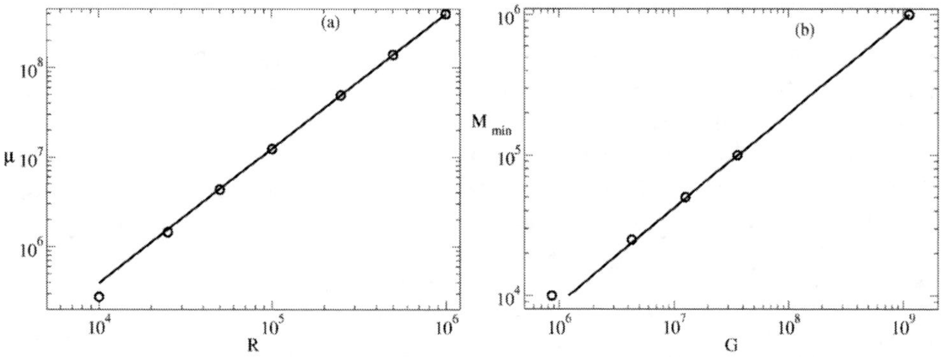

Figure 4. Figure (a): Dependence of $\mu = \langle r\sqrt{2}v_r v_x \rangle$ on the Reynolds number Re for the single-wavenumber solution of the variational problem. ——: theory of Busse. ○ : results from the numerical investigation . Figure (b): lower bound M_{min} on the mass transport as a function of the pressure gradient G for the single wavenumber solution of the variational problem. ——: Theory of Busse. ○ : results from the numerical simulation.

θ exist in larger interval of Taylor numbers and for an example their thickening for the case of single wavenumber solution (and when the Rayleigh number is fixed at $Ra = 10^9$) follows approximately the power law $\delta \approx 1.29 \times 10^{-10} \cdot Ta^{0.75}$.

3.4. First results on the bounds on in pipe flow

For the case of single-wavenumber solution the averaged product $\overline{w_1\theta_1}$ has value $\approx r\sqrt{2}$ in the interior of the layer and at the boundary $r = 1$ w_1 and θ_1 decrease in order to satisfy the boundary conditions. Fig. 4 presents comparison of the analytical results from the single-wavenumber theory of Busse [25] and the corresponding results from the numerical investigation. The homogeineity of the variational problem allows us to set $\mu = \langle r\sqrt{2}v_r v_x \rangle$. The analytical and numerical results for this quantity are shown in figure 4 (a). We note that the double - logarithmic scale suppresses the differences between the analytical asymptotic result and numerical results for the case of large Reynolds number. What we observe is that the numerical result slowly approaches from below the result from the analytical theory. Figure 4 (b) shows the lower bound on the mass transport as a function of the pressure gradient G. Again the results from the numerical investigation are close to the analytical results when the investigation enters the case of almost asymptotically large values of the pressure gradient. This can be expected as the analytical theory is strictly valid only for very large values of the Reynolds number. For intermediate values of the Reynolds number instead of the asymptotic version of the Euler-Lagrange equations one has to solve the complete equations and this leads to differences between the asymptotic theory and the numerical results as it can be seen at the bottom left corner of Fig. 4(b).

4. Concluding remarks

The study of the three variational problems above shows that the numerical investigation is very useful addition to the analytical asymptotic theories for the variational problems arising in the optimum theory of turbulence. We come to the following conclusion. For small and intermediate values of the Reynolds, Rayleigh or/and Taylor numbers one has to solve the complicated integro-differential Euler-Lagrange equations of the corresponding variational problem in order to obtain the bounds and the related profiles of the optimum fields. This solution can be only numerical. Our experience shows that the spectral methods (when applicable) are convenient and sufficient tool to solve this task. They give the solution up to the values of the characteristic numbers for which the Euler-Lagrange equations can be simplified and their analytical asymptotic solutions (given by the asymptotic boundary layer approach of the optimum theory of turbulence) are available. Thus one has bounds on the corresponding flow quantities from small and intermediate values of Reynolds, Rayleigh or/and Taylor numbers up to very large values of these values. And the bounds allow us to obtain impression about the maximum or minimum amount of mass, heat, momentum, etc which the corresponding turbulent flow is able to transport.

5. Acknowledgments

We thank to the COST Action MP0806 "Turbulence and particles" for the support of this research. This research was partially supported by the Grant DO 02/338-22.12.2008 of the National Fund for Scientific Researches of the Republic of Bulgaria.

References

[1] Howard L N 1963 *J. Fluid Mech.* **17** 405
[2] Busse F H 1969 *J. Fluid Mech.* **37** 457
[3] Vitanov N K and Busse F H 1997 *ZAMP* **48** 310
[4] Hoffmann N P and Vitanov N K 1999 *Phys. Lett* A **255** 277
[5] Vitanov N K and Busse F H 2001 *Phys. Rev* E **63** 016303
[6] Vitanov N K 2000 *Physica* D **136** 322
[7] Vitanov N K 2005 *Physics of Fluids* **17** 105106
[8] Vitanov N K 1998 *Phys. Lett.* A **248** 338
[9] Busse F H 1978 *Adv. Appl. Mech.* **18** 77
[10] Strauss J M 1976 *Dyn. Atm. Oceans* **1** 77
[11] Vitanov N K 2000 *Phys. Rev.* E **61** 956
[12] Vitanov N K 2003 *Phys. Rev.* E **67** 026322
[13] Doering C R and Constantin P 1992 *Phys. Rev. Lett* **69** 1648
[14] Doering C R and Constantin P 1994 *Phys. Rev.* E **49** 4087
[15] Doering C R and Constantin P 1996 *Phys. Rev.* E **53** 5957
[16] Plasting S C and Kerswell R R 2003 *J. Fluid Mech.* **477** 363
[17] Vitanov N K 2000 *Eur. Phys. J.* B **15** 349
[18] Vitanov N K 2000 *Phys. Rev.* E **62** 3581
[19] Vitanov N K 2001 *Eur. Phys. J.* B **23** 249
[20] Vitanov N K 2005 *Springer Proceedings in Physics* **101** 37
[21] Vitanov N K 2010 *Eur. Phys. J.* B **73** 265
[22] Chandrasekhar S 1981 *Hydrodynamic and Hydromegnetic Stability* (New York: Dover)
[23] Busse F H 1969 *ZAMP* **20** 1
[24] Busse F H 1970 *J. Fluid Mech.* **41** 219
[25] Busse F H 1972 *Symposia Mathematica* **9** 493

Droplet Coalescence in 2D Thermal Convection of a Thin Film

M. Winkler[1] **and M. Abel**[1]

E-mail: Michael.Winkler@uni-potsdam.de

E-mail: Markus.Abel@physik.uni-potsdam.de
[1] Institute for Physics and Astronomy, University of Potsdam

Abstract. We show first results on the dynamics of ultrathin regions, black film (BF) – bubbles, in a two-dimensional thermal convection experiment with already very thin aqueous films. The formation of such stable regions in thin films is a new feature which makes it possible to study the collision dynamics of BF-bubbles in detail. This is of high interest due to the analogy to bubbles in a fluid in 3D situations. For our experiment, we use a Rayleigh-Bénard setup and an extremely well–controlled chemistry: n-dodecyl-β-maltoside is used as the surfactant stabilizing the sheet of water making it possible to reach stable black films. We demonstrate how the formation and coalescence of bubbles can be tracked and give an overview about shape and size variation during this process.

1. Introduction

Liquid thin films are found often in nature, e.g. in cellular membranes, condensation layers on surfaces, and as interfaces between material domains. This contribution reports on experiments on a soap film captured in a frame, that combines three scientific areas: free-standing, thin, liquid films, thermal convection, and droplet coalescence. Thin films are not only beautiful as soap film, but highly important for process engineering and life sciences. Further, the limits of physical theories are tested when films become exceedingly thin, i.e., on the nanometer scale. Convection is often considered in the context of Rayleigh–Bénard (RB) convection as a model for large systems like the earth–mantle convection and atmospheric, thermally driven turbulence. Droplet coalescence, finally, is very important in the study of rain and cloud formation. In the presented study two approaches are combined: thin liquid films in their equilibrium regime and Rayleigh-Bénard convection. Due to perturbations in the film, stable bubbles are formed and their creation, convection and coalescence can be analyzed. Here, only the case of bubble coalescence will be presented.

When a fluid is subject to a temperature difference where the gradient is directed inverse to the gravitational force, above a critical value, the onset of turbulent convection is observed. This "cooling from above, heating from below" scenario leads to Rayleigh-Bénard convection which is well-studied [1, 2]. Yet some fundamental questions are still unsolved, regarding the scaling behavior of Nusselt vs. Prandtl for high Rayleigh numbers or the dependence on aspect ratio and the question of universality.

Thin liquid films are typically composed of a layer of amphiphilic molecules on each surface and a liquid core separating these layers [3]. Depending on the chemical details of the solution there exist two stable equilibrium thicknesses [4]. In contrast to previous convection experiments with thin liquid films in the transient state the presented setup makes it possible to study convection in films at their equilibrium thickness.

For the experimental investigation, a thin-film forming system was investigated in a special experimental setup. The scenario is very similar, but not identical to classic RB convection. In contrast to usual RB convection the system is necessarily open on both surfaces, and an exchange of matter through the surface happens. This means that heat exchange whith the surrounding atmosphere must be considered, in addition to the heat flux through the container, which is in our case just a metal frame.

2. Thin Film Convection

The thinning of a liquid film can terminate in two distinct equilibrium states. One being the Common Black Film (CBF) and the other one being the Newton Black Film (NBF) [5]. The latter is a bilayer formation of the surface stabilizing surfactant molecules without an aqueous core. This state can only be reached with an electrolyte concentration high enough to shield the long range molecular repulsion forces of the two surfaces. Thus, the final thickness is only determined by the balance between surface pressure and molecular short-range repulsion forces. Common Black Films on the other hand are achieved when there are only very few free ions present and the film is stabilized due to electrical double layer repulsion. Because of the dissimilar stabilizing mechanism it is expected that there is a distinct difference in the convection behavior. Typically a Newton Black Film reaches a thickness of roughly 5 nanometers whereas Common Black Films develop at a thickness of approximately 15 nanometers [6]. Since we are forcing the film it is necessary to keep in mind that these results are equilibrium values. Therefore the actual thickness in our experiment may vary and we refer to black film (BF), not distinguishing further the two possible states.

When the thin film is heated, convective motion sets in, which can be observed visually and in real time, hence video recording is used and the images subsequently characterized by statistical analysis and tracking of the fluid.

The temperature difference achievable by the experiment ranges from $\Delta T = 20K$ to $\Delta T = 50K$. The used frame has a height of $20\,mm$; this results in temperature gradients from $\frac{\partial T}{\partial y} = 10\,\mathrm{K/mm}$ to $\frac{\partial T}{\partial y} = 25\,\mathrm{K/mm}$ generating convection which is observable by the naked eye. The accessible Rayleigh numbers are $Ra \simeq 1 \cdot 10^6 - 1.5 \cdot 10^7$. The maximum temperature difference is limited by decreased film stability at high temperatures and increased condensation at the top border of the frame at temperatures close to the freezing point. A sketch of the experiment is given in Fig. 1. The temperature difference used in the analysis below was $\Delta T = 30K$ with a Rayleigh number of $Ra = 6 \cdot 10^6$.

The system is described by the Navier-Stokes equations, here, we use the 2D approximation with temperature in the Oberbeck Boussinesq approximation, the thickness is described in the lubrication approximation coupled to the fluid through the boundary conditions. The lubrication approximation assumes that scales in x and y are large compared to the typical film thickness (z-axis). An equation for the film thickness $h(x,y) = h_f(x,y) - h_b(x,y)$ is derived by multiscale expansion [7], assuming independence of the back and front surfaces h_b, h_f.

$$\begin{aligned} \mu\partial_t h \;&+\; \vec{\nabla}_1 \left[\left(\vec{\tau} + \vec{\nabla}_1 \sigma \right) \left(\frac{1}{2} h^2 + \eta h \right) \right] \\ &-\; \vec{\nabla}_1 \left[\left(\frac{1}{3} h^3 + \eta h^2 \right) \vec{\nabla}_1 \left(\Delta\Phi - \sigma \vec{\nabla}_1^2 h - \Pi \right) \right] = 0\,, \end{aligned} \tag{1}$$

with μ the (dynamic) fluid viscosity, $\vec{\nabla}_1 = (\partial_x, \partial_y)^T$, σ the surface tension, η the slip coefficient between surface and air, $\Delta\Phi = \Phi|_{z=h_l} - \Phi|_{z=h_r}$, Φ describes a body force, here this is gravity, and

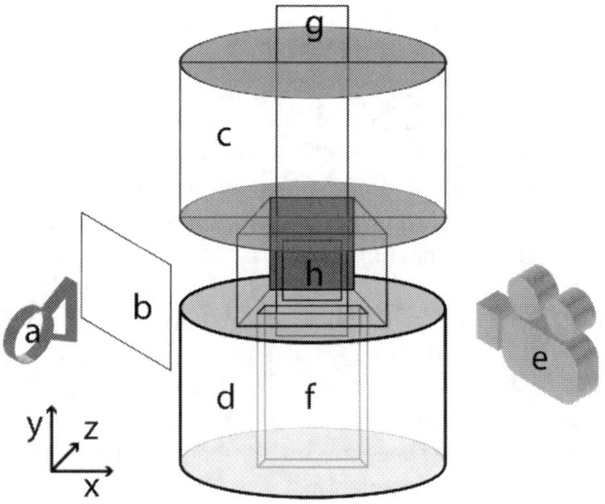

Figure 1. Experimental Setup: a) light source, b) diffuser, c) top, cool aluminum cylinder, temperature controlled, d) bottom, hot aluminum cylinder, temperature controlled, e) video camera, f) reservoir of thin film solution, located within the bottom aluminum cylinder (heated), g) sliding frame which is delved into the solution at the bottom, when lifted the thin film is produced, h) thin film inside the measurement cell. The measurement chamber is sealed and placed between the two slotted aluminum cylinders c) and d) with a stainless steel frame clamped in between them. The chamber is transparent with a black backside, the thin film area has a size of 20 x 20 mm^2.

at small scales, disjoining pressure, τ and Π are tangential and normal forcings at the surface.

In the bulk of the film we assume the Oberbeck-Boussinesq approximation [1]

$$\rho(\partial_t\vec{u} + \vec{u}\cdot\vec{\nabla}_1\vec{u}) = -\vec{\nabla}_1 p + \mu\Delta_1\vec{u} + \beta\vec{g}\theta, \tag{2}$$

$$\rho c(\partial_t\theta + \vec{u}\cdot\vec{\nabla}_1\theta) = \kappa_{th}\Delta_1\theta, \tag{3}$$

with ρ: density, β: thermal expansion coefficient, c: specific heat, κ_{th}: thermal conductivity. At the boundaries, h_b, and h_f, respectively, we have

$$\kappa_{th}\vec{\nabla}\theta\cdot\vec{n} + \alpha_{th}(\theta - \theta_\infty) = 0, \tag{4}$$

with α_{th}: heat transfer to ambient, θ_∞: outer temperature, $\vec{n}$: vector normal to the surface. For a two–dimensional thin film one can assume to 0th order that the thickness is advected passively. But the surface tension depends parametrically on the temperature, and thus certainly becomes important in the next higher order. Of course, in contrast to usual thermal convection, chemical potentials and surfactant forces matter.

3. Experimental Details

For the experimental investigation, a thin-film forming system was placed in a closed container to prevent evaporation and to ensure thermodynamic control. The setup is based on a measuring cell that was designed at the Chinese University of Hong Kong in the fluid physics group, similar to [8]. Very thin translucent films and thicker opaque films can both be prepared with a lifetime of several hours. The frame holding the thin liquid film was heated from below and cooled from above by two large aluminum cylinders, thus creating a thermal gradient.

A film consisting of pure water ruptures almost immediately due to the capillary forces generated by the surface tension. Adding a surfactant that covers the surface of the film reduces the surface tension and increases the surface elasticity, therefore stabilizing the sheet of water. Several further components help to maintain film stability over hours, as described below.

Macroscopic, quasi-two-dimensional liquid films are obtained by spanning a stainless steel or PVC frame with surfactant solution. The base solvent is deionized water, as it reduces contaminations with non-conducting carbon chain remnants or residual ion concentrations [6].

Glycerin is also required to achieve macroscopically large films with an area of several square centimeters or meters ([9, 10]), serving to adjust bulk and surface viscosity. Due to its high viscosity ($\eta = 945$ mPa s^{-1}, compared to ≈ 1 mPa s^{-1} for water) it reduces the thinning speed of vertical films and allows the film to remain stable for a longer period of time ([11]). Typically, a concentration of 25 $\%_{\mathrm{vol}}$ is used, yielding a net viscosity of ≈ 2 mPa s^{-1} (cf. [12]). Our experimental setup thus differs from the experiments on turbulence in soap films [13], and is quite similar to previous ones on convection [14].

The electrolyte concentration has a major effect on film stability and thickness [15]. The addition of ions shields the electrostatic double-layer repulsion as described by the DLVO (Derjaguin-Landau-Verwey-Overbeek) theory [3] hence permitting the appearance of a black film. Here, sodium chloride was chosen, since it has negligible influence on surface tension and critical micelle concentration (CMC). To eliminate any surface active contaminations, the laboratory grade $NaCl$ (Merck, Darmstadt, Germany) was roasted at $600°C$.

The chosen surfactant, n-dodecyl-β-maltoside (here after $\beta{-}C_{12}G_2$), belongs to a new generation of sugar-based surfactants which are non-toxic, biodegradable and of low cost. $\beta{-}C_{12}G_2$ is a non-ionic surfactant which consists of hydrophilic head group, made up of two glucose rings ($C_6H_{12}O_6$) connected by an ether bond, and a hydrophobic alkyl chain ($C_{12}H_{25}$). The surfactant is soluble in water so that the hydrophilic head group is submerged in the core liquid. The hydrophobic tail is oriented normal to the liquid-air-interface and is not in contact with film surface. All experiments were performed with a concentration above the CMC, to guarantee that the equilibrium surface tension remains constant. Above CMC, deviations in surface surfactant density are compensated by diffusion of surfactants from or to the bulk liquid.

4. Bubble Movement and Tracking

To capture the movement of the soap film turbulence we use a generic HD video camera which is able to record 1920 times 1080 pixel at 25 frames per second. The average angular velocity of the convection roll is approximately π/s and the maximum speed is $18\,{}^{\mathrm{mm}}/_{\mathrm{s}}$ so that the frame rate limited minimal resolution is roughly $0.5\,mm$. We use a demountable measuring cell that encloses a stainless steel or PVC frame holding the film and a reservoir of the solution. The film is quadratic with a side length of $20\,mm$, i.e. an area of $A = 400\,mm^2$.

It is vitally important for the long term stability of the film to keep the atmosphere saturated with vapor to prevent evaporation of solution from the film. This is achieved by a closed container with a glass front. The film is illuminated through this viewing window by an array of high power white LEDs which are placed behind a diffuser. To minimize reflections the rear panel of the measuring cell is covered by a sheet of black cellular rubber. Additionally, to avoid direct reflections of the light source from the viewing window the measuring cell is tilted with respect to the liquid film plane.

The diffused light is reflected by the front and back side of the film whereas the front reflection is shifted by $\pi/2$ due to reflecting of an optically thicker medium. If the film thickness is much smaller than $\lambda/4$ of the smallest emitted wavelength the light waves interfere $\pi/2$ out-of-phase and no light is reflected. Hence the name black film although more precisely it is transparent.

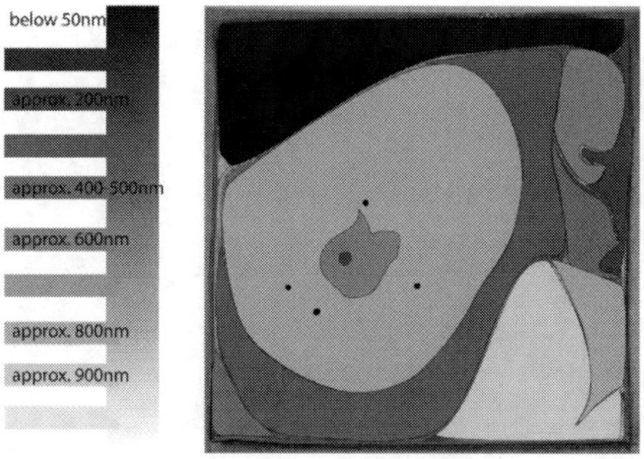

Figure 2. Sample frame of the convection in real color with manual thickness overlay information.

The reason why there is no smooth transition in the intensity of the reflected light is the abrupt change in thickness when the black film is formed. At $\lambda/4$ of the smallest wavelength, which is blue ($\sim 450nm$) in the case of the used white LEDs, the reflected waves interfere constructively. In addition, every other wavelength is reflected as well which sums up to a white color with a blue tint. Before that thickness is reached the film exhibits an orange tint as the film is approximately $200\,nm$ thick which corresponds to the cancellation of blue light. To get an absolute value of the thickness corresponding to a certain color displayed one needs to take into account that the refractive index of the solution is slightly higher than that of pure water.

We study domains of black film which are enclosed in thicker colored film, for convenience we call these spots BF-bubbles, cf. Fig. 2. One might consider calling them "holes" as they are stable spots of reduced thickness within the thicker, homogeneous liquid film. Due to the higher pressure in the black film and the vast thickness difference between the two phases, the system can be considered as a composition of two immiscible fluids - thicker film and the BF-bubbles. Since the BF-bubbles are stable, they grow due to the pressure gradient at their interface towards the colored film. This growth can be considered as a front motion, as known from evaporating films on substrates with interesting secondary instabilities [16]. However, this is not relevant here.

During motion the BF-bubbles experience a gravitational force, so while being convected they are lifted upwards. On the other hand, light media is drawn towards the center of rotational motion due to centrifugal forces. Obviously, these two forces are concurrent and which one wins basically depends on the size of the BF-bubble; in this experiment the BF-bubbles remain below the critical size which would drive them to escape towards the top of the experimental frame. Due to the disjoining pressure gradient across the interface line between black and colored film the shape of the BF-bubbles approaches a minimal circular shape when not disturbed by shear or adhesion of other bubbles.

It is worth noting that the BF-bubbles are not rigid and can deform due to shear stress or upon impact on one another. They have a line tension, analogous to surface tension of free droplets or bubbles in 3D. The exact value of it depends on the chemistry used. This makes them an ideal playground to study the collision of deformable bubbles in 2D in a controlled manner.

To track the paths of the domains of black film we employ a selective color tracking algorithm. A narrow range of colors matching the desired feature is selected and marked as black and the rest of the image is masked with white. These identified "clusters" are then tracked through

Journal of Physics: Conference Series **333** (2011) 012018 doi:10.1088/1742-6596/333/1/012018

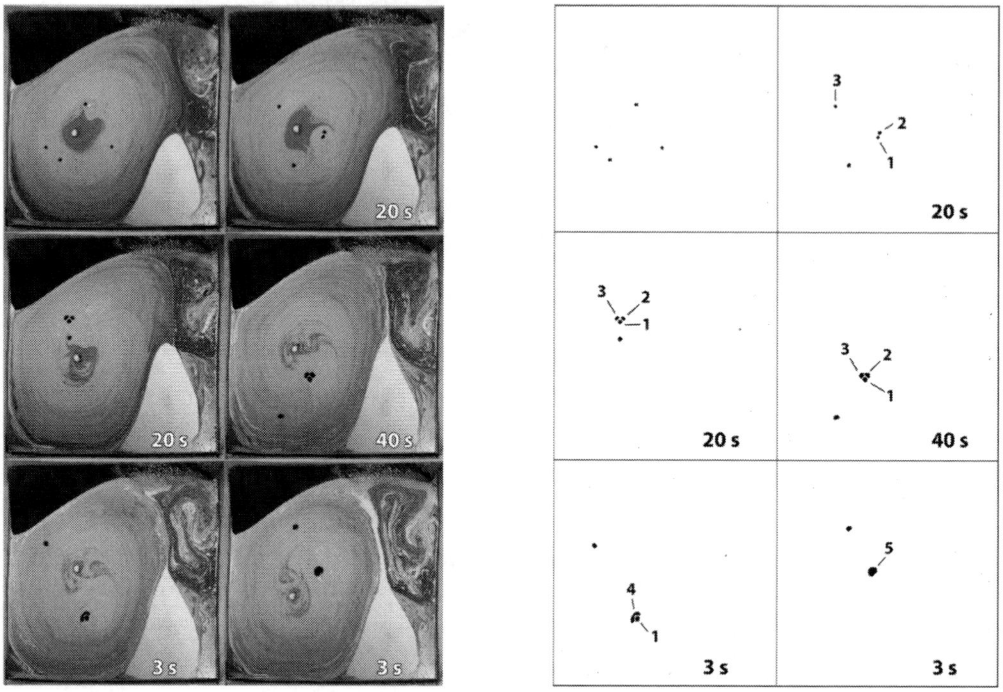

Figure 3. Left: Real color sequence of frames. The experiment has been run approximately 10 minutes at the upper left frame. Times shown are relative to the previous frame, numbering is left-right, and top-down.
Right: Same sequence as left, shown as a binary image identified by the tracking algorithm. Bubbles are black, the remainder is coded in white.

each frame. Here, we calculate for each instant of time the center of mass, the principle axes, the corresponding eigenvalues, giving the stretching/contraction rate, and the volume evolution. From these data we calculate the velocity by finite differences.

Due to the relatively low frame rate available during the experiment the data was interpolated from 25 frames per second to 200 frames per second, thus realizing a good connectivity of the cluster instances over time. A frame sequence, as recorded, is shown in Fig. 3. Although the calculated Rayleigh-number is in a relatively narrow range for the given bulk viscosity and temperature gradient the thin liquid film exhibits different turbulent motion depending on the atmospheric conditions (humidity and temperature of the surrounding gas). In this example one of the most common scenarios is presented with formation of a long term stable convection with one primary convection roll and several secondary eddies at the edge. The artifacts on the upper right side are condensation droplets on the front window. In the upper left corner a convection cell consisting of black film is present, this is not the topic of this publication and will be discussed elsewhere. The gravitationally driven thinning of the liquid layer leads to the formation of a stripe of black film where convection is present, too. However, due to the nature of the capturing technique this motion remains hidden.

In Fig. 3 the BF-bubble identification is illustrated. Real colors are transformed to black and white pictures which are then processed: To analyze the behavior of domains of the same thickness the video data is converted into a binary image where white corresponds to a single color or thickness, respectively. As the spectrum repeats continually this technique is only valid if the overall thickness deviation is smaller than a full period of the smallest wavelength. In

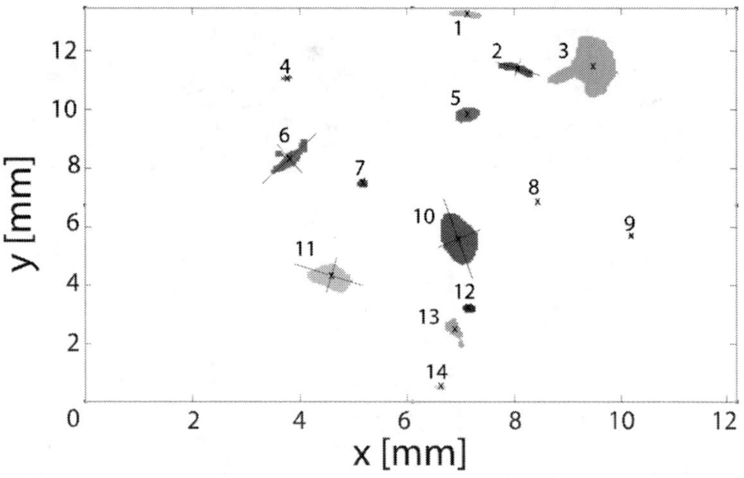

Figure 4. Example of the identification of clusters for one frame (time slice). The colors correspond to the unique indices of the domains of equal thickness, they have no physical meaning and serve for better visualization. The crossed lines correspond to the principle axes, their length to the unscaled weight, the crosspoint is the center of mass.

particular for the BF-bubble tracking, all black areas were extracted. Subsequently in each frame all spots are numbered and consecutively linked through the following frames (cf. Fig. 4). This enables us to track a BF-bubble uniquely with its volume, velocity, and deformation rate. Even the rotation rate can approximately be determined for non-circular bubbles by the rotation of the principle axes. For each bubble the velocity is calculated using the shift of its center of mass per frame. The deformation rate is calculated by determining the change of the scale of the principal components per frame. The algorithm is operating with a linear backwards memory of variable depth. The memory must be limited, because merging or separating clusters would be considered connected and a unique tracking is not possible. We set a memory of one frame in order to maintain connectivity of an individual cluster during its time evolution.

5. Bubble Coalescence

Let us have a closer look at the BF-bubbles in Fig. 3. The black film bubbles form due to local fluctuations, mainly at the boundary of the film which drive the thin liquid film to collapse to its equilibrium state. A preceding state is the formation of domains of very thin film ($\approx 100\,nm$) which appear as white spots in the convection. As shown in Fig. 3 the bubbles move along similar streamlines and change their relative position due to small perturbations in the flow. Once two bubbles reach a critical distance they stick together and move in the flow field as a conglomerate. The Young-Laplace-Equation $\Delta p = -\gamma\Delta \cdot \vec{n}$ gives an additional pressure term to the disjoining pressure balance and effectively reduces the pressure at the curved interface between black film and colored film. There remains an interstitial barrier of thicker film between the bubbles which will eventually break due to shear and lead to the coalescence of the bubbles as shown in frame 5 of Fig. 3. In this example a conglomerate of 3 bubbles is formed until the first coalescence occurs.

Initially, we observe (in Fig. 3 top left, i.e. after about 10 min.) 4 bubbles which rotate about the center. As explained above two bubbles approach inevitably due to small velocity differences (Fig. 3 top right), and the individual bubbles rotate about their mutual center of mass due to the

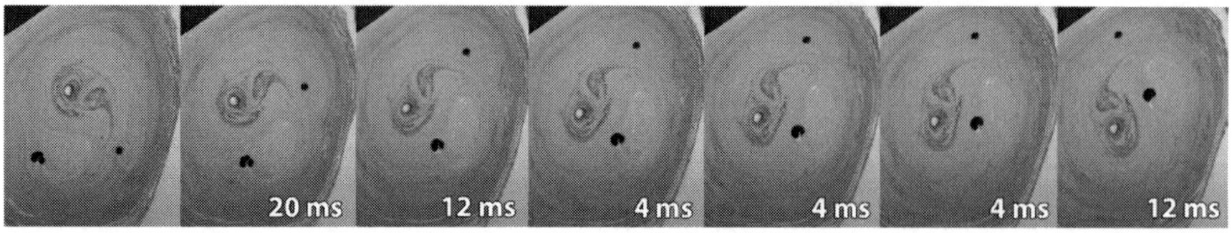

Figure 5. Sequence of 7 frames of the coalescence of two bubbles beginning just before the merge and ending with the complete withdrawal of the interstitial fluid. Time steps are given relative to the previous frame.

different velocities on their paths. This is like the common tumbling mechanism of non-circular particles in a shear flow. At their interface, they exert an attractive force onto each other. We suggest a fluid motion to cause this force, since the region between the bubbles is thicker as the surrounding, thus liquid is pushed outwards yielding an attraction. In addition, the black film wants to grow as the stable phase. All in all both bubbles stick together. A theory will have to involve a generic model for the line tension of the BF-bubbles and the repulsive and attractive forces of the BF-bubbles. With a third bubble, this game is repeated, cf. Fig. 3, middle.

The details of the coalescence of three bubbles are shown in Fig. 5. We see that first two bubbles merge to form a banana–shaped structure, which then unites with the third bubble. It is not clear in how far two bubbles can survive separately, rotating about a common center of mass. The rotation counteracts partially the attraction such that even a stable period orbit of the system might exist. Eventually, the third BF-bubble is swallowed by the now bigger and thus slower (relative to the surrounding fluid) "banana" and the resulting big bubble rotates about the center. One can nicely observe the vortex generated by the different rotation velocity of the bubbles with respect to the surrounding fluid: below the (global) center it is slower, above it is faster, resulting in a leading and trailing vortex respectively. It is seen in white since it is thicker fluid and originates on the rim which surrounds the bubbles.

Now, this spot is big enough to escape the rotation, because gravity is strong enough and the big BF-bubble escapes to the top of the experiment whereas the fourth stays trapped for quite a while. The volume evolution of the bubbles during the coalescence process is readily available from the cluster tracking algorithm. In Fig. 6 the volume of the bubbles involved in the coalescence is shown.

6. Conclusion

We have presented a prototype for an experiment on 2D BF-bubbles, where collision properties in rotating flows can be studied in detail. We suggest using the setup on a larger scale with dedicated bubble generation, e.g. by electrical forces. One drawback of the setup is that one can hardly generate bubbles with different mass, except one accepts changing the size.

However, under the given restrictions we can already state that the collision and sticking of the BF-bubbles is a very interesting phenomenon with parallels in other situations. Of course, the three dimensional world is different in that the contact forces are 2D instead of 1D, the motion has more degrees of freedom, and so on. Nevertheless, we think that principles can be tested on the playground of 2D to end up in dedicated experiments in 3D.

With respect to foam, our experiment has further implications: The coalescence of two or more bubbles is the intrinsic mechanism of foam collapse which can be studied in detail with this experiment as the time scales of the actual merging process are much larger than in the typical air and soap solution ensemble. The bubbles are very robust and can deform to a great extent until a permanent deformation, rupture or coalescence with surrounding bubbles occurs.

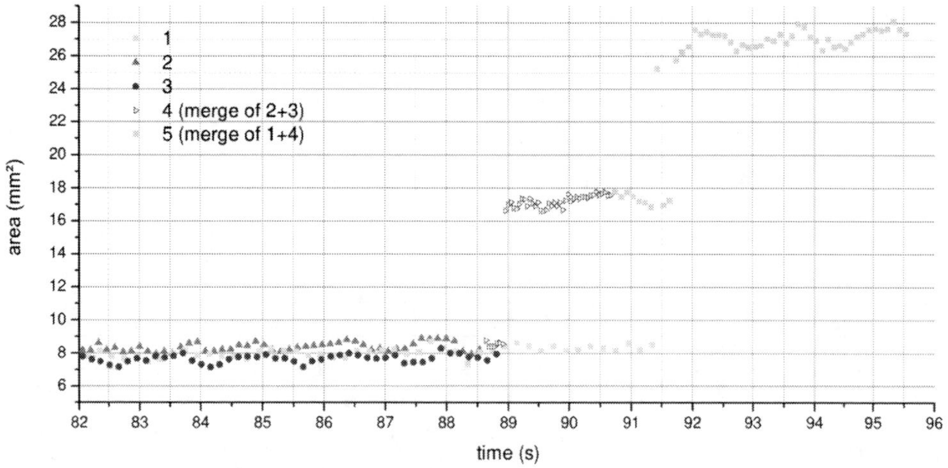

Figure 6. Volume evolution of the bubbles, as shown in Fig. 3 with according numbering. The two coalescences shown in Fig. 3 are equivalent to the merge of bubble 2+3 and 1+4 respectively.

Due to the pressure balance at the interface line of these black spots their area increases over time which is a unique feature and opposite to the usual volume decrease of bubbles in foam due to the higher inner pressure and gas permeability of the soap film.

The area tracking algorithm provides detailed information about the flow field and is adaptable to work on any data of deformable clusters with high enough contrast with respect to their surrounding.

Generally, we think that with our extremely simple experiment we can touch deep physical questions on the microscopic nature of thin film flows and treat at the same time very practice-oriented topics.

Acknowledgements

We acknowledge the initial idea by N.P. Abel. M. Winkler acknowledges support by UP Transfer, Potsdam.

References

[1] Ahlers G, Grossmann S and Lohse D 2009 *Reviews of Modern Physics* **81** 503
[2] Squires T and Quake S 2005 *Reviews of Modern Physics* **77** 977
[3] Derjaguin B 1989 *Theory of stability of colloids and thin films* (Consultants Bureau New York and London)
[4] Ivanov I 1988 *Thin liquid films: fundamentals and applications* vol 29 (Marcel Dekker, Inc.)
[5] Exerowa D and Kruglyakov P 1998 *Foam and foam films: theory, experiment, application* (Elsevier, New York)
[6] Muruganathan R, Krustev R, Mller H, Mhwald H, Kolaric B and Klitzing R 2004 *Langmuir: the ACS journal of surfaces and colloids* **20** 6352
[7] Oron A, Davis S and Bankoff S 1997 *Reviews of Modern Physics* **69** 931
[8] Sugiyama K, Ni R, Stevens R A, SChan T, Zhou S Q, Xi H D, Sun C, Grossmann S, Xia K Q and Lohse D 2010 *Phys. Rev. Lett.* **105** 034503–1–4
[9] Martin B, Wu X, Goldburg W and Rutgers M 1998 *Physical Review Letters* **80** 3964–3967
[10] Kellay H, Wu X and Goldburg W 1995 *Physical Review Letters* **74** 3975–3978
[11] Isenberg C 1992 *The science of soap films and soap bubbles* (Dover Pubns)
[12] Rutgers M, Wu X and Daniel W 2001 *Review of Scientific Instruments* **72** 3025

[13] Kellay H and Goldburg W 2002 *Reports on Progress in Physics* **65** 845–894
[14] Zhang J, Wu X and Xia K 2005 *Physical Review Letters* **94** 174503
[15] Zhang L, Somasundaran P and Maltesh C 1996 *Langmuir* **12** 2371–2373
[16] Diamant H and Agam O 2010 *Physical Review Letters* **104** 47801

Solid particle acceleration in a high Reynolds number channel flow: DNS and LES with stochastic modelling of subgrid acceleration

Rémi Zamansky, Ivana Vinkovic and Mikhael Gorokhovski
Laboratoire de Mécanique des Fluides et d'Acoustique CNRS UMR 5509
Ecole Centrale de Lyon, 36, av. Guy de Collongue, 69134 Ecully Cedex, France
E-mail: ivana.vinkovic@univ-lyon1.fr

Abstract. Inertial particle acceleration statistics are analyzed using DNS in the case of a turbulent channel flow. Along with effects recognized in homogeneous isotropic turbulence, an additional effect is observed due to high and low speed vortical structures aligned with the channel wall. In response to those structures, the inertial particles experience strong streamwise acceleration variations. DNS is also used in order to assess LES-SSAM (Subgrid Stochastic Acceleration Model), in which an approximation to the instantaneous non-filtered velocity field is given by simulation of both, filtered and residual, accelerations. Advantages of this approach in predicting particle dynamics in the channel flow at a high Reynolds number are shown.

1. Introduction

Understanding the Lagrangian behavior of inertial particles in turbulent channel flows has important implications for many environmental systems, from sediment transport to atmospheric dispersion of pollutants or solid deposition in marine flows. Previous experimental [1] and numerical [2] studies on particle-laden channel flows have examined particle deposition, trapping, segregation or the modification of particle velocity statistics due to the presence of coherent structures. It is recognized that inertial effects cause particle segregation and clustering [3]. In a high Reynolds number "free" turbulence [4, 5], these phenomena were linked to particle acceleration statistics. In our paper, such statistics are of interest in the case of a high Reynolds number channel flow. This may help in understanding complex interactions between particle dynamics and wall flow structures.

Lagrangian measurements [5, 6, 7] or computations [2, 8] provided insight into inertial particle accelerations and their statistics, including the acceleration probability density functions (PDF), in both homogeneous and inhomogeneous flows. These studies showed that, in the case of homogeneous isotropic turbulence, solid particles with low inertia exhibit highly non-Gaussian acceleration PDF with a high probability of intense acceleration events. These acceleration PDF tend to narrow and to Gaussianize as particle inertia increases [4]. In [4], the authors suggested that the trend of acceleration PDF to Gaussianity, as well as a monotonic decrease of acceleration variance, are both a consequence of preferential concentration (dominant at small Stokes numbers) and filtering (for particles with larger Stokes numbers).

One could expect the same tendencies in the presence of the wall: preferential concentration and filtering may decrease the variance of particle acceleration. However, close to the wall, the flow is very different from homogeneous isotropic turbulence. There are typical streaks, aligned with the wall, characterized by alternating high and low longitudinal speed regions. Low inertia particles do not see this alternation of regions; once such particles get trapped in the structure, they travel with it. Highly inertial particles do not respond to this intermittency neither, due to the filtering effect. Only particles with intermediate inertia may respond to the spanwise alternation of high and low speed regions; this may cause an additional agitation of those particles. In this paper, particle acceleration statistics in the channel flow are explored, in comparison with the effects described by [4] for the case of homogeneous isotropic turbulence.

For single phase flows, large-eddy simulation (LES) has become a suitable tool that produces acceptable results at much lower computational costs than direct numerical simulation (DNS). However, application of LES in turbulent channel flow becomes too expensive when the Reynolds number is high and the resolution of strong subgrid gradients is needed, specifically in the near-wall region, in the case of wall-bounded flows. An alternative to standard LES, referred to as LES-SSAM approach was proposed in [12, 13]. In this approach, the exact Navier-Stokes equation is replaced by a model equation in which the instantaneous total acceleration is viewed as a sum of two model accelerations. One represents the filtered acceleration closed by Smagorinsky model, the other one emulates the residual (subgrid) acceleration. The latter is done by two independent stochastic processes, for the norm of this acceleration and for its orientation. The new stochastic model for the subgrid acceleration was proposed in [14], and was assessed in the case of turbulent channel flow in the framework of LES-SSAM approach. The stochastic model for the norm of the fluid acceleration is based on the fragmentation process [15] in order to represent the long-range interactions across the channel. The fluid acceleration orientation is also simulated stochastically as a Brownian motion on a unit sphere, starting from random orientation on the wall plane, and then going toward stochastic isotropy away from the wall. Such a development of LES-SSAM approach [14] is applied to particle-laden channel flows in this paper. The accuracy of the method will be tested in regard of particle velocity and acceleration statistics. Results will be compared with DNS and classical LES. The tests will be run for different types of particles given by Stokes numbers from 1 to 125. The article is organized as follows. In the next section, the numerical method is described. Then, we present some results obtained by DNS. In the last section, the LES-SSAM approach is compared to DNS and classical LES. Finally, conclusions are stated.

2. Numerical method

2.1. Flow

The flow considered is an incompressible turbulent channel flow. The computational domain consisting of two infinite parallel walls is illustrated in Figure 1. Periodic boundary conditions are imposed on the fluid velocity field in x (streamwise) and z (spanwise) directions and no-slip boundary conditions are imposed at the walls.

Three different sets of numerical simulations for the fluid are treated here: direct numerical simulation of channel flow (DNS), standard large-eddy simulation (LES) and large-eddy simulation coupled with stochastic forcing of subgrid acceleration (LES-SSAM). For the case of LES the classical Smagorinsky model is used [11].

Details on the simulation characteristics are given in Table 1, where N_i and L_i are the number of grid points and the domain length in direction i. The size of the grid is also given. The superscript "+" denotes quantities expressed in wall units, normalized by the friction velocity u_τ and the viscosity ν. For all simulations, the Reynolds number based on the u_τ, the channel half height h and ν, is $Re_\tau \sim 587$. In the case of DNS, this corresponds to a Reynolds number based on the mean velocity U at the center of the channel of $Re \sim 12500$. The dimensions of the computational domain have been changed in LES with respect to DNS in order to match previous LES of turbulent channel flow. It has been checked that the difference in size does not

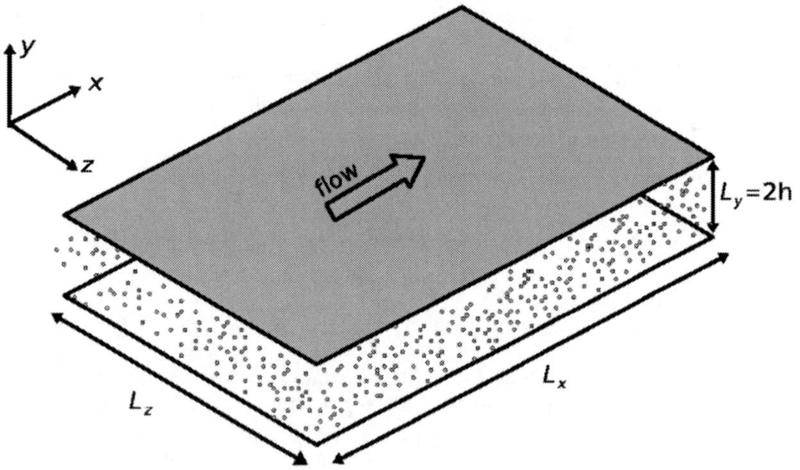

Figure 1. Channel flow. x, y, z represent the streamwise, the wall-normal and the spanwise directions respectively.

affect the particle statistics presented here.

Table 1. Simulation parameters.

Type	Re_τ	Re	$N_x \times N_y \times N_z$	$L_x \times L_y \times L_z$	$\Delta x^+ \times \Delta y^+ \times \Delta z^+$
DNS	587	12490	$384 \times 257 \times 384$	$3/2\pi h \times 2h \times 3/4\pi h$	$7.2 \times (0.04 \sim 7.2) \times 3.6$
LES	587	14160	$64 \times 65 \times 64$	$3\pi h \times 2h \times \pi h$	$87 \times (0.71 \sim 29) \times 29$
LES-SSAM	587	12760	$64 \times 65 \times 64$	$3\pi h \times 2h \times \pi h$	$87 \times (0.71 \sim 29) \times 29$

The incompressible Navier-Stokes equations in a turbulent channel flow are solved using a Galerkin spectral approximation (Fourier Chebyshev) and a variationnal projection method on a divergence free space [16]. Steady state fluid statistics have been compared with other DNS in [14].

2.2. Particles

Particles are injected into the flow at low concentrations in order to consider dilute systems. Particle-particle interactions are neglected as well as gravity or the influence of particles on the carrier fluid. Particle-wall collisions are assumed to be elastic. Furthermore, particles are considered to be point-wise, spherical, rigid and to obey the following Lagrangian dimensionless equation of motion :

$$St\frac{d\vec{v}_p}{dt} = (\vec{u} - \vec{v}_p)\, f(Re_p) \qquad (1)$$
$$\frac{d\vec{x}_p}{dt} = \vec{v}_p$$

Here, $\vec{v}_p$ and $\vec{x}_p$ are the dimensionless particle velocity and position. The solid particle - fluid interaction is modelled by a drag force with the correction term $f(Re_p) = 1 + 0.15Re_p^{0.687}$ suggested by [17]. Re_p is the local and instantaneous particle Reynolds number based on the local relative velocity, the particle diameter d_p and the fluid viscosity.

St is the Stokes number given by :

$$St = \frac{\tau_p}{\tau_f} \tag{2}$$

with $\tau_p = \dfrac{\rho_p d_p^2}{18\rho\nu}$ and $\tau_f = \dfrac{\nu}{u_\tau^2}$, ρ_p and ρ being respectively the particle and fluid density. The Stokes number characterizes the response time of a particle to fluid solicitation.

Once the steady state for the fluid is obtained, 200000 particles are released at randomly chosen locations within the channel, then tracked at each time step. Five sets of particles are considered with $St = 1$, $St = 5$, $St = 15$, $St = 25$ and $St = 125$.

The initial velocities of the particles are set equal to the interpolated fluid velocities at each particle location. A high order three dimensional Hermite interpolation is used for computing the fluid velocity $\vec{u}(\vec{x}_p, t)$ at the particle position. The time-integration of the particle motion Equation (1) is performed using a second-order Adams-Bashforth method with the same time step as the DNS. Once the particles released, the simulations are run over several particle timescales τ_p. Particle statistics are sampled starting from $t^+ \sim 1000$, counted from particle release. For all simulations, velocity statistics for the solid phase are at stationary state.

In this study, the acceleration is evaluated using the velocity time derivative along particle trajectories. Even though we use a three dimensional Hermite interpolation for computing the fluid velocity at particle position, as suggested in [18] numerical errors are generated when a particle crosses a grid point. Therefore, as in [19], the acceleration is estimated by a convolution of the Lagrangian velocity with the derivative of a Gaussian kernel. This ensures both time derivation and filtering. The filter width, of the order of τ_f, is such that there is agreement between fluid Lagrangian and Eulerian acceleration statistics.

In the case of DNS, for a lower Reynolds number, particle velocity statistics have been compared with the benchmark [20]. For higher Reynolds numbers, velocity statistics have also been compared to experiments [21]. Details about these comparisons and other results based on particle velocity statistics may be found in [22].

3. DNS acceleration statistics

In this section we present some results obtained by DNS.

3.1. Acceleration variance

Figure 2 shows the streamwise acceleration variance of solid particles as a function of $y^+ = yu_\tau/\nu$. For comparison, other DNS [23, 8] and experiments are also illustrated [7]. In [23], the fluid acceleration in a turbulent channel flow at $Re_\tau = 180 \sim 600$ without gravity is studied. From Figure 2, it is seen that the peak of acceleration variance obtained in our DNS is lower than the one computed by [23]. The measurements reported in [7] for inertial particles in the turbulent boundary layer at $Re_{\lambda_0} = 100$ and $St_0 = 0.035$ are plotted, where Re_{λ_0} is the Reynolds number based on the Taylor-scale and the streamwise velocity root-mean-square (RMS), and where the subscript zero refers to the free-stream conditions. This corresponds to $Re_\tau \sim 470$ and $St \sim 0.7$ and is closest to our $St = 1$ and $Re_\tau \sim 600$ simulation. Although gravity was not taken into account here, the DNS results are close to the measurements [7]. The DNS results of [8] are also shown in Figure 2 for the case without gravity, $St = 0.87$ and $Re_\tau = 300$. It can be seen that the present DNS is close to the DNS reported in [8] except close to the wall, where we slightly underestimate the peak of inertial particle acceleration variance.

From Figure 2 it is seen that as the distance to the wall increases, the acceleration variance decreases. In addition to this, as particle inertia increases the acceleration variance departs from the fluid. This is in accordance with previous studies in homogeneous isotropic flows [4, 5, 24] and results from the simultaneous effect of preferential concentration and filtering. For $St = 5$, the streamwise acceleration variance presents the highest peak close to the wall. This peak is even higher than the fluid streamwise acceleration variance. To gain insight on this behavior, we analyze the acceleration statistics of the fluid seen by the solid particles.

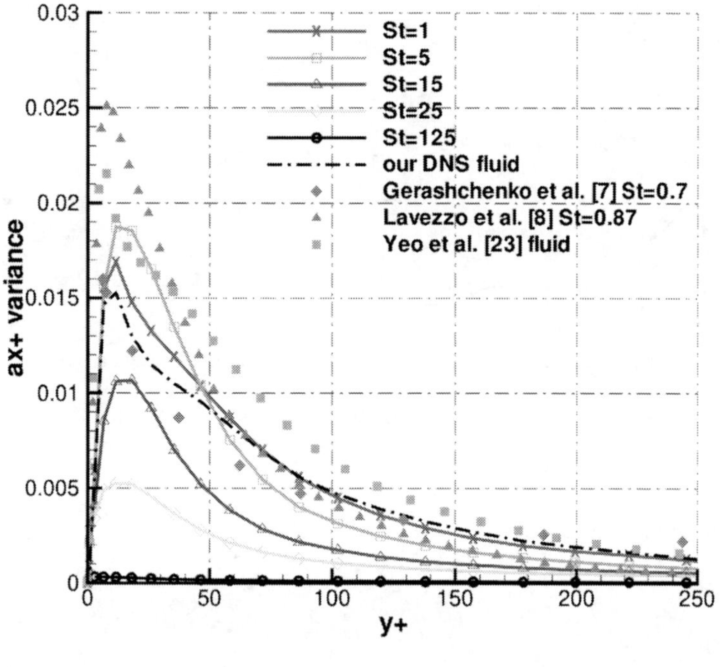

Figure 2. Streamwise acceleration variance profile for solid particles with different Stokes numbers and for the fluid (dashed line). [7] for $St \sim 0.7$ and $Re_\tau \sim 470$ (full diamonds), [8] for $St = 0.87$ and $Re_\tau = 300$ (full triangles), [23] for fluid particles at $Re_\tau = 600$ (full squares).

Figure 3 illustrates the streamwise acceleration variance of the fluid seen by the solid particles. The variance for the fluid seen by the solid particles is higher than the variance of the non-conditional fluid. This is especially the case close to the wall, at the position of the peak of streamwise acceleration variance. The peak of fluid streamwise acceleration at the solid particle position increases as the Stokes number increases. Whereas, the peak of solid particle acceleration first increases from $St = 1$ to $St = 5$ and then decreases as the Stokes number increases.

These observations suggest that in the case of wall bounded flow, solid particles are entrained preferentially by regions with relatively high streamwise acceleration variance. The contrary is observed in homogeneous isotropic turbulence [4, 5, 24], where previous studies found that inertial particles tend to cluster in regions of the fluid experiencing relatively low fluid accelerations. This behavior of the solid particle streamwise acceleration variance suggests that because of inertia particles respond less and less to the increasingly varying fluid solicitations. This results in a net decrease of solid particle streamwise acceleration variance.

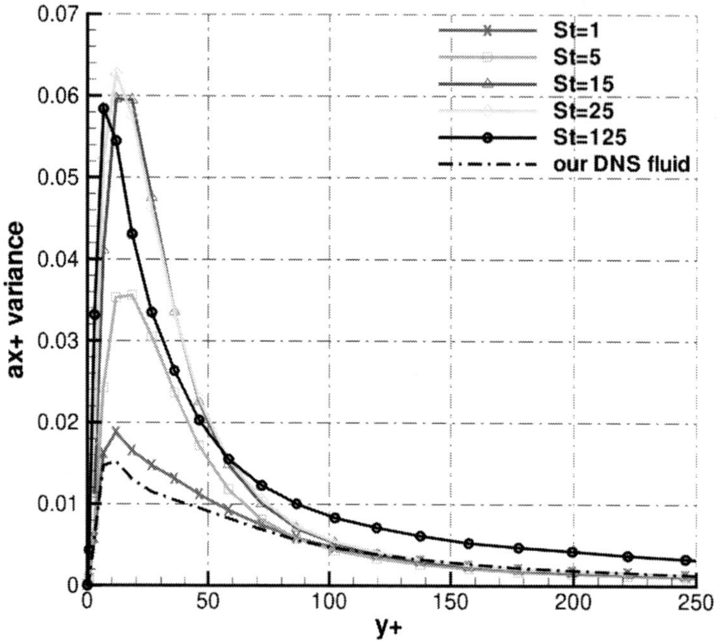

Figure 3. Streamwise acceleration variance profile for the fluid seen by the solid particles with different Stokes numbers and for the fluid ($St = 0$).

3.2. Acceleration PDF

In Figures 4 and 5 we compare the normalized PDFs of streamwise and wall-normal acceleration, respectively, at $y^+ \sim 100$ for different Stokes numbers with the normalized PDFs obtained by using the fluid acceleration on the particle position. For reference, the normalized PDF for the non-conditional fluid acceleration is plotted as well. As expected, when the Stokes number increases, the tails of the normalized solid particle PDFs become narrower. For all Stokes numbers, the normalized PDFs of the fluid seen by the solid particles overlap almost perfectly with the normalized PDFs of the non-conditional fluid. The same conclusions can be drawn for other distances to the wall and for the spanwise component of the acceleration. For simplicity reasons, these plots are not shown here. The overlap suggests similarity of structure in terms of statistical distribution of fluid acceleration seen by the solid particles. The scaling factor is given by the acceleration RMS, which is different for each Stokes number.

These results show that in wall bounded flows, effects predicted in [4] are present with a new aspect which is additional agitation of particles by alternation of high and low speed streaks. Wall bounded flows present spatial alternation of high and low speed vortical structures aligned with the channel wall. As in homogeneous isotropic turbulence, inertial particles are ejected from high vorticity regions toward high strain (high dissipation rate) regions. Inertial particles are swept by these regions and due to the random alternation of high and low speed streaks, experience strong streamwise velocity fluctuations. Thereby, in these regions, inertial particles see fluid with high streamwise acceleration variance. Slightly inertial particles (with a response time τ_p similar to the characteristic fluid timescale) may well respond to the fluid solicitations and therefore experience an increase of the streamwise acceleration variance relatively to the

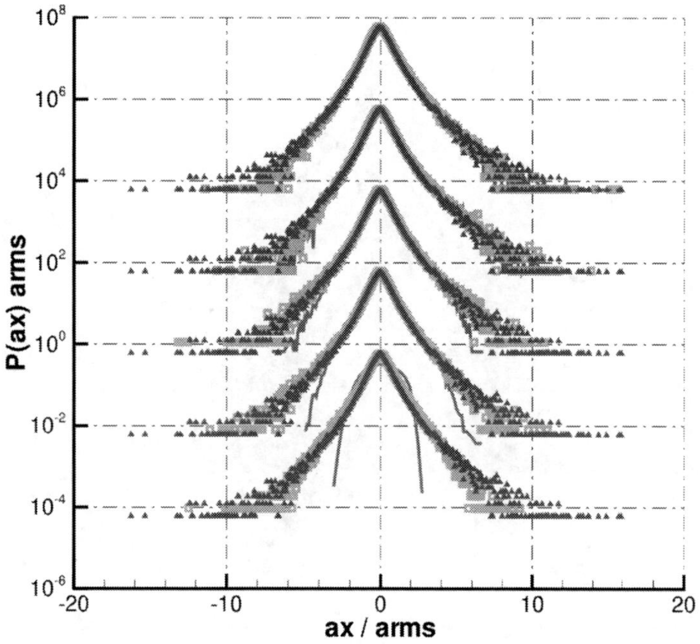

Figure 4. Streamwise acceleration PDF at $y^+ = 100$, for (from top to bottom) $St = 1, 5, 15, 25, 125$. Solid particles (line), fluid particles (triangle), fluid seen by solid particles (square).

fluid. Due to the filtering effect of inertia, very inertial particles ignore the wall turbulent structures and therefore present a more homogeneous concentration.

Here, the significance of wall structures on particle acceleration statistics has been shown. In the frame of LES, wall structures are under-resolved. Therefore, in the next section, classical LES is coupled with a stochastic model for subgrid acceleration in order to introduce flow intermittency at subgrid scales.

4. LES-SSAM applied to particles

Here we briefly describe the stochastic subgrid model for the acceleration used in the LES-SSAM approach. The LES-SSAM is then compared to classical LES and DNS in the case of particle laden turbulent channel flows.

4.1. LES-SSAM approach

In order to take into account the non-resolved acceleration in standard LES, the approach proposed in [12] and [13] is used here. In this approach, the total instantaneous acceleration is given by two contributions: $a_i = \bar{a}_i + a'_i$. The first component is the filtered total acceleration where the overbar denotes the filtering operation. The second component represents the total acceleration in the residual field. When both parts are modeled, their sum gives an approximation to the non-filtered velocity field.

In the case of channel flow, a stochastic model for the subgrid acceleration was proposed in [14]. This model is based on the norm/orientation decomposition of the fluid acceleration. The

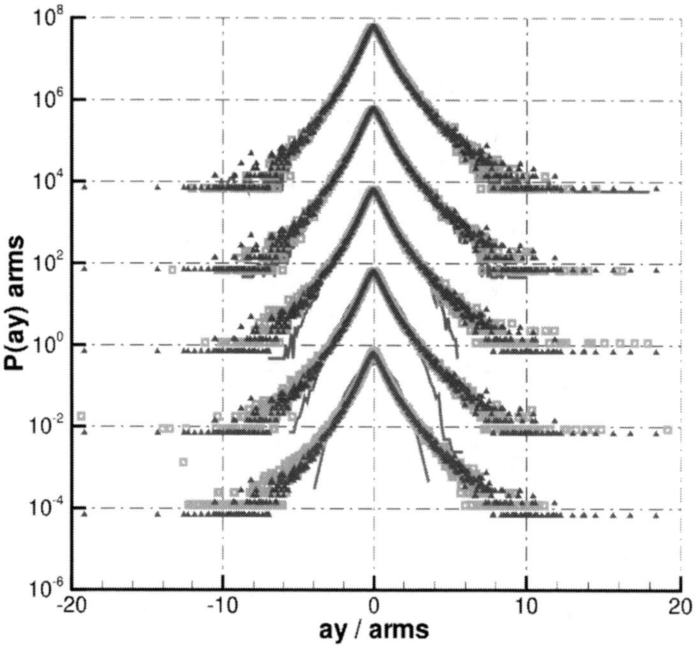

Figure 5. Wall-normal acceleration PDF at $y^+ = 100$, for (from top to bottom) $St = 1$, 5, 15, 25, 125. Solid particles (line), fluid particles (triangle), fluid seen by solid particles (square).

stochastic model for the norm of the fluid acceleration is based on the fragmentation process in order to represent the long-range interactions across the channel. The fluid acceleration orientation is simulated by a random walk on a sphere in order to reproduce the relaxation towards isotropy with increasing wall distance. In comparison with standard LES, it was showed that the LES-SSAM approach gives better prediction of velocity spectra on small spatial scales and better statistics of fluid acceleration [14]. In the following section, we apply this model to the case of particle laden channel flow. We assume that the modelled subgrid acceleration can be attributed to the acceleration of the fluid seen by the solid particle.

4.2. Results and comparisons

Figures 6 and 7 show the solid particle mean and RMS velocity profiles obtained from the LES-SSAM model and compared to the DNS and LES results. The LES-SSAM model gives a better estimation of the mean solid particle velocity than classical LES (Figure 6). This is illustrated by the average relative distance of LES and LES-SSAM to the DNS, given in the inset of Figure 6. In addition to this, there is also an improvement of the prediction of the wall-normal (Figure 7) velocity RMS profile.

A quantitative indicator of the improvements made by LES-SSAM is the average streamwise particle flux:

$$< u_p > = \frac{1}{2h} \int_0^{2h} u_p(y) dy$$

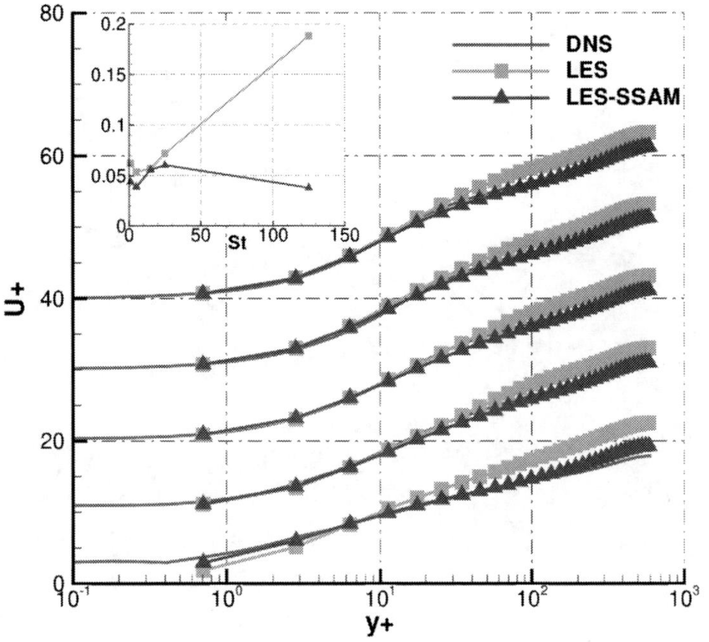

Figure 6. Mean streamwise velocity profile of solid particles for (from top to bottom) $St = 1$, 5, 15, 25, 125. DNS (line), LES (squares), LES-SSAM (triangles). The average relative distance of LES (squares) and LES-SSAM (triangles) to DNS is given in the inset.

This flux is given in Figure 8 as a function of the Stokes number. The inset shows the relative distance to DNS. A better global estimation of the flux is achieved by LES-SSAM.

Figures 9 and 10 illustrate the RMS and PDF of the solid particle acceleration. The profile of the RMS is shown for the spanwise component, whereas the PDFs are given for the wall-normal component at $y^+ \sim 100$. As described in the previous section, the PDFs obtained by the DNS display stretched tails. As the Stokes number increases the acceleration PDFs tend to Gaussianize and the RMS of solid particle acceleration decreases. This evolution is well reproduced by LES-SSAM, while standard LES predicts much higher tails for the normalized PDF and much lower values of the acceleration RMS. The overall improvement achieved by the model is illustrated by the average relative distance of LES and LES-SSAM to DNS, given in the inset of Figure 10. For other distances to the wall or other components of the acceleration we observe either the same improvement either no difference with classical LES. These results point out the importance of predicting subgrid scale acceleration for estimating solid particle velocity and acceleration statistics.

5. Conclusion

In this study, numerical simulations of particle laden turbulent channel flow are performed for five different Stokes numbers and a high Reynolds number. Acceleration statistics obtained by DNS show that in wall bounded flows, effects predicted in homogeneous isotropic turbulence [4] are completed by a new aspect which is interaction of particles with high and low speed

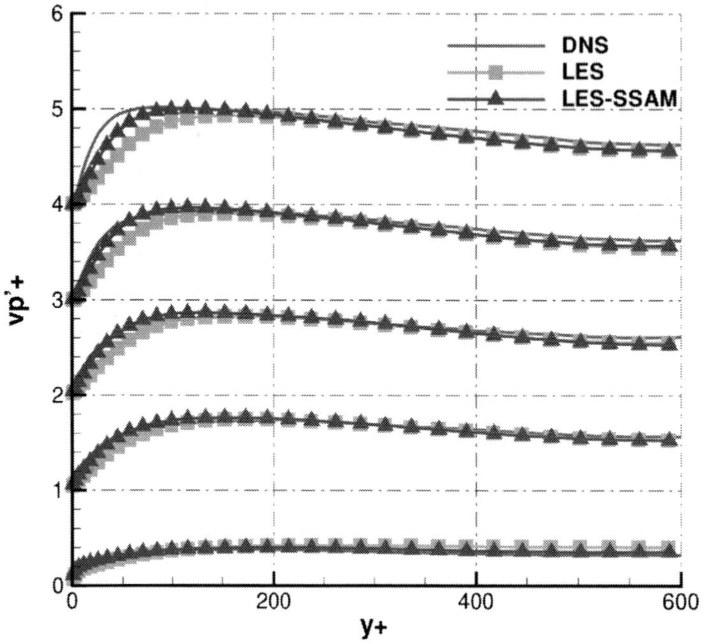

Figure 7. Wall-normal velocity RMS profile of solid particles for (from top to bottom) $St = 1$, 5, 15, 25, 125. DNS (line), LES (squares), LES-SSAM (triangles).

streaks. As in homogeneous isotropic flows, inertial particles are ejected from high vorticity regions toward high strain (high dissipation rate) regions. Due to the proximity of high and low speed streaks, particles experience strong streamwise acceleration variations. Depending on their inertia particles may well respond to those fluid solicitations (experiencing an increase of the streamwise acceleration variance) or ignore the wall turbulent structures (presenting in that case a more homogeneous concentration).

In order to take the intermittency of subgrid flow into account, a model for the subgrid acceleration is introduced according to [14]. The capability of LES-SSAM for particle-laden turbulent channel flow is assessed by comparing the results with DNS and classical LES. The LES-SSAM model gives a better estimation of the solid particle velocity and acceleration statistics. These improvements are due to two factors. First, particle velocity statistics are mainly influenced by the large scale flow which, as shown in [14], are better simulated. Second, with LES-SSAM the influence of small scales is taken into account, which is important for particle acceleration statistics.

Along with fundamental interest, the LES-SSAM approach has a practical relevance, when significant physics take place on subgrid scales. Examples include atomization, droplet evaporation and combustion.

6. Acknowledgments

M. Buffat is acknowledged for the development of the computational code. The authors express their gratitude to F. Laadhari who kindly provided his initial fields for the DNS. This work

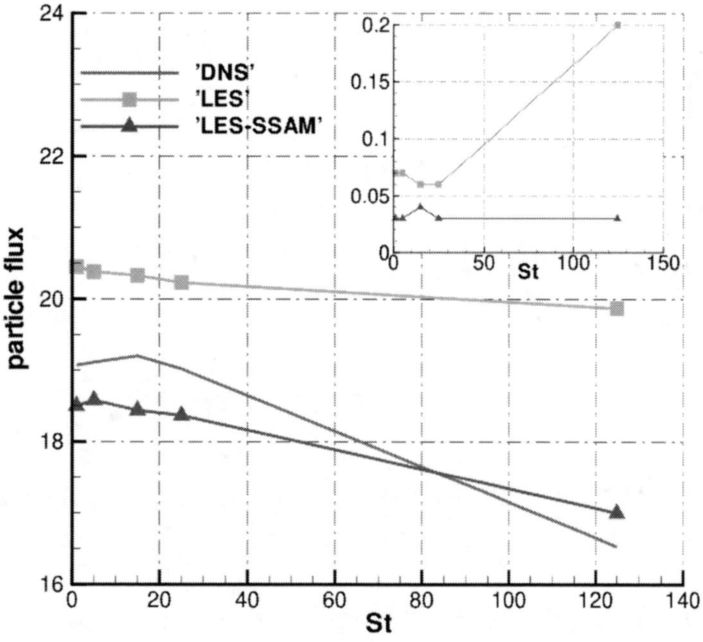

Figure 8. Streamwise particle flux $< u_p >$ obtained by DNS (line), LES (squares), LES-SSAM (triangles). The average relative distance of LES (squares) and LES-SSAM (triangles) to DNS is given in the inset.

was granted access to the HPC resources of CINES under the allocation 2009-c200902560 made by GENCI (Grand Equipement National de Calcul Intensif). Numerical simulations were also performed on the P2CHPD parallel cluster. The authors are therefore grateful to C. Pera for the administration of the computational tool.

References

[1] Kaftori D, Hestroni G and Banerjee S 1995 *Phys. Fluids* **7** 1095
[2] Marchioli C and Soldati A 2002 *J. Fluid Mech.* **468** 283
[3] Kiger K T and Pan C 2002 *J. of Turbulence* **3** 1
[4] Bec J, Biferale L, Boffetta G, Celani A, Cencini M, Lanotte A, Musacchio S and Toschi F 2006 *J. Fluid Mech.* **550** 349
[5] Ayyalasomayajula S, Warhaft Z and Collins L R 2008 *Phys. Fluids* **20** 095104
[6] Qureshi N M, Arrieta U, Baudet C, Cartellier A, Gagne Y and Bourgoin M 2008 *Eur. Phys. J. B* **66** 531
[7] Gerashchenko S, Sharp N S, Neuscamman S and Warhaft Z 2008 *J. Fluid Mech.* **617** 255
[8] Lavezzo V, Soldati A, Gerashchenko S, Warhaft A and Collins L R 2010 *J. Fluid Mech.* **658** 229
[9] Meneveau C and Katz J 2000 *Annu. Rev. Fluid Mech.* **32** 1
[10] Piomelli U and Balaras E 2002 *Annu. Rev. Fluid Mech.* **34** 349
[11] Sagaut P 2002 *Large Eddy Simulation for Incompressible Flows: An introduction* 2nd edition (Springer Verlag)
[12] Sabel´nikov V, Chtab A and Gorokhovski M 2007 *Advances in Turbulence XI* **117** 209
[13] Sabel´nikov V, Chtab-Desportes A and Gorokhovski M 2011 *Eur. Phys. J. B* **80** 177
[14] Zamansky R, Vinkovic I and Gorokhovski M 2010 *J. Turbulence* **11** 1
[15] Gorokhovski M and Saveliev V 2008 *J. Phys. D: Appl. Phys.* **41** 085405
[16] Buffat M, Le Penven L and Cadiou A 2011 *Comput. Fluids* **42** 62

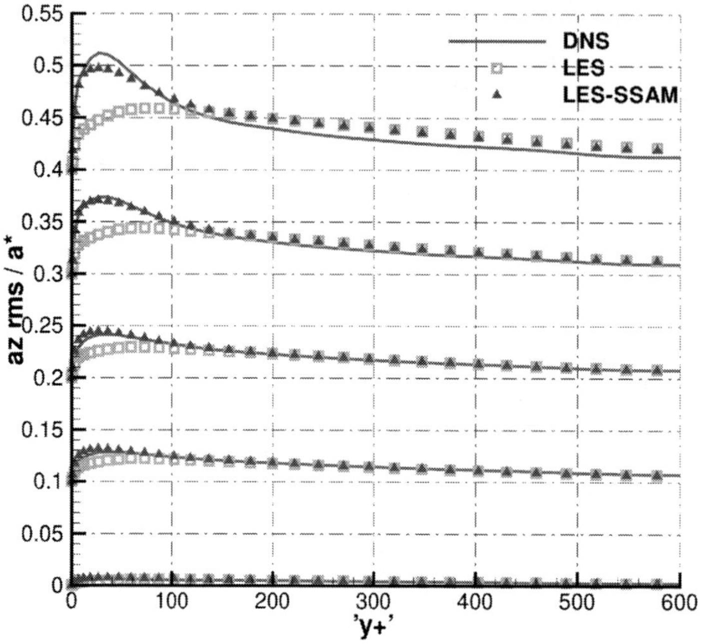

Figure 9. Spanwise acceleration RMS profile of solid particles for (from top to bottom) $St = 1$, 5, 15, 25, 125. DNS (line), LES (squares), LES-SSAM (triangles).

[17] Schiller L, Naumann A 1933 *Vereines Deutscher Ingenieure* **77** 318
[18] Choi J-I, Yeo K and Lee C 2004 *Phys. Fluids* **16** 779
[19] Mordant N, Crawford A M and Bodenschatz E 2004 *Physica D* **193** 245
[20] Marchioli C, Soldati A, Kuerten J G M, Arcen B, Tanière A, Goldensoph G, Squires K D, Cargnelutti M F and Portela L M 2008 *Int. J. Multiphase Flow* **34** 879
[21] Lelouvetel J, Bigillon F, Doppler D, Vinkovic I and Champagne J-Y 2009 *Water Resour. Res* **45** doi:10.1029/2007WR006520
[22] Vinkovic I, Doppler D, Le Louvetel J and Buffat M 2010 *Int. J. Multiphase Flow* **37** 187
[23] Yeo K, Kim B-G and Lee C 2010 *J. Fluid Mech.* **659** 405
[24] Calzavarini E, Volk R, Bourgoin M, Lévêque E, Pinton J-F and Toschi F 2009 *J. Fluid Mech.* **630** 179

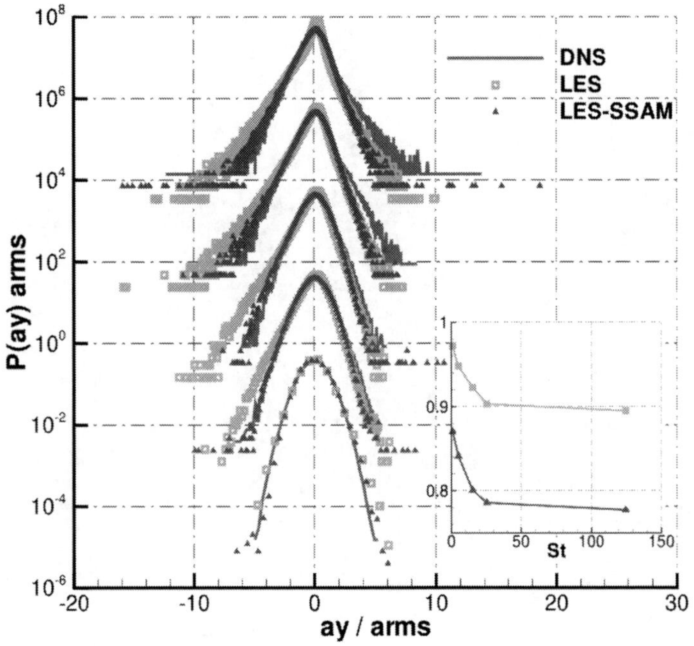

Figure 10. Wall-normal acceleration PDF of solid particles, at $y^+ = 100$, for (from top to bottom) $St = 1, 5, 15, 25, 125$. DNS (line), LES (squares), LES-SSAM (triangles). The average relative distance of LES (squares) and LES-SSAM (triangles) to DNS is given in the inset.

AUTHOR INDEX